feedback and
control systems

A Boeing 747 flight simulator designed and built by CAE Industries, Ltd.; a most sophisticated combination of engineering disciplines.
By kind permission of CAE Industries, Ltd., Toronto, Canada.

feedback and control systems

A.C. McDonald, Lecturer
sir sandford fleming college
of applied arts and technology
peterborough, ontario, canada

H. Lowe, Ph.D
senior control engineer
spectrum engineering corporation, ltd.
peterborough, ontario, canada

RESTON PUBLISHING COMPANY, INC.
A PRENTICE-HALL COMPANY
RESTON, VIRGINIA

for christine, nick, andy, theo, paul and liam
for deborah, helen and andrew

Library of Congress Cataloging in Publication Data

McDonald, Anthony C.
   Feedback and control systems.

   Includes index.
   1. Feedback control systems. I. Lowe, H.
II. Title.
TJ216.M36   629.8'3   81-4538
ISBN 0-8359-1898-X   AACR2

10   9   8   7   6   5   4   3   2   1

Printed in the United States of America

contents

preface

The frontespiece picture shows a state of the art flight simulator, a most complex piece of machinery, that requires the harmonious coordination of many engineering disciplines to achieve a properly operating system. It is our intention, in this book, to draw back the veil of mystery surrounding the design and operation of many multidisciplined control systems.

Within the text we present a fairly concise exposition on feedback control systems and highlight those topics which, as practicing engineers and teachers, we feel are most relevant to the requirements of technology programs. The topics dealt with represent the basic tools and yardsticks which the technologist and junior engineer will need when working with control systems, whether it be for the purpose of design, analysis, or even discussing a control system with another colleague. We also feel that the student should know something of mathematical modelling with application to analog circuits. This practice in itself can give insight not only into feedback systems but also into the problems attendant when "breadboarding" electronic feedback circuits.

Finally, we believe that a knowledge of feedback control systems should be supported by a level of mathematics well within the capability of the reader. We are convinced that this math must not only be easily understood in the lecture room but must be simply demonstrated in the laboratory.

The text is essentially in five parts.

Part 1: Chapters 1 to 5 inclusive form the foundation or the working tools in the study of control systems and explain what a feedback system is and how to describe such a system. In these chapters all the mathematics required for this text is discussed, explained, and demonstrated by means of examples. The authors share the opinion that students find math forbidding only by dint of its presentation. We wish to downplay its emphasis so that it may be considered as the "workhorse of engineering" rather than the "queen of sciences."

Part 2: Chapters 6 to 8 inclusive cover the central theme, the analysis of systems in both the time domain and the frequency domain. They concentrate on the response when the input changes with respect to time or frequency. Methods of analysis are demonstrated with a step-by-step approach which forces the student to follow a logical, systematic procedure.

Part 3: Chapters 9 and 10 concern themselves with the modification and design of real life feedback systems. Once again using a step-by-step method of procedure, we take the reader through the task of compensating a number of real systems so that they meet specific engineering limitations.

Part 4: Chapters 11 and 12 deal, in very practical terms, with the design of op amp circuitry and modeling of control systems on an analog computer. "Rule of thumb" techniques are described so that the student can easily understand and construct simple analog computer models for himself. Examples showing how to construct analog models are given in Chapter 12, where practical control systems analyzed in earlier chapters are used. Chapter 12 concludes with a discussion of designing and optimizing a control system using an analog computer.

Part 5: Finally, Chapters 13 and 14 address what perhaps is the most recent development in control systems, namely digital control. In Chapter 13 the discussion centers on the components and techniques used at the output end of the digital servo system with emphasis on simple methods used to determine load position, accuracy, and resolution. To complete a general coverage of these systems, Chapter 14 examines the circuits used in a hybrid system to convert analog signals to digital signals and vice versa.

In conclusion, we would like to add that we were also concerned with literary style, in that whenever possible we have tried to convey a sense of conversation with the reader rather than using the dry inflexible style cultivated in most engineering texts. We felt that such intimacy would enhance comprehension of a sometimes thought-provoking subject.

A. C. McDonald
H. Lowe

author's note

It is not by chance that the words of Lewis Carroll have been used to introduce each chapter. Although the Reverand Charles Lutwidge Dodgson was considered to be a mediocre mathematician who taught at Oxford for twenty-seven years without producing anything of value in his subject, it is patently obvious that as Lewis Carroll he invested his stories of Alice with a most astonishing insight into mathematical logic.

He lived, as we all do, in an analog world and, as his writings demonstrate, was preoccupied by games, puzzles, mnemonics, algorithms, and logical paradoxes, so it is perhaps not improper to suggest that had he lived into the latter half of the twentieth century he would have delighted himself in the world of computers and control systems.

Before his death in January 1898 he wrote to a friend in the USA, "words mean more than we mean to express when we use them; so a whole book ought to mean a great deal more than the writer means." It is with this thought in mind that the authors wish to instruct and excite the minds of their readers.

A. C. McDonald
H. Lowe

part 1

learning
the basic tools

introduction to control systems

"Would you tell me, please, which way I ought to go from here?"
"That depends a good deal on where you want to get to," said the Cat."
"I don't much care where-," said Alice.
"Then it doesn't matter which way you go," said the Cat.

Lewis Carroll, *Alice in Wonderland*

1-1 WHAT ARE CONTROL SYSTEMS AND WHY DO WE NEED THEM?

Control systems, of one type or another, are present in just about every facet of our everyday life, although we are often on such intimate terms with these systems that we take them very much for granted and tend not to be aware of them. For example, our body temperatures are controlled with excellent accuracy by our cardiovascular systems; insect and animal populations are controlled by very delicately balanced predator-prey relationships. (Unfortunately, just how delicately balanced this particular control system is becomes more and more apparent as the result of the indiscriminate damage caused by the recent use of highly toxic pesticides).

3

Control systems such as these two examples are provided for us by nature. However, whether a control system is natural or man-made, they all share a common aim, which is to control, or regulate, a particular variable within certain operating limits.

Natural control systems are usually concerned with regulating or sustaining the operating environment of some larger system or process. Similarly, when we design control systems, we do so because we want to improve the performance, quality, or accuracy of some other system or process. Control systems can be designed to perform such tasks much more speedily, efficiently, and accurately than is possible with manual control and, in most cases, they are designed to permit process operations that are otherwise impossible.

Apart from sharing a common aim, control systems tend to have a common structure. This structure can be broadly broken down into:

1. A reference signal (input) which dictates the desired value of the controlled variable.

2. A dynamic element or actuator which acts upon the controlled variable to change it in accordance with the value of the input signal.

3. The controlled variable (or output), which was body temperature or animal population in the above two examples.

At this stage, we have what is called an open loop control system. It is called open loop control because the value of the output variable has no way of influencing the input signal. Thus if the output of the actuator were to drift, the value of the controlled variable would become incorrect and no corrective action would be forthcoming from the input signal, which would remain unchanged. This is obviously not the condition with the control systems of the above two examples. For example, if our body temperature rises slightly (perhaps due to the environment) then sensors in the brain note the change and the brain instructs the cardiovascular system to increase the cooling action of the blood flow; artery walls are expanded (to increase flow); body pores are opened, and moisture from the body is allowed to escape and evaporate, providing further cooling action. In short, the rise in temperature is measured and the control action is adjusted to reduce the body temperature. Consequently, the final elements needed to convert our open loop control system into a closed loop control system are

4. The feedback loop, where we measure the size or behavior of the controlled variable in order to feed the signal back to the input summing junction.

5. The input summing junction, where the value of the controlled variable is subtracted from the reference value to generate an error signal (or actuating signal) which causes the control action of the actuator to be reduced as the controlled variable approaches the reference value (see Figure 1-1).

A control system whose action takes account of the value of the controlled variable in this way is termed a feedback control system.

1-2 FEEDBACK CONTROL SYSTEMS

The interconnection of the five elements of a general feedback control system are shown in block diagram form in Figure 1-1.

In the diagram the summing junction has two inputs, one being the reference input, often designated by the letter R, and the second being the output feedback signal. The arrow connecting the summing junction to the actuator is called the actuating signal and represents the difference between the reference input and the feedback signal. It is this error actuating signal that causes the system to operate.

The actuator includes an optional amplifying device, which may be needed in some processes to bring the low-power actuating signal up to a level where it can be fed directly (as a drive) to the system dynamics. The dynamic equations describing the actuator represent the system dynamics; i.e., they may describe how the velocity of a dc motor changes with input voltage or how a diaphragm-operated control valve opens with changing diaphragm pressure. In cases such as these, the output controlled variable would be motor speed and valve stem position, re-

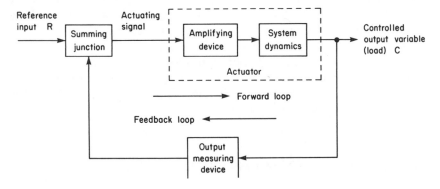

figure 1-1 the general form of the block diagram for a single input-single output feedback control system

spectively. The controlled variable, designated by the letter C, is then measured by some transducer and fed back to the input summing junction.

The block diagram of Figure 1-1 shows the general form of a single input-single output feedback control system. Of course, all control systems are not this simple and straightforward. Control systems can have multi-inputs, multi-outputs, multifeedback loops, feedforward loops, or any combination of these. Some control systems change their behavior with time, whereas others are nonlinear and difficult to analyze. However, our intention in this text is to introduce the reader to the basics of feedback control systems, and we will confine ourselves to dealing with time invariant linear systems that can be described by constant parameter, linear, differential equations of no greater than second order.

1-3 SYSTEM LINEARITY

To understand feedback systems and to be able to perform simple analysis demands a basic knowledge of the laws that govern their operation. Some of these laws and their associated relationships are covered in Chapter 2, so we will not dwell upon them at this time. It is important to realize at the outset that all real systems, regardless of the laws they obey, are nonlinear to a greater or lesser extent, although for the purposes of analysis we often assume that they are linear systems. Why do we need to be at odds with reality and is it valid to assume linearity in this way? The answer is that it is far easier to analyze an approximate linear system than a nonlinear one; by considering that the system in question operates within a restricted linear range (which is generally the case, when one is designing control systems), we can apply simple linear methods of analysis to the system. Operation within a linear working range will often be attained in a control system if the system variables are restricted in magnitude to avoid the nonlinear regions of otherwise linear components, e.g., amplifier saturation, plastic deformation of a spring, breakdown of a capacitor; these are all examples of the nonlinearities that occur in otherwise fairly linear components when their normal working ranges are exceeded.

1-4 CONTROL SYSTEM CHARACTERISTICS

Feedback system characteristics are discussed at great length in the following chapters, but as an introduction let us look, in a discursive way, at one or two advantages (and disadvantages) of applying feedback control.

1-4.1 accuracy.

The primary object of feedback is to increase output accuracy by means of comparison of the output with an input reference. However, absolute accuracy is not always achieved, and some systems, usually because of their simplicity, always exhibit an error. This error can usually be reduced to within acceptable limits, but penalties for this reduction are often incurred in the form of reduced stability.

1-4.2 stability.

If a system is unstable, its output will be uncontrolled; i.e., it will increase without bounds.

Open loop control systems can never go unstable; however, when feedback is introduced, certain control systems can become unstable if care is not exercised. Consequently, eliminating instability is the first concern of the control system designer.

1-4.3 sensitivity.

It will be shown that the introduction of feedback drastically reduces the sensitivity of a control system to system disturbances and changes occurring in fixed component values (due to aging, etc.).

1-4.4 noise.

All systems are subjected to unwanted signals. If any signal, even though it carries intelligence, is unwanted, it is considered to be noise. Feedback control can be used to reduce the effects of noise within a control system.

1-4.5 input and output impedance.

Impedance can be generally described as an impediment or restriction to the flow of energy. Although energy is necessary for the operation of a system, it is very undesirable to obtain that energy by means of the input signal because of loading considerations. Ideally the input and output impedance of a system should be infinite and zero, respectively. Feedback never achieves this ideal but often vastly improves the impedance characteristics.

1-4.6 cost.

Since the process of comparison which takes place within a feedback system requires extra components to accomplish this task, closed loop control systems will always be more expensive than open loop systems. Fortunately, we can be assured that if cost permits, then feedback control will offer a much better control system than would exist if the system were left in the open loop condition.

1-5 SUMMARY

The foregoing general comments set the stage for what will be encountered in the ensuing chapters. This book sets seven tasks for the student, which, if carried out sequentially, will provide the student with a basic understanding of feedback control systems.

1. We must reacquaint ourselves with some fundamental engineering concepts based upon the laws of physics.
2. We have to learn to describe feedback control systems in a simple mathematical form.
3. We shall examine control system responses, as a function of time, when various input disturbances are applied to the system.
4. We shall also examine the control system response, as a function of frequency, when sinusoidal input disturbances are applied.
5. We will learn to combine the knowledge of the preceding four points so that we can optimize the system performance in accordance with what is demanded of the system.
6. We will learn to design and build analog models that will simulate actual control systems.
7. Finally we will compare different techniques used to condition and process signals digitally within a control system.

basic relationships —a review

2

2-1 INTRODUCTION

Before designing a control system we must construct a mathematical model of the process to be controlled. For the types of electrical, mechanical, and chemical processes considered in this text we can construct such mathematical models quite simply from a consideration of the physical laws and basic relationships governing the behavior of the process. It is assumed that the reader has previously confronted most of these laws and subsequent equations, and that the brief review in this chapter together with the references at the end of the text is all that will be necessary to consolidate these ideas and refresh the memory.

In general, this text is only concerned with first- or second-order, linear, time-invariant processes and control systems—that is to say, only with those systems that can be described by a first- or second-order differential equation and with equation parameters that are unchanging as time progresses. In this chapter we will examine only the more important laws and relationships of some mechanical, thermal, hydraulic,

9

and electrical systems. Most of the symbols used are in preferred SI [International System of Units] form; however, reserve SI symbols are used when the preferred symbol is already being used with a different significance.

2-2 MECHANICS

2-2.1 equations of motion

Linear Motion: Let us first consider the dynamics of a body moving with a varying velocity, v. If the body travels an arbitrary distance x in a given time, which we shall call t, the average velocity of the body can be simply expressed by the equation

$$v = \frac{x}{t}$$

If, while the body is traveling, the movement of the body could be examined over a very small period of time, Δt, we would see that the velocity of the body has little opportunity to change within this period and the equation

$$v = \frac{\Delta x}{\Delta t}$$

is a very close approximation of the instantaneous velocity of the body when moving over the small distance Δx. As we continue to reduce the size of the period we are examining, we find that as Δt approaches zero the above equation becomes an exact expression of the instantaneous velocity of the body at any instant of time, say t'. The mathematical way of stating this is to write

$$v(t') = \lim_{\Delta t \to 0} \frac{\Delta x}{\Delta t} \bigg|_{t=t'}$$

More usually this expression is written thus for convenience:

$$v(t) = \frac{dx(t)}{dt} \qquad (2.1)$$

which says that both v and x are functions of time t and that v is a measure of the instantaneous rate of change of the distance x with respect to time t; i.e., v is the first derivative of distance with respect to time.

Now consider a change of velocity with respect to time; as time passes, velocity increases or decreases. This is the case when a car speeds up or slows down, and on these occasions the vehicle is said to accelerate or decelerate. The motion can then be described by the relationship

$$a(t) = \frac{dv(t)}{dt} \qquad (2.2)$$

where $a(t)$ is the acceleration and is a function of time.

Equation (2.1) can be substituted into equation (2.2) to obtain

$$a(t) = \frac{d\left(\dfrac{dx(t)}{dt}\right)}{dt}$$

or

$$a(t) = \frac{d^2x(t)}{dt^2} \qquad (2.3)$$

which is the second derivative of distance with respect to time.

Rotary Motion: Rotational motion can be explained in precisely the same terms as motion in a straight line. For instance, if θ is considered to be angular distance in radians, then angular velocity ω is described by

$$\omega(t) = \frac{d\theta(t)}{dt} \qquad (2.4)$$

and angular acceleration α is described by

$$\alpha(t) = \frac{d\omega(t)}{dt} \qquad (2.5)$$

or

$$\alpha(t) = \frac{d^2\theta(t)}{dt^2} \qquad (2.6)$$

The symbol t in parentheses again indicates that θ, ω, and α are functions of time. Symbols θ, ω, and α are analogous to x, v, and a described in the previous section.

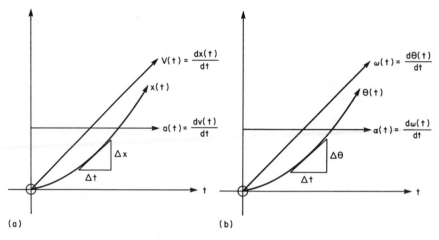

figure 2-1 linear and angular motion of a moving body

These relationships can also be shown graphically, as illustrated in Figure 2-1.

EXAMPLE 2.1

A body moves from rest a distance x such that

$$x(t) = kt^2$$

where k is a constant valued at 1/2 m/s².

(a)

Plot a graph of $x(t)$, $\dfrac{dx(t)}{dt}$, and $\dfrac{d^2x(t)}{dt^2}$.

(b) Determine from the graph
    (1) The acceleration of the body.
    (2) The velocity of the body at $t = 2$ seconds.
    (3) The distance moved after 4 seconds.

solution

(a)

Since
$$x(t) = kt^2$$

then
$$x(t) = \tfrac{1}{2}t^2$$

and
$$\frac{dx(t)}{dt} = t$$

$$\frac{d^2x(t)}{dt^2} = 1$$

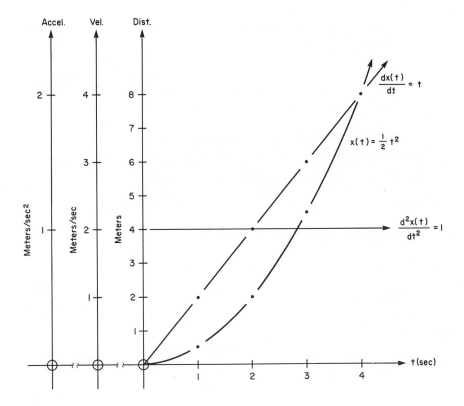

(b)

(1) From the graph the acceleration is found to be a constant of 1 m/sec².

(2) At $t = 2$ seconds the graph indicates the velocity to be 2 m/sec or

since
$$\frac{dx(t)}{dt} = t$$

then
$$\frac{dx(t)}{dt} = 2 \text{ m/sec}$$

(3) From the graph the distance moved after 4 seconds 8 metres, or

since

$$x(t) = \tfrac{1}{2}t^2$$

then

$$x(t) = (\tfrac{1}{2})(4)^2 = 8 \text{ metres}$$

EXAMPLE 2.2

A body is rotating such that its angular displacement in radians at any time can be described by

$$\theta(t) = 3t^2 + 2t$$

Calculate

(a) The angular displacement when $t = 4$ seconds.

(b) The angular velocity in rad/s and rpm when $t = 10$ seconds.

(c) The angular acceleration at $t = 12$ seconds.

solution

(a) Since the displacement is given by

$$\theta(t) = 3t^2 + 2t$$

then when $t = 4$ s

$$\theta = (3)(4)^2 + (2)(4)$$
$$\theta = (48) + (8)$$
$$\theta = 56 \text{ radians}$$

(b) The angular velocity is given by the equation

$$\omega(t) = \frac{d\theta(t)}{dt}$$

therefore $\quad \omega(t) = 6t + 2$

when $\quad\quad\quad t = 10$

$$\omega = (6)(10) + (2) = 62 \text{ rad/sec}$$

or

$$\omega = \frac{62 \times 60}{2\pi} = 592 \text{ rpm}$$

(c) Angular acceleration is given by the equations

$$\alpha(t) = \frac{d^2\theta(t)}{dt^2} = \frac{d\omega(t)}{dt}$$

Therefore, $\qquad \alpha(t) = \frac{d}{dt}(6t + 2) = 6$

Since 6 is a constant, then the angular acceleration is the same at any value of t, namely 6 rad/s^2.

2-2.2 dynamics of motion

Force and Mass: It is now convenient to introduce the concept of mass into these ideas concerning motion, and to do this we make use of Newton's second law of motion, which states, in its most simple form, that the rate of change of momentum of a body is proportional to all the forces acting upon the body.

Because momentum is defined as the product of mass and velocity, this law may be expressed mathematically by the equation

$$F(t) = K \frac{d}{dt}[m \times v(t)]$$

where F is the net force acting on the body and may or may not be constant with time, m is the mass of the body, and K is a constant of proportionality. Because the mass m is constant (at least at speeds much less than the speed of light) the equation may be rewritten thus:

$$F(t) = K m \frac{dv(t)}{dt} \qquad (2.7)$$

which from equation (2.2) can be reduced to

$$F(t) = K m \, a(t)$$

If SI units are used such that F is expressed in newtons, m in kilograms, and a in metres per second per second, then the value of K is equivalent to unity and our equation becomes

$$F(t) = m \, a(t) \qquad (2.8)$$

or from equation (2.3)

$$F(t) = m \frac{d^2x(t)}{dt^2} \qquad (2.9)$$

This law is very easily observed in everyday life; for instance, when a car accelerates, the mass of our body is "forced" back into the car seat, and when the brake is applied, we are "forced" forward onto the seat belt. In fact, many seat belts will lock into position when the car makes a sudden stop because the seat belt's locking device is "forced" forward due to the sudden deceleration. Actually, the forward motion is only relative to the car body; what is really happening is that all the loose masses in the quickly decelerating car are trying to continue on their original path at their original velocity.

Torque and Inertia: In Section 2-2.1 we said that the equations of rotational motion were analogous to the equations of motion in a straight line; in the same fashion we can say that the dynamics of a rotating mass are analogous to the dynamics of a mass moving in a straight line. However, before considering the dynamics of a rotating body we must introduce another concept, the concept of inertia, more properly known as the moment of inertia when one is talking of rotating bodies.

When we speak of inertia we are usually referring to the inertial mass of an object. If we say that a body has a large amount of inertia, then we mean that it is difficult to start the body moving or to accelerate movement of the body and that once the body is moving it is equally difficult to bring the body to rest again. When we talk of inertia in this way, the body could be moving in a straight line, could be rotating on a shaft, or could be a combination of both motions, as in the wheel of a car or bicycle.

However, when we speak of the moment of inertia of a rotating body, we refer to a specific property of the body and its manner of rotation. The moment of inertia is a measure of the body's ability to resist any attempt to change its speed of rotation and is dependent not only upon the mass of the body but also upon how that mass is distributed about the axis around which the body rotates. This property is illustrated in Figure 2-2, where Jxx, Jyy, and Jzz are the moments of inertia of a flat disc, of mass m, about the axes xx, yy, and zz, respectively. Jxx is referred to as the *polar moment of inertia* and is quite different in magnitude from Jyy and Jzz.

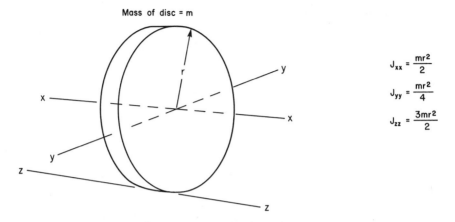

figure 2-2 moments of inertia of a flat disc about three perpendicular axes

The rotary force required to start or stop a body spinning is called torque, and Newton's second law of motion when applied to rotating bodies can be simply stated thus: The rate of change of angular momentum of a body is directly proportional to the torque acting on the body. Because the angular momentum of a body is defined as the product of its moment of inertia and its angular velocity, the law may be expressed thus:

$$T(t) = K\frac{d}{dt}[J\omega(t)]$$

where T is the net torque acting on the body and may vary with time or have a constant value, J is the moment of inertia of the body about the rotating axis, and ω is the angular velocity of the body. By analogy with equations (2.7) to (2.9) describing the dynamics of linear motion we may rewrite the above expression in the following ways:

$$T(t) = J\alpha(t) \tag{2.10}$$

$$T(t) = J\frac{d\omega(t)}{dt} \tag{2.11}$$

$$T(t) = J\frac{d^2\theta(t)}{dt^2} \tag{2.12}$$

where T has the units of newton-meters, J is expressed in kilogram-(metres)^2, α, the angular acceleration, has the units radians per second per second, ω is in radians per second, and the angular displacement θ is expressed in radians.

EXAMPLE 2.3

The moment of inertia of a motor rotor is 3.3×10^{-7} kg-m² and the starting torque is 1.076×10^{-2} Nm. What is the acceleration in rad/s² at the instant of starting?

solution

Since $$T = J\alpha$$

then $$\alpha = \frac{T}{J}$$

$$\alpha = \frac{1.076 \times 10^{-2}}{3.3 \times 10^{-7}}$$

$$\alpha = 32,600 \text{ rad/s}^2$$

2-2.3 linear springs

In the seventeenth century during Sir Isaac Newton's lifetime, another Englishman, Robert Hooke, performed many experiments with springs and discovered what is now called Hooke's Law, which states that over a particular linear range the distortion of a body is directly proportional to the distorting force. This law as applied to a spring is expressed as

$$F_s = K_s x \qquad (2.13)$$

where F_s is the force on the spring in newtons, x is the extension or compression distance in metres, and K_s is the spring proportionality constant in newtons per meter. Figure 2-3 illustrates such a spring together with graphs showing the relationship between the applied force and the spring distortion and the resulting restoration force exerted by the spring.

Furthermore, if a force F_s is used to extend or compress a linear spring a distance x, the spring will then be exerting a force F_K, equal in magnitude to F_s, but directly opposing F_s. This example illustrates the principle of Newton's third law of motion, which states that action and reaction are always equal and opposing. Hence,

$$F_K = -F_s$$

and from equation (2.13) we have

$$F_K = -K_s x \qquad (2.14)$$

This equation is illustrated in Figure 2.3(c).

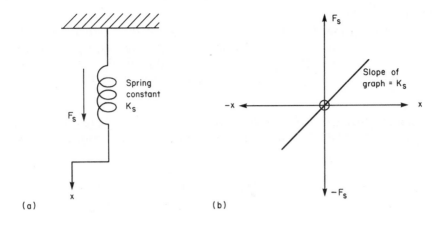

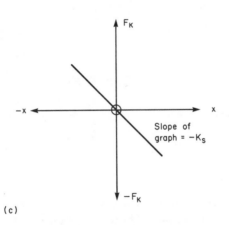

figure 2-3 force versus distortion relationships for a linear spring

If a spring is twisted rather than extended or compressed the angular distortion is similarly proportional to the applied torque. The spring also supplies a restoring torque and can be expressed as

$$T_K = -K_s'\theta \qquad\qquad (2.15)$$

where T_K is the restoring torque in newton-metres, θ is the angular distortion in radians, and K_s' is the torsional proportionality constant of the spring in newton-metres/rad.

Both K_s and K_s' are often referred to as the spring stiffness constants even though they are expressed in different units.

EXAMPLE 2.4

A linear spring deflects through an angle of 0.15 radian when 20 Nm of torque is applied to it. What is the restoring torque if the angle of deflection is 0.24 radian? Also, what is the angular stiffness constant of the spring measured in newton-metres per radian?

solution

Since restoring torque will always be equal and opposite to the applied torque over the linear range of the spring, then this restoring torque T_K is given by

$$T_K = -K_s\theta$$

Therefore,

$$T_{K_1} = -K_s\theta_1$$

and

$$T_{K_2} = -K_s\theta_2$$

so that

$$\frac{T_{K_1}}{T_{K_2}} = \frac{\theta_1}{\theta_2}$$

or

$$T_{K_2} = \frac{T_{K_1}\theta_2}{\theta_1} = \frac{20 \times 0.24}{0.15}$$

which yields

$$T_{K_2} = 32 \text{ Nm}$$

Angular stiffness constant,

$$K_s = \frac{T_{K_1}}{\theta_1} = 133.333 \text{ Nm/rad}$$

2-2.4 friction

Friction is a force that always opposes the relative motion between any two bodies that are in contact. There are essentially three kinds of friction: static, coulomb, and viscous. The first two are nonlinear and difficult to describe mathematically because they depend upon many varying factors and characteristics of the two bodies in contact. Viscous friction, however, is reasonably linear over a particular range and is used with great effect to dampen various kinds of motion both oscillating and nonoscillating. Since designers go to a great deal of trouble to get rid of the first two and often introduce large amounts of the third type, we shall examine only the effects of viscous friction.

Consider an oil dashpot which is essentially a piston in a cylinder which contains oil and is shown diagramatically in Figure 2-4. When the piston moves in either direction, oil is forced either around the piston or through an orifice in the piston in the opposite direction to the motion. When this occurs the oil molecules rub on one another, and, de-

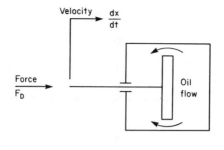

(a)

(b)

figure 2-4 (a) oil dashpot, (b) force/velocity characteristic of an oil dashpot

pending upon the oil's viscosity and the size of the orifice, it will flow slowly or quickly through the orifice. It is obvious that as the piston "flows" down the cylinder in one direction the volume of oil is displaced in the opposite direction. The oil and hence the piston flow at a rate proportional to the force applied to the piston. This relationship is expressed by

$$F_D = B \frac{dx(t)}{dt} \qquad (2.16)$$

where F_D is the applied force to the damper and is expressed in newtons, $dx(t)/dt$ is the piston velocity in metres/second, and B is the damper proportionality constant with units newtons/(metres/second).

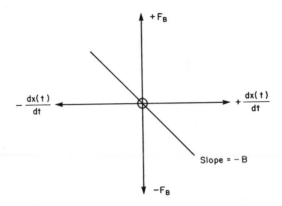

figure 2-5 restraining force/velocity characteristic of a viscous damper

From Newton's third law of motion the restraining force exerted by the damper is equal and opposite to the applied force, as shown in Figure 2-5, and this relationship is expressed by the equation

$$F_B = -B \frac{dx(t)}{dt} \tag{2.17}$$

EXAMPLE 2.5

At what steady velocity does the piston of a viscous damper move if the force applied is 3 N and the damper proportionality constant is 10 Ns/cm?

solution

If the piston is moving at a steady velocity then the viscous reaction is equal to the applied force

$$F_D = B \frac{dx(t)}{dt}$$

$$B = 10 \text{ Ns/cm}$$

or

$$B = 1000 \text{ Ns/m}$$

Since

$$\frac{dx(t)}{dt} = \frac{F_D}{B}$$

then

$$\frac{dx(t)}{dt} = \frac{3}{10^3} = 0.003 \text{ m/s}$$

or

$$\text{Velocity} = 3 \text{ mm/s}$$

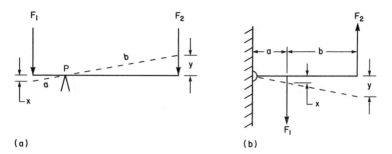

figure 2-6 (a) simple first-order lever pivoted at P, (b) simple second-order lever

2-2.5 levers and gears

Levers: Figure 2-6(a) shows a simple first-order lever and assumes a frictionless pivot or fulcrum point at P. A mass m_1 with weight F_1 acts on the lever at a distance a from P. By definition the counterclockwise moment of force acting about the point P is $F_1 \times a$. In order to balance the lever a clockwise moment of equal magnitude must be produced. If a downward force F_2 is applied to the lever, at a distance b from P on the opposite side of the fulcrum, the moments about P will be balanced if

$$F_2 b = F_1 a \qquad (2.18)$$

That is,

$$F_2 = \frac{F_1 a}{b} \qquad (2.18a)$$

From an inspection of Figure 2.6(a) we can see that if the left-hand side of the lever is moved downwards a distance x, then the right-hand side of the lever is raised a distance y, where

$$y = \frac{b}{a} x \qquad (2.19)$$

By differentiating this equation we also see that if the weight F_1 moves downwards with velocity x then the force F_2 is moving upwards with a velocity $\dot{y}$, such that

$$\frac{\ddot{y}}{\ddot{x}} = \frac{\dot{y}}{\dot{x}} = \frac{y}{x} = \frac{b}{a} \qquad (2.19a)$$

where
$$\dot{x} = \frac{dx}{dt}$$

and
$$\ddot{x} = \frac{d^2x}{dt^2}$$

Multiplying y by F_2 in equation (2.19), we obtain

$$yF_2 = x\frac{F_2 b}{a}$$

which from equation (2.18) yields

$$yF_2 = xF_1 \tag{2.20}$$

The term on the right-hand side of equation (2.20) will be seen to be the amount of work or energy required to lift the mass of weight F_1 newtons a distance of x metres, whereas the term on the left-hand side of the equation denotes the amount of work expended on the lever when a force of F_2 newtons moves the lever downwards a distance of y metres. Because these two terms must be equal in order to balance the system, we see that "work in" (effort supplied by F_2) equals "work out" (lifting the weight F_1). The example used in deriving this statement provides a simple demonstration of the principle of the conservation of energy, which states that energy cannot be destroyed but only converted into another form. Hence, if the fulcrum cannot be assumed to be frictionless, a certain amount of the mechanical effort would be dissipated as heat in the bearing and F_2 would have to be slightly greater than the value implied by equation (2.18a) in order to supply the extra effort needed to overcome this friction.

The second-order lever shown in Figure 2-6(b) is a lever pivoted at one end rather than somewhere under the beam, but once again the moments of force about the fulcrum must be equal in order to balance the lever; i.e.,

$$F_1 a = F_2(a + b) \tag{2.21}$$

$$\frac{F_2}{F_1} = \frac{a}{a + b} \tag{2.21a}$$

Also, by reasoning similar to that used with the first-order lever example, the ratios of displacement, velocity, and acceleration are given by

$$\frac{\ddot{y}}{\ddot{x}} = \frac{\dot{y}}{\dot{x}} = \frac{y}{x} = \frac{a + b}{a} \tag{2.22}$$

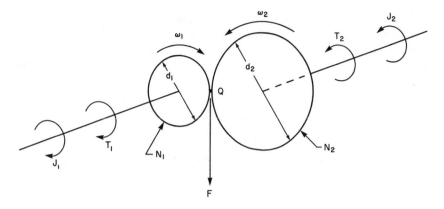

figure 2-7   simple spur gear system

Gears: The equations describing the dynamics of a gear train are very closely related to the equations we have just examined in the section dealing with levers. Consider Figure 2-7, in which the small gear of pitch diameter d_1 is driving a larger gear of pitch diameter d_2.

An external torque T_1 is applied to the small gear, causing it to rotate with angular velocity ω_1. The gear exerts a downward force F at point Q, on the pitch circumference of the larger gear. By definition

$$T_1 = F\frac{d_1}{2}$$

hence

$$F = \frac{2T_1}{d_1}$$

Similarly, the torque T_2 exerted on the larger gear will be

$$T_2 = F\frac{d_2}{2}$$

Substituting for F in this equation yields

$$T_2 = T_1\frac{d_2}{d_1} \tag{2.23}$$

For proper meshing of the gears to take place, all gear teeth must be the same size. It can, therefore, be seen that the number of teeth on each gear wheel is proportional to the pitch diameter of the relevant gear. The gear ratio is therefore defined as the ratio of the number of teeth on the second gear to the number of teeth on the first gear or more simply the ratio of gear teeth pitch diameters, which leads to the relationship

$$\frac{d_2}{d_1} = \frac{N_2}{N_1}$$

Therefore, equation (2.23) may be rewritten thus:

$$\frac{T_2}{T_1} = \frac{d_2}{d_1} = \frac{N_2}{N_1} \tag{2.24}$$

where N_1 is the number of teeth on the small gear and N_2 the number of teeth on the large gear.

Let us now examine how the gear ratio affects the relative speed of the two gears. One revolution of the small gear will cause only N_1 of the N_2 teeth of the large gear to pass the point Q. Thus, if the first gear is rotated by an angle θ_1, then the second gear will be rotated by an angle θ_2, where

$$\theta_2 = \frac{N_1 \theta_1}{N_2}$$

Successive differentiation of both sides of this expression yields the following equation:

$$\frac{\alpha_2}{\alpha_1} = \frac{\omega_2}{\omega_1} = \frac{\theta_2}{\theta_1} = \frac{N_1}{N_2} \tag{2.25}$$

where α_1 and α_2 are the angular accelerations of the first and second gears, respectively. Thus the ratio of angular acceleration or velocity or rotation of one gear to the other is inversely proportional to the ratio of their number of gear teeth.

If the output torque T_2 of the second gear wheel is used to rotate a mass with inertial load J_2, then from equation (2.10) the mass will accelerate such that

$$\alpha_2 = \frac{T_2}{J_2}$$

Also from equation (2.25) the acceleration of the first gear will be

$$\alpha_1 = \alpha_2 \frac{N_2}{N_1} = \frac{T_2}{J_2} \cdot \frac{N_2}{N_1} \qquad (2.26)$$

Because the driving force is supplying a torque T_1 to the first gear, the apparent inertial load seen by the driving force is J_1, where, from equation (2.10),

$$J_1 = \frac{T_1}{\alpha_1}$$

Substituting equation (2.26) into this expression yields

$$J_1 = J_2 \cdot \frac{T_1}{T_2} \cdot \frac{N_1}{N_2}$$

which from equation (2.24) results in the expression

$$J_1 = J_2 \left(\frac{N_1}{N_2}\right)^2 \qquad (2.27)$$

Thus by using a step-down gear train, the apparent inertia seen by the driving force is reduced by a factor equal to the square of the gear ratio.

EXAMPLE 2.6

A rotating load has a polar moment of inertia of 250 kg-m², and we desire to drive this load using a motor whose maximum torque capability is 159.25 Nm.

(a) What is the load torque if the desired load acceleration is 25 rad/s²?

(b) What is the optimum gear ratio of the drive system?

(c) What is the inertial load on the motor drive shaft? (Assume that the inertial load of the gearbox can be neglected.)

solution

(a) Now,

$$T = J\alpha$$

Therefore the load inertia T_L is given by

$$T_L = J_L \alpha = 250 \times 25 \text{ Nm}$$

or $\qquad T_L = 6250 \text{ Nm}$

(b) The motor torque $T_m = 159.25$ Nm and since the gear ratio

$$\frac{N_1}{N_2} = \frac{T_m}{T_L}$$

then $\qquad \dfrac{N_1}{N_2} = \dfrac{159.25}{6250} = \dfrac{1}{40}$

i.e., a 40:1 step down gear box.

(c) Since

$$\frac{J_m}{J_L} = \frac{J_1}{J_2} = \left(\frac{N_1}{N_2}\right)^2$$

then $\qquad J_m = \dfrac{250}{40^2} = 0.1563 \text{ kg-m}^2$

2.3 THERMODYNAMICS

Heat transfer is a complex topic, but there are a few relationships that are very useful in analyzing simple thermal systems.

For example, by considering one-dimensional heat conduction through a homogenous wall of uniform section, we can illustrate the basic heat transfer equations. However, even though these equations describe only a simple example they are often used to obtain approximate solutions to more complex heat transfer problems.

If k is defined as the thermal conductivity of a material expressed in the dimensions of the rate of heat flowing through a sample of the material having a unity cross-sectional area and with a unity temperature gradient in the direction of the heat flow, then, from Figure 2-8,

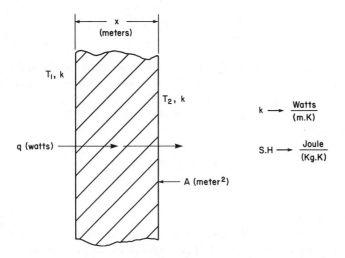

figure 2-8 heat conduction through a wall

$$q = kA \frac{\Delta T}{x} \text{ watts} \qquad (2.28)$$

where q is the heat flow rate, A is the area of the wall, x is the wall thickness, and ΔT is the temperature drop across the wall; i.e., $T_1 - T_2$, where T_1 and T_2 are in degrees Kelvin.

Let
$$R = \frac{x}{kA} \text{ (K/watts)} \qquad (2.29)$$

then
$$q = \frac{\Delta T}{R} \text{ watts} \qquad (2.30)$$

which is Ohm's law of heat transfer. R is termed the thermal resistance of the material, and the reciprocal of R is termed the thermal conductance of the material.

As heat flows into a body (solid or fluid) the temperature of the body rises due to the heat being stored in the body. Now, by definition, the amount of heat required to raise a body by T degrees Kelvin is Q, where

$$Q = M \times \text{S.H.} \times T \text{ joules} \qquad (2.31)$$

where M is the mass of the body and S.H. the coefficient of specific heat of the body material. By differentiating equation (2.31), we obtain

$$q = \frac{dQ}{dt} = M \times \text{S.H.} \times \frac{dT}{dt} \text{ watts} \qquad (2.32)$$

Thus we see that the rate of change of temperature of the body varies directly as q, the rate of heat flow into the body, and varies indirectly as the magnitude of the product $M \times \text{S.H.}$ This product is termed the thermal capacitance of the body, C, and equation (2.32) is more often written thus:

$$q = C \frac{dT}{dt} \text{ watts} \qquad (2.33)$$

EXAMPLE 2.7a

An electrically heated hot water tank, measuring 2 m × 1 m × 1 m is used to store water at 65°C. The heating element is rated at 15 kW, and the tank is covered on all sides with 8 cm of insulation having a thermal conductivity $k = 0.42$ W/m²/(°K/m). If the surface temperature of the insulation is 25°C, what is the rate of heat loss from the tank? (Assume that heat is lost uniformly from all sides of the insulation and that the effects of ends and corners can be neglected.)

As we are neglecting corner and end effects, the total surface area of the tank is A, where

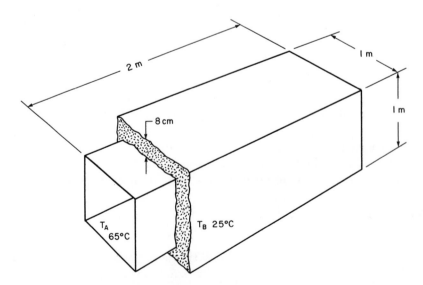

sketch of the insulated hot water tank

$$A = 4 \times 2 \times 1 + 2 \times 1^2$$

i.e.,
$$A = 10 \text{ m}^2$$

The thickness of the insulation x is 0.08 m; therefore, the rate of heat loss q through the insulation is given by

$$q = \frac{kA}{x}(T_A - T_B) \text{ watts}$$

or
$$q = \frac{0.42 \times 10 \times (65 - 25)}{0.08} \qquad \text{(i)}$$

$$= 2100 \text{ watts}$$

EXAMPLE 2.7b

If the tank contains 1900 kg of water when heat to the tank is turned off, how long will it take for the water to cool to 40°C? (Assume that tank temperature and water temperature are the same.)

When the heat is turned off, the water is 65°C and begins to lose heat according to expression (i) above; i.e., if T_A is the tank or water temperature, then the heat loss q is

$$q = \frac{0.4 \times 10 \times (T_A - 25)}{0.08}$$

It is more convenient for solving this problem to replace the term in brackets by a new variable $T_{A'}$ such that

$$T_{A'} = T_A - 25$$

∴ when $T_A = 65°C$

$$T_{A'} = 65 - 25 = 40°C$$

and when $T_A = 40°C$

$$T_{A'} = 40 - 25 = 15°C$$

Also
$$\frac{dT_{A'}}{dt} = \frac{dT_A}{dt}$$

Therefore, from the above,
$$q = 50T_{A'} \text{ watts} \qquad \text{(ii)}$$

Also, we know that heat loss is given by

$$q = -M \times \text{S.H.} \times \frac{dT_A}{dt} \quad \text{(negative because } T_A \text{ is decreasing)}$$

or $\qquad q = -M \times \text{S.H.} \times \frac{dT_{A'}}{dt}$

where M is the mass of the water (i.e., 1900 kg) and S.H. is the specific heat of water, which is 4200 joules/kg/°K. Therefore,

$$q = -1900 \times 4200 \times \frac{dT_{A'}}{dt} \qquad \text{(iii)}$$

Equating equations (ii) and (iii) yields

$$\frac{-dT_{A'}}{dt} = \frac{50}{1900 \times 4200} T_{A'} = 0.000006 T_{A'} \, °\text{K/s}$$

or $\qquad \dfrac{1}{T_{A'}} dT_{A'} = -0.000006 \, dt$

Integrating both sides of this equation gives us

$$\int \frac{dT_{A'}}{T_{A'}} = -\int \frac{6}{10^6} \, dt$$

which has the solution

$$\text{Ln} \, (T_{A'}) = -0.000006t + C' \qquad \text{(iv)}$$

where C' is a constant of integration which we can determine from our knowledge of the initial conditions; i.e., when $t = 0, T_{A'} = 40°\text{C}$; therefore from equation (iv)

$$C' = \text{ln} \, (40)$$

and equation (iv) can be rewritten thus:

$$\text{Ln} \, (T_{A'}) = -0.000006t + \text{Ln} \, (40)$$

Thus the time taken for T_A to cool to 40°C or for $T_{A'}$ to reach 15°C is given by

$$t = \frac{\text{Ln} \, (40) - \text{Ln} \, (15)}{0.000006}$$

i.e., $\qquad t = 163,471 \text{ secs}$

or $\qquad t = 45 \text{ hours}$

2.4 ELECTRICAL RELATIONSHIPS

In the year 1854, at the age of thirty, Gustav Robert Kirchhoff was appointed a professor of physics at Heidelberg University in Germany. It was at this juncture in his life that he started his rise to fame, but even before this he had researched into heat and electricity and had discovered two fundamental laws which today are called Kirchhoff's Voltage Law and Kirchhoff's Current Law. Simply stated, the voltage law says that the sum of all the voltage rises and voltage drops around a closed loop is equal to zero; the current law states that the sum of all the currents entering and leaving a node is equal to zero. Figure 2-9(a) shows a simple closed loop; if Kirchhoff's voltage is applied, we can say that

$$E - V_1 - V_2 - V_3 = 0 \qquad (2.34)$$

where all the voltage rises are positive and all the voltage drops are negative.

Figure 2.9(b) shows a circuit node where currents entering the node are considered positive and currents leaving the node are considered negative. If Kirchhoff's current law is applied, we can say

$$i_1 + i_2 - i_3 - i_4 = 0 \qquad (2.35)$$

These two laws are so important that they are the very heart of any electrical circuit analysis or design.

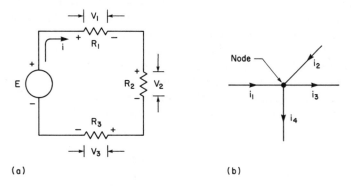

(a)   (b)

figure 2-9   (a) A closed electrical loop, (b) an electrical circuit node

figure 2-10 (a) resistance, (b) inductance, (c) capacitance

Figure 2-10 shows the symbols commonly used to describe the three major passive electric circuit components:

1. A resistor R.
2. An inductance coil L.
3. A capacitor C.

Resistance: Consider a resistance R with a current of i amps passing through it due to the voltage difference between E_1 and E_2. It can be said that the relationship between current and voltage over the linear range range of the resistor is such that

$$R = \frac{E_1 - E_2}{i} \qquad (2.36)$$

where the units of resistance are expressed in ohms (Ω).

Inductance: If we try to alter the amount of current flowing in a coil, an electromagnetic field is set up around the coil and this field causes a voltage or emf (electromotive force) to be induced across the coil in such a direction as to try to stop the current changing. This rule, which is the basis of electromagnetism, was first stated by another German, Heinrich Lenz, in 1834 and is today known as Lenz's Law. The property of the coil to resist such changes in current is termed the coil's inductance L and is measured in henrys (H). If the current flowing in the coil is i amps and the rate of change of current is di/dt amps per second, then an emf e is induced across the coil such that

$$e = -L\frac{di}{dt} \text{ volts} \qquad (2.37)$$

where the negative sign indicates that the emf opposes the change in the current i. Thus in Figure 2-10(b) if E_1 increases, thereby attempting to cause i to increase, then an emf e will be induced in the coil in the direction shown so as to oppose the increase in i. Similarly, if E_1 is reduced, then the direction of e will be reversed, thereby attempting to maintain i at its present value.

Capacitance: Both inductors and capacitors have reactance; i.e., they display an ability to resist changes in their electrical status. We have seen that an inductive coil reacts to any attempt to change the current flowing through the coil, and the coil displays a greater reactance to rapid changes of current than it does to slow changes in the current flow. The reactance displayed by a capacitor, however, is quite different. If we change the voltage across a capacitor rapidly, we cause a current to flow through the capacitor. On increasing the rate of change of voltage across the capacitor, we would find that the magnitude of the current flowing through the capacitor increases. Conversely, if capacitor voltage is changed gradually, then hardly any current will flow. Thus in the case of a capacitor the reactance of the capacitor is such as to oppose the passage of low frequency current. The current i flowing through the capacitor varies directly as the rate of change of the voltage V across the capacitor; thus

$$i = C\frac{dV}{dt} \text{ amps} \qquad (2.38)$$

where $V = E_1 - E_2$ in Figure 2-10(c) and C is termed the capacitance of the capacitor and is expressed in farads or, as is usually the case, microfarads (μF) or picofarads (pF).

EXAMPLE 2.8a
If the voltage across a capacitor changes by 2000 V in 40 ms when a current of 50 mA passes into it, what is the value of the capacitor?

Since
$$I = \frac{CV}{t}$$

then
$$C = \frac{It}{V} = \frac{50}{10^3} \times \frac{40}{10^3} \times \frac{1}{2000}$$

or
$$C = 1\ \mu F$$

EXAMPLE 2.8b

If 2000 volts is generated across a pure inductance of 500 mH, what is the rate of change of current through the inductance?

solution

Since
$$E = -L\frac{di}{dt}$$

then
$$\frac{di}{dt} = \frac{-E}{L}$$

$$= 2000 \times \frac{10^3}{500}$$

or
$$= 4000 \text{ A/s}$$

2.5 HYDRAULICS

The topic of hydraulics is concerned with using fluids to do work. The science of hydraulics is very old, and, therefore, examples are numerous. Reservoirs and irrigation schemes were used in early Mesopotamia, Egypt, China, and South America. Lake Moesis was constructed over 4000 years ago to contain the flood waters of the River Nile. Waterwheels, clepsydras (water clocks), and rotary lift pumps were used by the ancient Greeks, while hot water central heating was an early convenience in many of the villas of imperial Rome.

In the fifteenth century Leonardo da Vinci commenced the study of hydraulics as we understand it today when he composed an amazing treatise on the flow of water and flow measurement through various restrictions such as orifices and weirs. This work was continued by Torricelli in the seventeenth century, but it was not until the nineteenth century and the industrial age that the study and application of hydraulics began in earnest.

When we think of hydraulics today, we probably think of such examples as the hydraulic hoist used for lifting automobiles, or the hydraulic rams used to drive the digging tool on a backhoe. Hydroelectric power stations and electricity pump storage schemes are also familiar examples in our modern world, and an example of hydraulics that we may see often in the near future is the coastal wave power generating station, where the energy of the waves is converted into electrical energy.

In all these examples, although it may not be immediately obvious, energy that has been stored in the fluid is given up and converted into work. In Section 2-2.5 we spoke about how energy cannot be created or destroyed but is converted from one type of energy to another. In 1738, Daniel Bernouilli in his thesis *Hydrodynamica* proposed perhaps the most famous law in hydraulics, namely, "In a steady stream of fluid in which there is no loss by friction or other causes, the sum of the potential energy, the compressive (or stored) energy, and the kinetic energy is constant at all points along the stream." Thus in Figure 2-11, a fluid of density ρ kg/m³ is used to propel a hydraulic hoist, where a force of F newtons is applied at point E to raise a mass of M kg sitting on the hoist at point A. If it is assumed that we can neglect friction in the system, then if the piston at E moves downwards with a velocity of V_E m/s, thereby lifting the hoist at a velocity V_a m/s, we can invoke Bernouilli's expression and state that at any point in the system the sum

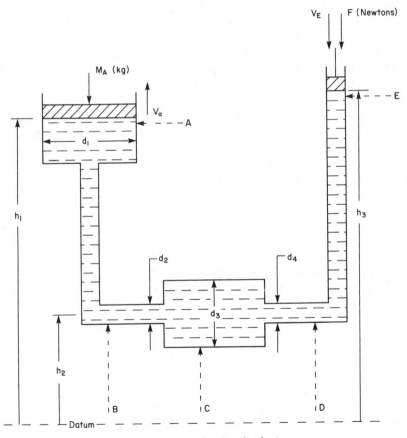

figure 2-11 a hydraulic hoist

$$gh + \frac{P}{\rho} + \frac{V^2}{2} \qquad (2.39)$$

is constant, where P is the fluid pressure in pascals, and V is the fluid velocity in m/s, at the point being considered; g is the gravitational constant.

Examining expression (2.39), we see that if we consider a small amount of the fluid of mass M, at position D, for example, so that we can substitute values for h, P, and V into expression (2.39) and we multiply this equation by M, we obtain

$$gMh + \frac{MP}{\rho} + \frac{MV^2}{2} \qquad (2.40)$$

The first term of expression (2.40) will be recognized as the potential energy of a mass M; the second term is the energy stored in a fluid of volume M/ρ m³ under a pressure of P pascals, and the third term is the well-known expression describing the kinetic energy of a mass M moving with a velocity V. Thus as h, P, and V change as the fluid moves along the pipe, we see from expression (2.40) that the energy within the system is being converted from one form to another.

The velocity of the fluid as it passes through various sections of the pipe will obviously change as the diameter of the pipe changes. As the area of the pipe reduces, fluid velocity must increase to maintain a constant flow rate. Fluid velocity varies inversely as the cross-sectional area of the pipe; hence we can say

$$\frac{V_E}{V_A} = \frac{\pi d_1^2}{4} \cdot \frac{4}{\pi d_4^2} = \left(\frac{d_1}{d_4}\right)^2 \qquad (2.41)$$

Similar expressions can be derived for the velocities V_B and V_C. The pressure at point E, P_E, is determined by the force F acting on the fluid over the full area of the pipe such that

$$P_E = \frac{4F}{\pi d_4^2} + P_a \qquad (2.42)$$

where P_a is the atmospheric pressure outside the system.

Static Behavior: Under static conditions when the piston force F_s just supports the mass M on the hoist, the pressure through the system is determined from equation (2.42) and expression (2.39) when V equals zero. Thus

$$gh + \frac{P}{\rho} = gh_3 + \frac{1}{\rho}\left(\frac{4F}{\pi d_4^2} + P_a\right) \qquad (2.43)$$

Therefore at point A the fluid pressure is given by

$$P_A = \rho g(h_3 - h_1) + \frac{4F}{\pi d_4^2} + P_a \qquad (2.44)$$

and the net force acting upwards on the hoist is $P_A - P_a$. Hence the mass that can be supported on the hoist is given by the product of the hoist area and fluid pressure:

$$M_A \text{ (static)} = \frac{\pi d_1^2}{4g} \cdot \left[\rho g(h_3 - h_1) + \frac{4F}{\pi d_4^2}\right] \text{ kg} \qquad (2.45)$$

If points A and E are of the same height, then

$$M_A \text{ (static)} = \frac{F}{g} \cdot \left(\frac{d_1}{d_4}\right)^2 \text{ kg} \qquad (2.46)$$

It can be seen that by making d_1 much larger than d_4 we can obtain a large mechanical advantage to lift heavy objects by the use of only modest forces.

Dynamic Behavior: When, however, the piston is moving downwards with a velocity V_E, the fluid is forced through the vertical section of the pipe at the same velocity and at velocities V_C, V_B, and V_A at other sections of the pipe according to equation (2.41) such that:

$$gh_3 + \frac{P_E}{\rho} + \frac{V_E^2}{2} = gh_2 + \frac{P_D}{\rho} + \frac{VD^2}{2}$$

$$= gh_2 + \frac{P_C}{\rho} + \frac{V_C^2}{2}$$

$$= gh_2 + \frac{P_B}{\rho} + \frac{V_B^2}{2}$$

$$= gh_1 + \frac{P_A}{\rho} + \frac{V_A^2}{2} \qquad (2.47)$$

Suppose now that h_1 is equal to h_3; then

$$\frac{P_E}{\rho} + \frac{V_E^2}{2} = \frac{P_A}{\rho} + \frac{V_A^2}{2} \qquad (2.48)$$

Substituting equation (2.41) into equation (2.48) and rearranging, we obtain

$$P_A = P_E - \frac{V_E^2\rho}{2}\left[\frac{d_1^4 - d_4^4}{d_1^4}\right] \qquad (2.49)$$

From equation (2.49) we see that when the piston is moving the fluid pressure at P_A can never equal the applied piston pressure, and the faster the piston moves, the greater the pressure loss becomes at point A. This pressure loss is, of course, because part of the applied energy has been converted into the kinetic energy of the fluid. The pressure reduction at the hoist means that the maximum load that can be lifted by the hoist is determined not only by the fluid pressure but also by the velocity of the hoist and hence the fluid through the pipes, so that

$$M_A \text{ (max)} = \frac{\pi d_1^2}{4g} \cdot (P_A - P_a) \text{ kg} \qquad (2.50)$$

or $$M_A \text{ (max)} = \frac{F_s}{g} \cdot \left(\frac{d_1}{d_4}\right)^2 - \frac{V_E^2\rho\pi}{8g}\left[\frac{d_1^4 - d_4^4}{d_1^2}\right] \text{ kg} \qquad (2.50)$$

EXAMPLE 2.9

In the hydraulic hoist shown in Figure 2-11, F is a force of 100 N moving at a velocity of 15 m/s. (i) What is the pressure at point C? (ii) What is the maximum mass that the hoist will lift?

Where: atmospheric pressure $= 10^5$ pascals

$$g = 9.81 \text{ m/s}^2$$

$$d_4 = 3 \text{ cm}$$

$$d_3 = 5 \text{ cm}$$

$$d_1 = 20 \text{ cm}$$

$$h_1 = h_3 = 2 \text{ m}$$

$$h_2 = 1 \text{ m}$$

neglect the effects of pipe friction. Fluid used is water; i.e., $\rho = 1000$ kg/m³.

Case (i)
The pressure exerted on the fluid at point E is

$$P_E = \frac{4F}{\pi d_4^2} + 10^5 \text{ pascals}$$

$$= \frac{4 \times 100}{\pi \times 0.03^2} + 10^5$$

$$= 141,471 + 10^5$$

$$= 241,471 \text{ pascals} \qquad (A)$$

We know that

$$gh_3 + \frac{P_E}{\rho} + \frac{V_E^2}{2} = gh_2 + \frac{P_C}{\rho} + \frac{V_C^2}{2} \qquad (B)$$

and that

$$\frac{V_C}{V_E} = \left(\frac{d_4}{d_3}\right)^2$$

$$V_C = 15 \times \left(\frac{3}{5}\right)^2$$

$$V_C = 5.4 \text{ m/s} \qquad (C)$$

Therefore, from equations (B) and (C)

$$P_C = \left[9.81(2-1) + \frac{241,471}{1000} + \frac{15^2}{2} - \frac{5.4^2}{2}\right] \times 1000$$

$$P_C = 349,201 \text{ pascals} \qquad (D)$$

Case (ii)

$$V_A = V_E \cdot \left(\frac{d_4}{d_1}\right)^2$$

i.e. $\qquad V_A = 0.338 \text{ m/s} \qquad (E)$

From equations (E) and (B)

$$P_A = \left[9.81(2-2) + \frac{241,471}{1000} + \frac{15^2}{2} - \frac{0.338^2}{2}\right] \times 1000$$

$$P_A = 353,914 \text{ pascals} \qquad (F)$$

the net upward force on the hoist is given by F_A, where

$$F_A = \frac{\pi d_1^2}{4} \times (P_A - P_a) \text{ N}$$

Therefore, the maximum mass that can be lifted by hoist is

$$M_A \text{ (max)} = \frac{\pi d_1^2}{4g} \times (P_A - P_a) \text{ N}$$

$$= \frac{\pi \times 0.2^2}{4 \times 9.81} \times (353{,}914 - 100{,}000)$$

$$M_A \text{ (max)} = 813 \text{ kg} \qquad\qquad (G)$$

EXAMPLE 2.10

A fluid flows steadily at 20 m/s through a 3.568-cm diameter pipe. The pressure at the inlet to the pipe is 20 P_a and at the outlet is 14 P_a. What is the restriction resistance of the pipe expressed in $P_a/(m^3/s)$?

solution

The cross-sectional area $\quad A = \dfrac{\pi d^2}{4}$

or $\qquad\qquad\qquad A = \dfrac{(\pi)(0.03568)^2}{4} \text{ m}^2$

$$A = 0.001 \text{ m}^2$$

Flow rate $\qquad\quad Q = Av$

$$Q = 0.001 \times 20 = 0.02 \text{ m}^3/s$$

The pressure difference across the pipe $= \Delta P$

and $\qquad\qquad\qquad \Delta P = P_1 - P_2$

or $\qquad\qquad\qquad \Delta P = 20 - 14 = 6P_a$

R, the restriction resistance, is equal to the flow rate divided by the pressure drop. Therefore,

$$R = \frac{\Delta P}{Q} = \frac{6}{0.02}$$

$$R = 300 \text{ } P_a/(m^3/s)$$

2-6 SUMMARY

In this chapter we have reviewed and discussed some expressions and physical laws commonly found in the areas of mechanics, thermodynamics, electrical science, and hydraulics. This review should serve to refresh the memory of the reader and provide the necessary level of engineering knowledge to deal with the work in the following chapter.

building a mathematical model 3

Twas brillig, and the slithy toves
Did gyre and gimble in the wabe;
All mimsy were the borogoves,
And the mome raths outgrabe.

Lewis Carroll, *Through the Looking Glass*

3-1 INTRODUCTION

The quotation from the "Jabberwocky" almost seems to make sense, although if we take a closer look we can see that it is mostly made up of nonsense words. It does, however, convey a feeling to the reader that if he were to try just a little harder he could, in fact, understand every word, but try as he may, understanding is always out of reach.

This same frustrating situation often besets students studying control systems when it becomes necessary for them to use mathematics to analyze the operation of a system. For many people mathematics seems to bear no relation to reality, and this conception can be a barrier to understanding when carried into those areas of science and engineering that need mathematics to a considerable degree. Often this veil of mystery can be drawn back by approaching the topic boldly and in as simple a way as possible.

Block diagrams were first introduced in Chapter 1 but only in a qualitative sense; i.e., it was necessary to show how parts of systems were tied together physically but without the complication of shafts, gears, wires, etc. We are now at the point in our study of systems where we have to use block diagrams in a quantitative sense. By this we mean to use block diagrams in such a way that they will show the mathematical operations that take place within a system. It is very necessary to understand these operations, because they are used to describe the laws of physics that govern most control systems. We reviewed some of these laws and relationships in Chapter 2.

It is not by chance that block diagrams and control systems mathematics are intimately connected in this manner; a block diagram is an elegantly visual way of describing differential equations. Seeing a problem, or visualizing it in some fashion, usually is necessary for most of us before understanding can be achieved, hence the expression, "Oh yes! I see," when what is really meant is "Oh yes! I understand."

In this chapter we shall explain the methods of handling block diagrams quantitatively; we shall examine the relationship between the input and output of a single block and then a combination of blocks. We shall construct first- and second-order differential equations and explore ways of finding solutions to our equations. Finally, we shall demonstrate how block diagrams are used to interconnect the individual terms of a differential equation and thereby form a mathematical model such as may be used to describe a control system.

3-2 BLOCK DIAGRAMS

Introduction: Any engineering system and for that matter any biological, social, or economic system can be represented by a combination of blocks. Each block has a single line into it and a single line out. Within the block is a statement of its operation; that is, the block indicates what happens to the input information before that information is passed on at the output. Figure 3-1(a) shows this process. In the illustration a simple amplifier is shown as a block; if, for example, the amplifier has a gain of 200 and a 1-mv signal is fed into it, then the amplifier will produce 200 mv at its output terminals.

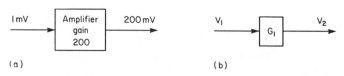

figure 3-1   a block diagram of an amplifier (a) precise representation, (b) general representation

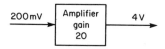

figure 3-2   another precise block diagram representation of an amplifier

Now if the input voltage, output voltage, and gain of the amplifier are unknown but are said to equal V_1, V_2, and G_1, respectively, then a more general, but just as precise, diagram can be drawn, as in Figure 3-1(b), showing the operation of the amplifier.

A statement can be made about the latter diagram; it can be said that G_1 *operates* on V_1 to give V_2. In fact, if V_2 is divided by V_1, the result will always equal the gain (provided that the amplifier does not saturate). This statement can be expressed:

$$\frac{V_2}{V_1} = G_1 = \text{a constant} \qquad (3.1)$$

Thus the block diagram illustrates that a mathematical operation is performed on the input voltage V_1; i.e., V_1 is multiplied by the constant, G_1. Therefore, for any new and arbitrary value of input voltage, say V_1', the new output voltage will be given by

$$V_2' = G_1 V_1' \qquad (3.1a)$$

Now consider the effect of feeding the output of this amplifier into the input of a second amplifier which has a gain of 20. The 200-mv signal will now be amplified to 4000 mv or 4 V, as shown in Figure 3.2.

If the gain of the second amplifier is designated as G_2 and its output as V_3, the entire system can be shown in block diagram fashion, as in Figure 3-3.

By similar reasoning to the above we note that

$$\frac{V_3}{V_2} = G_2 \qquad (3.2)$$

or $$V_3 = V_2 G_2 \qquad (3.2a)$$

figure 3-3   cascaded blocks

However, from equation (3.1a)

$$V_2 = V_1 G_1$$

Therefore, if $V_1 G_1$ is substituted for V_2 in equation (3.2a), we find

$$V_3 = V_1 G_1 G_2 \qquad (3.3)$$

$$\frac{V_3}{V_1} = G_1 G_2 \qquad (3.3a)$$

Equations (3.3) and (3.3a), which are different forms of the same equation, demonstrate two ideas. First, that G_1 multiplied by G_2 operates on V_1 to give V_3. Second, that simple blocks strung together in cascade (or tandem) fashion can be multiplied together, a fact which allows us to reduce such a string to a single block, as shown in Figure 3-4.

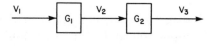

figure 3-4 reduction of cascaded blocks

EXAMPLE 3.1
Reduce the following series of elements to a single element, using the input and output shown.

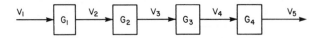

solution
All the elements are multiplied together to give the overall gain; thus

$$\frac{V_5}{V_1} = G_1 \times G_2 \times G_2 \times G_3$$

and the reduced block diagram is as follows:

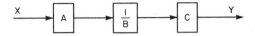

EXAMPLE 3.2

Reduce the following block diagram to a single block, leaving the input and output unchanged.

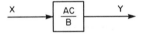

solution

All the elements are multiplied together to obtain the overall gain; thus

$$\frac{Y}{X} = A \times \frac{1}{B} \times C$$

Thus the block diagram becomes

X ⟶ $\boxed{\dfrac{AC}{B}}$ ⟶ Y

Notice that the second block contains a reciprocal term, $\left(\dfrac{1}{B}\right)$; this in effect constitutes a division, that is to say the product AC is divided by B. However, this is nothing new, since we know that division is only multiplication by a reciprocal value.

Summing Junctions: It is sometimes necessary to feed more than one signal into an amplifier at the same time, and this is done by a summing network in an actual circuit. However, in block diagram symbols it is done by means of a summing junction, as shown in Figure 3-5.

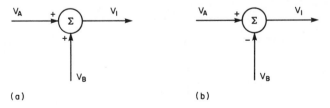

(a)   (b)

figure 3-5 summing junctions (a) positive summing, (b) negative summing

A summing junction is always shown as a circle with arrows into and out of the symbol. The arrows into the symbol are always identified with a plus or minus sign, indicating either a positive or a negative signal. Inside the circle is sometimes shown a Greek letter sigma Σ, which means *sum*. In Figure 3-5 (a) V_B is added to V_A to give V_1 (summing positively) and in Figure 3-5 (b) V_B is subtracted from V_A to give V_1 (summing negatively). These two operations can be stated thus:

For Figure 3-5(a) $\qquad\qquad V_A + V_B = V_1 \qquad\qquad\qquad\qquad (3.4)$

and for Figure 3-5(b) $\qquad V_A - V_B = V_1 \qquad\qquad\qquad\qquad (3.5)$

Now if both V_A and V_B are simultaneously fed into the first amplifier, whose gain was 200, the output is the amplified sum of the two inputs, as shown below:

From equation (3.4) $\qquad\qquad V_1 = V_A + V_B$

and from equation (3.1a) $\qquad V_2 = G_1 V_1$

Therefore, $\qquad\qquad\qquad\qquad V_2 = G_1(V_A + V_B) \qquad\qquad\qquad (3.6)$

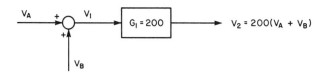

figure 3-6 summing two voltages and multiplying by a constant

The block diagram in Figure 3-6 is not the only way of representing the summing and multiplying operations of equation (3.6). These operations may be rearranged in a new block diagram, as shown in Figure 3-7. Although apparently more cumbersome, this new arrangement can be very useful, as we shall see later in this chapter.

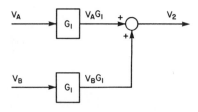

figure 3-7 another method of summing two voltages and multiplying by a constant

From Figure 3.7

$$V_2 = V_A G_1 + V_B G_1$$

Hence $\qquad V_2 = G_1(V_A + V_B)$

This equation is identical to equation (3.6).

Notice that the summing junction in Figure 3-7 has been moved to the right of the blocks, and to maintain the same validity, as in the block diagram in Figure 3-6, each input has to be separately multiplied or operated on by G_1. This is a good example of superposition, which was first introduced in Chapter 1.

EXAMPLE 3.3

Given the diagram below, (a) find V_2 in terms of the inputs and gain; (b) show the block diagram in two other forms.

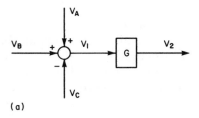

(a)

solution

(a)
$$V_1 = V_A + V_B - V_C$$
$$V_2 = V_1 G$$
$$V_2 = G(V_A + V_B - V_C)$$

(b)

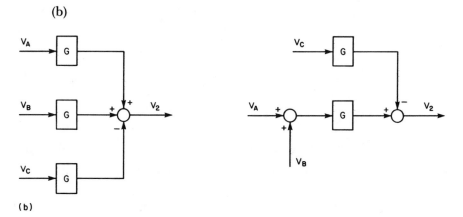

(b)

Four arithmetic operations have now been covered: multiplication, division, addition, and subtraction, and two further operations remain to be discussed. These operations are not, however, arithmetic; they are in fact the calculus operations of differentiation and integration.

The D Operator: It is important to pause for a moment before continuing with differentiation and integration and consider what has taken place. So far it has been seen that the output of any single block is obtained by multiplying the input by the term appearing in the block. It will be remembered, however, that the term in the block is really an operator which describes the action that the block has on the input. In the examples so far, all the operations have been algebraic products; hence, a simple rule of multiplication has been possible. However, not all the operations that we shall be considering are pure products (e.g., differentiation and integration), and we must find a way to deal with these new operations that will allow us not only to describe a differential equation, by means of a block diagram, but also using only the rules of algebra.

In order that the operations of differentiation and integration can be handled as easily and as uniformly as the algebraic products in the previous examples, the operations must be transformed. This transformation is done by means of a simple code which replaces the operations of differentiation and integration by operations that involve only straightforward multiplication or division, respectively. For example, if we consider velocity, it can be expressed by the relationship

$$v(t) = \frac{dx(t)}{dt} \tag{3.7}$$

where the velocity v equals the rate change of the distance x with respect to time t. It could also be written as:

$$v(t) = \frac{d}{dt} \cdot x(t) \tag{3.8}$$

This does not mean that $v(t)$ equals $\frac{d}{dt}$ multiplied by $x(t)$, but rather $v(t)$ equals $x(t)$ when operated on by $\frac{d}{dt}$.

We will now replace the differential operator $\frac{d}{dt}$ by an algebraic operator D, thus transforming $v(t)$ in equation (3.8) into $v(D)$, where

$$v(D) = D \cdot x(D) \qquad (3.9)$$

i.e., $v(D)$ equals D multiplied by $x(D)$.

The symbol D is termed the D operator. Equation (3.9) can now be combined with similar terms; such combinations are governed solely by the rules of algebra. For example, if $f(t)$ is the acceleration of a body whose velocity is $v(t)$, then

$$f(t) = \frac{d}{dt} \cdot v(t) \qquad (3.10)$$

Using the D operator, we find that the D transformation of equation (3.10) is

$$f(D) = D \cdot v(D) \qquad (3.11)$$

But from equation (3.9) we know that

$$v(D) = D \cdot x(D)$$

Therefore,

$$f(D) = D \cdot D \cdot x(D) = D^2 \cdot x(D) \qquad (3.12)$$

Alternatively, $\qquad x(D) = \frac{1}{D^2} \cdot f(D) \qquad (3.12a)$

The result of such manipulations is usually the D transform of the output variable, which when decoded will yield an answer that describes the behavior of the output variable with respect to time. In our example the input variable is the acceleration of the body $f(t)$; e.g., depressing the gas pedal causes our car to accelerate. The output variable is, of course, $x(t)$, the distance the car travels in time t. Equation (3.12a) can be illustrated by means of a block diagram, as shown in Figure 3-8; note how the terms appearing inside the blocks can now be multiplied by each other and by the input variable. To decode the D transform described in equation (3.12a), let us examine equation (3.9). Equation (3.9) could also be written as

$$x(D) = \frac{1}{D} \cdot v(D) \qquad (3.9a)$$

$$f(D) \rightarrow \boxed{\frac{1}{D}} \xrightarrow{v(D)} \boxed{\frac{1}{D}} \xrightarrow{x(D)}$$

figure 3-8 block diagram of equation (3.12a) $f(t) = \dfrac{d^2x(t)}{dt^2}$ when transformed

to appear as $x(D) = \dfrac{1}{D_2} \cdot f(D)$

From equation (3.7) we know that the distance $x(t)$ is the integral of the velocity $v(t)$, with respect to time t; i.e.,

$$x(t) = \int v(t) \cdot dt \qquad (3.13)$$

Comparing equations (3.9a) and (3.13), we see that $\dfrac{1}{D}$ is the D transformation of the integral operator. Thus decoding the transformation in equation (3.12a) yields

$$x(t) = \int \int f(t) \, dt \cdot dt \qquad (3.14)$$

The following list shows the equivalent D operator transformation for some typical calculus operations:

OPERATION	D TRANSFORM
d/dt	D
d^2/dt^2	D^2
d^3/dt^3	D^3
$\int \cdot dt$	$1/D$
$\int\int \cdot dt \, dt$	$1/D^2$

Tie Points: In block diagrams it is often necessary to have the signal flow in more than one direction. To effect this symbolically a tie point is used (see Figure 3-9).

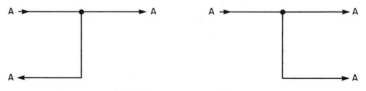

figure 3-9 tie points

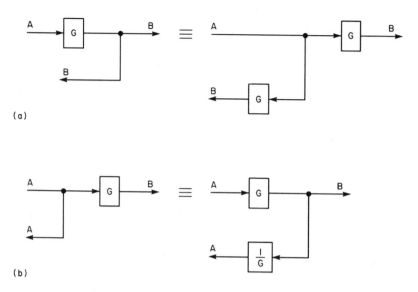

figure 3-10 transferring tie points across a block (a) from the output side to the input side, (b) from the input side to the output side

A tie point can be thought of as a node or junction where no summing (positively or negatively) is done. It is just a convenient method of providing extra paths along which the signal may flow without affecting its flow along its original path. Tie points, like summing junctions, may be transferred across blocks, the same rules being used as applied to the movement of summing junctions. These rules are illustrated in Figure 3-10.

Block Diagram Reduction: Block diagram reduction is an important step in control system analysis. It is a technique which results in a diagram consisting usually of only one block whose input and output is that of the original blocks, but it is much more useful.

Since we are now considering more general systems, it is appropriate to dispense with the use of an amplifier as an example. One reason is that when using an amplifier we tend to use the usual symbols associated with amplifiers rather than the generally accepted control system notation. Second, we restrict our conception of control systems to that defined by one of the more simple electronic devices.

Consider the illustration in Figure 3-11 for which the following system notation is used.

R = the reference input

C = the output controlled variable

ε = the error signal

G = the forward loop gain

H = the feedback loop gain

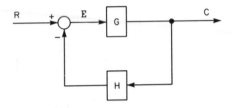

figure 3-11   a simple servo-system

The block diagram in Figure 3-11 shows the simplest form of feedback control system, and it is incorporated into so many block diagrams that it has a reduced form that should be remembered. The reduction is based upon the following block diagram algebra. From Figure 3-11 we see that

$$\varepsilon G = C \qquad (3.15)$$

also

$$R - CH = \varepsilon \qquad (3.16)$$

Substituting equation (3.16) into equation (3.15) yields

$$(R - CH)G = C$$

or,

$$RG - CGH = C$$

$$RG = C + CGH$$

hence,

$$RG = C(1 + GH)$$

and, The Control Ratio,

$$\frac{C}{R} = \frac{G}{1 + GH} \qquad (3.17)$$

Figure 3-11 can now be reduced to a "single block" block diagram, as shown in Figure 3-12, having the same input and output as the original, more complicated diagram.

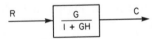

figure 3-12   reduced form of the simple servo-system

EXAMPLE 3.4

Reduce the following unity feedback system to one block and hence derive the control ratio $\dfrac{C}{R}$.

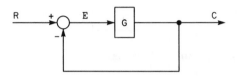

solution

The feedback loop should be considered as having a block in it whose operation is $H = 1$, since all the output C is fed back to the summing junction.

The reduction is, therefore, the same as that shown in Figure 3-12.

$$\frac{C}{R} = \frac{G}{1 + G \times 1}$$

$$\frac{C}{R} = \frac{G}{1 + G}$$

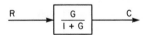

EXAMPLE 3.5

Reduce the following block diagram to one block and hence derive the control ratio $\dfrac{C}{R}$.

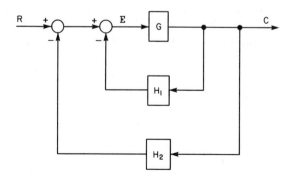

solution

This is a multiloop system, and the simplest approach to a solution would be to reduce the minor (or inner) loop first.

The minor loop reduces to

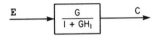

and the block diagram now becomes

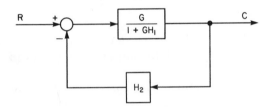

which is reduced to one block by the standard algebra of equation (3.17).

$$\frac{C}{R} = \frac{\dfrac{G}{1 + GH_1}}{1 + \dfrac{G}{1 + GH_1} \cdot H_2}$$

$$\frac{C}{R} = \frac{G}{1 + G(H_1 + H_2)}$$

Example 3.5 is rather interesting in that it shows that two feedback loops which have adjacent tie points and adjacent summing junctions are additive and could have been reduced to one feedback loop with a block whose operation would have been $H_1 + H_2$.

EXAMPLE 3.6

Reduce the following system to one block and in so doing derive the control ratio $\dfrac{C}{R}$.

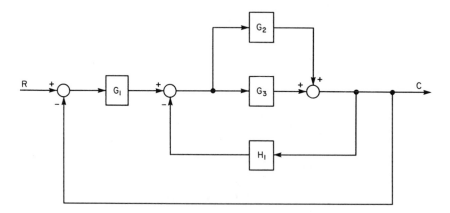

solution

Step 1 is to reduce the minor feed forward loop thus:

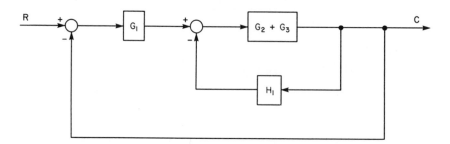

When the next minor loop is reduced we obtain:

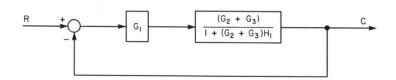

The forward loop is then reduced to one block as follows:

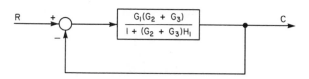

We now have a unity feedback system which can be reduced in the normal way:

$$\frac{C}{R} = \frac{\dfrac{G_1(G_2 + G_3)}{1 + (G_2 + G_3)H_1}}{1 + \dfrac{G_1(G_2 + G_3)}{1 + (G_2 + G_3)H_1} \cdot 1}$$

$$\frac{C}{R} = \frac{G_1(G_2 + G_3)}{1 + (G_2 + G_3)H_1 + G_1(G_2 + G_3)1}$$

$$\frac{C}{R} = \frac{G_1(G_2 + G_3)}{1 + (G_2 + G_3)(H_1 + G_1)}$$

The reduced block diagram thus becomes

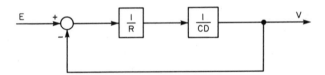

EXAMPLE 3.7

Reduce the following simple system to one block and derive the control ratio $\dfrac{V}{E}$.

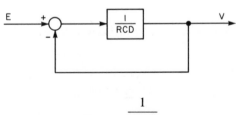

solution

Although this contains the operator D it is handled completely algebraically. Thus the system becomes:

$$\frac{V}{E} = \frac{\dfrac{1}{RCD}}{1 + \dfrac{1}{RCD} \cdot 1}$$

$$\frac{V}{E} = \frac{1}{RCD + 1}$$

and the reduced block diagram is

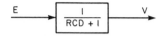

3-3 THE TRANSFER FUNCTION

In the analysis and design of linear systems, the transfer function and block diagrams comprise two of the most important tools. Block diagram reduction or simplification was discussed at some length in Section 3-2. The reduction, as we saw, resulted in a single block containing a mathematical expression. This expression, for the moment, will be referred to as the transfer function of the system (we shall see in Chapter 5 that the term "transfer function" is usually reserved to describe a function which is similar in appearance to our expression but more rigorously defined).

Let us consider, therefore, that the transfer function of a system is simply the ratio of the D transform of the output divided by the D transform of the input. It is then possible to list the properties of our transfer function thus:

1. The transfer function is independent of the input to the system; i.e., the characteristics of the system are not modified by the input signal (at least within the working range). An example of this property is the constant gain of the amplifier considered in Section 3-2.

2. All initial conditions are assumed to be zero; i.e., the system is at rest initially.

3. The transfer function describes only a time-invariant linear system, i.e., a linear system whose parameters do not change, or change only a little, during operation.

The first property is self-explanatory; the remaining two properties, however, deserve a word of explanation. The assumption of zero initial conditions is a property always associated with the use of D operator transfer functions; it is the primary difference between the transfer functions that we are using at present and those that we shall be using later, in Chapter 5. However, the D operator type of transfer function is still very useful, as many systems that concern us can be considered to be at rest initially. The third property is really a statement of two important generalizations, usually made about any "real" system that we may want to analyze. We assume that the system is both linear and time-invariant, a situation which is rarely ever true in practice. For example, we use Ohm's Law to describe how the current through a resistor varies in a constant, linear manner directly with the voltage across the resistor. If, however, the voltage is too high, the resulting high current will cause heating in the resistor, thereby causing an increase in resistance, and eventually the resistor will melt. We see that this "system," if taken out of its working range, becomes both time-variant and non-linear and eventually discontinuous. The resistor is designed to have a working range of voltage and current, and if it is operated within its working range, its behavior can be assumed to be linear and time-invariant. This argument applies to most systems that concern us and so allows us to use transfer functions as a general way of describing such systems.

A number of linear systems are examined in Sections 3-5 and 3-6, and the transfer function for each system is derived. However, before we can derive these functions we must determine the differential equation that describes the behavior of each system.

3-4 DIFFERENTIAL EQUATIONS

The dynamics of linear systems can be fully described only by the use of differential equations. Differential equations are part of calculus and calculus is the mathematics of change. The output of a system will change whenever the input is changed, either in magnitude or frequency. Change can be mathematically expressed in a number of ways; for instance, a tank changes its volume as the height of the tank changes. It is not this kind of change which is of interest in systems analysis and design, however, but rather the change of the system variables with the passing of time.

Differential equations are to be found in many degrees of complexity and difficulty, but fortunately we are interested only in linear differential equations with constant coefficients, because these are the equations that describe systems central to our study. They also happen to be the easiest differential equations to solve. The solution of these equations describes the initial short-term (transient) behavior and the long-term (steady state) behavior of the system as it responds to a change in the input variable.

When talking about differential equations, we often talk about the *order* of the equation; sometimes we also refer to the *degree* of the equation.

1. *Order:* The order of the equation is the maximum number of times that the dependent variable is differentiated in the equation; e.g., consider the following equation:

$$\frac{d^2y}{dt^2} + A\frac{dy}{dt} + By = \frac{dx}{dt} + Cx$$

This equation is linear and of second order, because y has been differentiated twice.

2. *Degree:* The degree of an equation is the maximum number of times the dependent variable or its derivations are multiplied by themselves or each other; e.g., in the equation

$$\left(\frac{dy}{dt}\right)^2 + y^3 = x$$

the equation is of third degree and first order.

All the differential equations to be considered in this text are linear and therefore of first degree.

We shall now examine some first-order and second-order systems, so called because the equations that we shall derive, describing these systems, are of first order and second order, respectively. Solutions to the equations derived will be reserved until Chapter 4, where each solution will be examined to show the significance of each of the parameters of the equation.

3-5 FIRST-ORDER DIFFERENTIAL EQUATIONS

Obtaining the system equation is reasonably straightforward but can be somewhat lengthy at times. The method requires that the dynamics of each component in the system be described by a mathematical statement and each statement be linked together by a common term in the equation. Finally, the complete equation is arranged and ordered with the terms of the dependent variable on one side and those of the independent variable on the other.

A Mechanical Example: The illustration in Figure 3-13 shows a large blower fan such as those used in wind tunnels or mine ventilation systems.

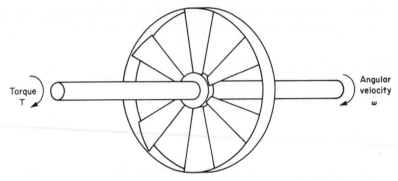

figure 3-13   a large blower fan—an entirely mechanical system

When torque is applied to the drive shaft, the fan begins to rotate, accelerating at a rate dependent upon the inertial load of the fan. As the angular speed of the fan increases, the wind resistance becomes more significant, and the opposing torque due to this drag can be assumed to vary directly with the speed of the fan. When the drag torque equals the applied torque, the fan can no longer accelerate and runs at a steady-state speed.

The parameters of this system are

Applied torque	$= T(t)$ Nm
Drag torque	$= T_D(t)$ Nm
Polar moment of inertia of fan and shaft	$= J$ kg-m^2
Drag coefficient	$= B$ Nm/(rad/s)
Angular velocity of fan	$= \omega(t)$ rad/s
Angular accel. of fan	$= \alpha(t)$ rad/s^2
	$\alpha(t) = d\omega(t)/dt$ rad/s^2

Now the net torque available to accelerate the fan is the difference between the applied torque and the drag torque; i.e., $T(t) - T_D(t)$; therefore, from Newton's second law of motion

$$T(t) - T_D(t) = J\frac{d\omega(t)}{dt} \qquad (3.20)$$

but

$$T_D(t) = B\,\omega(t) \qquad (3.21)$$

Combining these two equations yields

$$T(t) - B\omega(t) = J\frac{d\omega(t)}{dt} \qquad (3.22)$$

which may be written

$$J\frac{d\omega(t)}{dt} + B\omega(t) = T(t) \qquad (3.23)$$

Equation (3.23) is a first-order differential equation with constant coefficients J and B. This equation expresses how the angular velocity, ω, changes when a torque T is applied to the fan shaft. The form of equation (3.23) is one that we will encounter many times in our studies of control systems, as it describes the most common type of feedback system.

Although initially our fan does not appear to be a feedback system, we shall see upon examination that it is. In a feedback system the output variable ω must first be transformed into the same units as the input variable. We have seen that this is done by multiplying ω by B to obtain the drag torque, T_D. Thus the output is constantly being compared to the input by means of the drag torque. As long as the input exceeds the drag, the fan will continue to accelerate, thus increasing ω and hence the drag. As the drag torque approaches the magnitude of the applied torque, the acceleration slowly decreases and the angular velocity levels off at a steady value, whereupon both torques completely oppose each other. The magnitude of this steady state value is directly proportional to the applied torque; i.e.,

$$\omega_{max} = T/B$$

Equation (3.23) can be transformed by means of the D operator. The resulting expression can then be manipulated to produce a transfer function describing the relationship between the D transforms of the input and output variables: thus:

$$J \cdot D\omega(D) + B\omega(D) = T(D)$$

Therefore,

$$\left(\frac{J}{B}D + 1\right)\omega(D) = \frac{T(D)}{B} \qquad (3.24)$$

Hence

$$\frac{\omega(D)}{T(D)} = \frac{1/B}{1 + \frac{J}{B} \cdot D} \qquad (3.25)$$

The numerator of equation (3.25) is a constant and will be replaced by K. The term J/B can be seen to have the units of time and will be replaced by τ. Rewriting equation (3.25), we find

$$\frac{\omega(D)}{T(D)} = \frac{K}{1 + D\tau} \qquad (3.25a)$$

The form of the transfer function in equation (3.25a) can be considered a basic building block in the construction of block diagrams.

A number of systems that we may commonly encounter have similar dynamics to this one, and similar transfer functions may be derived for them. We will examine two such systems briefly.

An Electrical Example: Consider the electrical circuit shown in Figure 3-14, consisting of a resistor R in series with a capacitor C. An emf, $e(t)$, is applied across the input terminals, causing a current, $i(t)$, to flow in

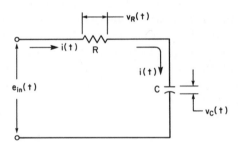

figure 3-14   a simple *RC* circuit

the circuit. Now from Kirchhoff's Law we know that the sum of the voltages across each component is equal to the applied voltage; i.e.,

$$e(t) = v_R(t) + v_C(t)$$

Also from Ohm's Law we know that

$$v_R(t) = i(t)R$$

Therefore,

$$e(t) = i(t)R + v_C(t) \qquad (3.26)$$

and
$$i(t) = \frac{e(t) - v_C(t)}{R} \qquad (3.26a)$$

Now if the capacitor is initially uncharged, then the voltage v_C varies as the magnitude of the charging current and the length of charging time; thus

$$v_C(t) = \frac{1}{C} \int^t i(t) \cdot dt \qquad (3.27)$$

However, the most common way of expressing this information is to differentiate equation (3.27) so that

$$i(t) = C \frac{dv_C(t)}{dt} \qquad (3.28)$$

If we now substitute equation (3.28) into equation (3.26), we obtain

$$e(t) = RC \frac{dv_C(t)}{dt} + v_C(t) \qquad (3.29)$$

Equation (3.29) is a first-order differential equation expressing how the output v_C responds to an input signal e. The system dynamics are very similar to those describing the fan examined previously. Let us compare the two differential equations, first for the fan:

$$T(t) = J \frac{d\omega(t)}{dt} + B\omega(t) \qquad (3.23)$$

or

$$\frac{T(t)}{B} = \frac{J}{B} \frac{d\omega(t)}{dt} + \omega(t) \qquad (3.23a)$$

and for the electrical circuit

$$e(t) = RC \frac{dv_C(t)}{dt} + v_C(t) \qquad (3.29)$$

It can be easily seen that if we could choose the values of R and C such that

$$RC = \frac{J}{B}$$

and use an input voltage of magnitude such that

$$e(t) = \frac{T(t)}{B}$$

then the magnitude and behavior of v_C would exactly coincide with that of ω, and the electrical circuit could be said to be an electrical analogue of the fan system. In fact, by suitable use of operational amplifiers and scale adjustments, analogue computers are used in just this fashion.

Like the fan example, initially the value of v_C (ω) is zero and the full voltage is applied across the resistor, causing maximum current (acceleration) to flow in the circuit. This current begins to charge the capacitor (rotate the fan), which causes the voltage across the resistor to be reduced, thereby reducing the current and the rate of charging of the capacitor. We thus see, as we did in the example using the fan, that the circuit has feedback associated with its behavior. The output is continually compared to the input, and as long as the input voltage exceeds the output voltage, current will flow and v_C will continue to grow, although at a steadily reducing rate. When v_C equals the input, e, no more current can flow and the steady state condition is obtained.

As in the previous example, the differential equation (3.29) can be transformed by use of the D operator to make the equation suitable for using either in block diagrams or for further manipulation.

$$e(t) = RC \cdot \frac{dv_C(t)}{dt} + v_C \qquad (3.29)$$

Using the D operator, we obtain

$$e(D) = RC \cdot Dv_C(D) + v_C(D) \qquad (3.30)$$

The units of RC are time units and so we can replace RC by τ; therefore

$$e(D) = (D\tau + 1)v_C(D) \qquad (3.31)$$

or
$$\frac{v_C(D)}{e(D)} = \frac{1}{1 + D\tau} \qquad (3.31a)$$

The transfer function in equation (3.31a) is, not surprisingly, of the same form as the transfer function for the fan expressed in equation (3.25a).

A Thermal Example: A heat transfer system is now considered; a simplified version of a hot water cistern is shown in Figure 3-15. Some

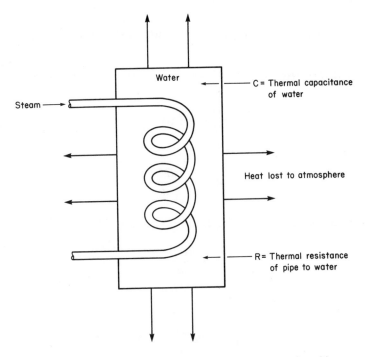

figure 3-15 a steam-heated hot water system—an example of heat transfer

assumptions have been made and some details neglected because the system is not quite as simple as is suggested. For example, the source temperature cannot be applied as suddenly as the voltage or torque in the previous examples; also, the heat flow is limited in practice and is influenced too much by the condition of the steam: the water is also moving through the tank at a nonlinear rate. However, for the purposes of preliminary studies we often make such approximations; therefore, let us assume that the tank has a thermal capacitance of C joules/°C and thermal resistance of pipe to water of R °C·sec/joule. The temperatures of the steam and water are T_1 and T, respectively, and just before steam heat is applied the water temperature is T_0.

Heat flows through the thermal resistance because of the difference in temperature between the steam and the water. The relationship is

$$\frac{T_1(t) - T(t)}{R} = q(t) \quad \text{(Ohm's Law analogy)} \qquad (3.32)$$

where q = heat flow in joules/sec. This heat is stored in the water, causing the water temperature to rise such that

$$T(t) - T_0 = \frac{1}{C}\int^t q(t)\, dt \quad \text{(Faraday's Law analogy)} \qquad (3.33)$$

If both sides of equation (3.33) are differentiated with respect to time, we find

$$\frac{dT(t)}{dt} = \frac{q}{C}(t)$$

therefore,

$$q(t) = C\frac{dT(t)}{dt} \qquad (3.34)$$

Equation (3.34) can be substituted into equation (3.32) to give

$$\frac{T_1(t) - T(t)}{R} = C\frac{dT(t)}{dt}$$

or

$$T_1(t) - T(t) = RC\frac{dT(t)}{dt}$$

and

$$T_1(t) = RC\frac{dT(t)}{dt} + T(t) \qquad (3.35)$$

This is the system differential equation, and the dynamics are precisely the same as those of equations (3.23) and (3.29). As before, (3.35) can be transformed by use of the D operator such that

$$T_1(D) = RCDT(D) + T(D) \qquad (3.36)$$

Therefore,

$$T_1(D) = (RCD + 1)T(D)$$

so that

$$\frac{T(D)}{T_1(D)} = \frac{1}{(RCD + 1)} \qquad (3.37)$$

The coefficient of D in the denominator of the right-hand side of equation (3.37) has units of time; i.e.,

$$R \times C = \frac{°C \cdot \sec}{\text{joules}} \times \frac{\text{joules}}{°C} = \text{seconds}$$

and the symbol τ is substituted for RC to give

$$\frac{T(D)}{T_1(D)} = \frac{1}{\tau D + 1} \qquad (3.37a)$$

which is the same form as equations (3.25a) and (3.31a).

Three very different examples have been used to illustrate first-order systems and in each case the same method was applied to obtain the system equation. It should be noted that with only minor algebraic manipulations the system transfer functions were also obtained and were identical in form.

3-6 SECOND-ORDER DIFFERENTIAL EQUATIONS

Feedback control systems are not usually so simple that they can be described by first-order differential equations. It is true that many components can be of the first order, but sometimes it is impossible to prevent loading of one component by another; if this occurs, the two components must be considered as acting in concert with one another. This leads to the synthesis of higher-order equations. It goes without saying, therefore, that if the component or system has any complexity to it at all, it will be second order or higher. This being the case, let us examine the operation of some such systems and derive their differential equations.

A Mechanical Example: An automobile and a mechanical schematic diagram of one of its shock absorber/spring systems are shown in Figure 3-16.

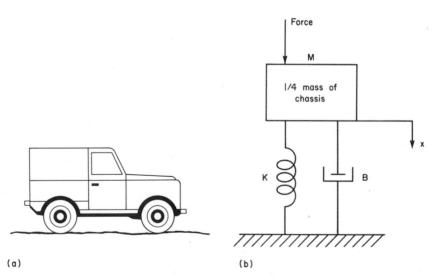

(a) (b)

figure 3-16 a more complicated mechanical system (a) a motor vehicle, (b) a shock absorber/spring system attached to one wheel

Since there are four springs supporting the vehicle, it is assumed that each one carries one-fourth of the mass of the vehicle. It has also been assumed that the pneumatic tire plays no part in the dynamics of the system, nor does the mass of the wheel itself; this, of course, is not true. For simplicity of reference let the mass be subjected to displacement and the reference surface be rigid. The system equation is derived as follows:

The mass of the system	$= M$ kg
The spring stiffness factor	$= K$ N/m
The shock absorber coefficient	$= B$ N/(m/sec)
The applied force	$= F(t)$ N
The displacement due to force	$= x(t)$ m

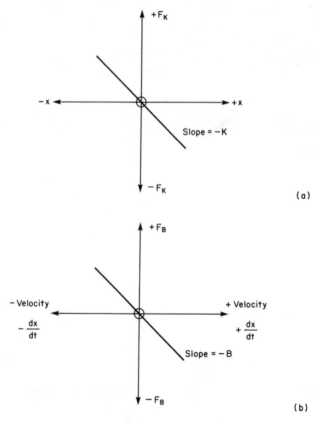

(a)

(b)

figure 3-17 a graphical representation of the proportionality constants, of the shock absorber/spring system (a) the spring constant K, (b) the damper constant B

When a force F is applied to the mass in a downward direction, it accelerates the mass. Opposing the motion are the restoring force of the spring, which resists being downpressed, and the restraining force of the shock absorber, which resists any movement. The restoring force due to the spring is $F_k(t)$, where

$$F_k(t) = -Kx(t) \tag{3.38}$$

Equation (3.38) is illustrated in Figure 3-17(a).

The restraining force due to the damper is $F_B(t)$, where

$$F_B(t) = -B\frac{dx(t)}{dt} \tag{3.39}$$

Equation (3.39) is illustrated in Figure 3-17(b).

From Newton's second law of motion (see Chapter 2) we know that the acceleration of a body of mass M is directly proportional to the algebraic sum of all the forces acting on the body; i.e.,

$$M\frac{d^2x(t)}{dt^2} = F(t) + F_k(t) + F_B(t) \tag{3.40}$$

Substituting equations (3.38) and (3.39) into (3.40), we have

$$M\frac{d^2x(t)}{dt^2} = F(t) - Kx(t) - B\frac{dx(t)}{dt} \tag{3.41}$$

or

$$M\frac{d^2x(t)}{dt^2} + B\frac{dx(t)}{dt} + Kx(t) = F(t) \tag{3.42}$$

Equation (3.42) is a second-order differential equation and expresses the dynamic behavior of the position of the mass when acted upon by an external force $F(t)$. If the mass has been displaced from its rest position to a distance x_m and the external force is suddenly released, equation (3.42) would become

$$M\frac{d^2x(t)}{dt^2} + B\frac{dx(t)}{dt} + Kx(t) = 0 \tag{3.42a}$$

and the mass M would be accelerated back toward its rest position by the restoring force of the spring, F_k. Just before the spring is released, the energy stored in the spring is $Kx_m^2/2$. After the spring is released, the mass is accelerated towards its rest position and the potential energy stored in the spring is converted into the kinetic energy of the mass and heat, which is generated within the shock absorber as it resists the movements of the mass. If the shock absorber was not present and we had an ideal situation of no damping forces at all, then the kinetic energy of the mass, as it passes through its rest position, would have a magnitude exactly equal to $\dfrac{Kx_m^2}{2}$, the original stored energy. That this condition is not true is obvious, and the net loss of energy is equal to the heat dissipated in the shock absorber; i.e.,

$$\frac{Kx_m^2}{2} - \frac{Mv_0^2}{2} = \int_{x_m}^{0} F_B \, dx$$

where v_0 is the velocity of the chassis as it passes through its rest position. Unless the shock absorber is sized to completely absorb all the spring's energy, as the chassis returns to its normal rest position, then the residual momentum of the chassis will carry it past this point, causing the spring to be extended, setting up a new but opposite restoring force. Although the spring extends, it can not extend as much as x_m, i.e., the amount of original compression, because of the energy lost within the shock absorber. If the shock absorber has only a small damping action, then the chassis will oscillate about the rest position and the oscillations will take a long time to die away, giving a very soft ride; this condition is known as underdamping. If overdamping occurs, then no oscillation takes place, but the chassis will only return slowly to its rest position, giving a stiff and bumpy ride. Ideally, the damping should be adjusted to return the chassis to its rest position quickly with little overshoot and no oscillation.

Equation (3.42) can be further simplified, by use of the D operator, to show the relationship between the D transforms of the input and output variables. Thus the D transform of equation (3.42) is

$$MD^2x(D) + BDx(D) + Kx(D) = F(D) \qquad (3.43)$$

therefore, $\qquad \dfrac{M}{K}D^2x(D) + \dfrac{B}{K}Dx(D) + x(D) = \dfrac{F(D)}{K}$

Hence

$$\frac{x(D)}{F(D)} = \frac{\dfrac{1}{K}}{\dfrac{M}{K}D^2 + \dfrac{B}{K}D + 1} \qquad (3.44)$$

If a unit analysis is performed upon equation (3.44), we find that it is a balanced equation in that the left-hand side has units of metres/newton, as does the right-hand side. In the denominator of the right-hand side the term M/K has units of (seconds)^2 and B/K has units of seconds; therefore, if the symbols τ_1^2 and τ_2 are substituted for these two terms respectively, equation (3.44) simplifies to

$$\frac{x(D)}{F(D)} = \frac{1/K}{\tau_1^2 D^2 + \tau_2 D + 1} \qquad (3.44a)$$

An Electromechanical Example: A dc armature-controlled motor is shown schematically in Figure 3-18. The running speed of the motor is controlled by varying the armature voltage $V_a(t)$. However, as the load on the motor is increased, the speed of the motor reduces. The speed can be increased again, within the limits of the motor, by increasing the armature voltage. In fact, within the operating range of the motor, the speed varies directly with the armature voltage and inversely to the motor load.

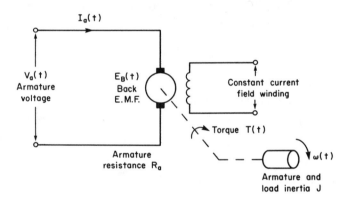

figure 3-18 diagram of a dc motor with armature voltage control

We can derive the differential equation, expressing the motor dynamics, from an examination of some of the basic laws of electromagnetism. When an armature conductor passes one of the magnetic poles within the motor, an emf, e, is induced in the conductor such that

$$e(t) = -B\ell v(t) \text{ (Lenz's Law)}$$

where B is the flux density at the pole in Webers per square metres (Teslas)

      ℓ is the length of the conductor in metres

      v is the velocity at which the conductor passes the pole in meters per second

If the radius of the armature is r, then

$$v(t) = \omega(t)r$$

where ω is the angular velocity of the motor. Now B, ℓ, and r are constants of the motor; thus the back emf, E_B, induced in the motor winding varies directly as the angular velocity, ω, i.e.,

$$E_B(t) = -K_1' \omega(t) \tag{3.45}$$

(where K_1' is a motor constant and relates to $B\ell r$). Also the force acting on the conductor is F, where

$$F = B\ell I_a$$

and I_a is the armature current. Now the torque acting on the conductor is Fr; therefore, by similar reasoning to the above the torque will vary directly as the armature current I_a; i.e.,

$$T(t) = K_2' \cdot I_a(t) \tag{3.46}$$

(where K_2' is also a motor constant, again relating to $B\ell r$). The armature resistance is R_a; hence, from Ohm's Law,

$$I_a(t) = \frac{V_a(t) + E_B(t)}{R_a} \tag{3.47}$$

Substituting equations (3.45) and (3.47) into equation (3.46), we find:

$$T(t) = \frac{K_2'}{R_a} \{V_a(t) - K_1'\omega(t)\} \tag{3.48}$$

If the load is purely inertial, then

$$T(t) = J\frac{d\omega(t)}{dt} = J\frac{d^2\theta(t)}{dt^2} \tag{3.49}$$

where J is the polar moment of inertia of the motor plus the load, and θ is the angle of rotation of the shaft at time t; i.e.,

$$\omega(t) = \frac{d\theta(t)}{dt}$$

Substituting equation (3.49) into equation (3.48), we obtain

$$J\frac{d\omega(t)}{dt} = \frac{K_2'}{R_a}\{V_a(t) - K_1'\omega(t)\}$$

Combining K_1', K_2', and R_a into two new constants K_1 and K_2, we may rewrite the equation thus:

$$J\frac{d\omega(t)}{dt} + K_1\omega(t) = K_2V_a(t) \qquad (3.50)$$

Equation (3.50) will be recognized as a first-order equation with input $V_a(t)$ and an output of $\omega(t)$. To convert this equation to a second-order differential equation, we replace $\omega(t)$ by

$$\omega(t) = \frac{d\theta(t)}{dt}$$

to obtain

$$J\frac{d^2\theta(t)}{dt^2} + K_1\frac{d\theta(t)}{dt} = K_2V_a(t) \qquad (3.51)$$

Although equation (3.51) is a second-order differential equation, the quantity $\theta(t)$ is usually of little interest to us if the motor is continuously rotating, and in this case equation (3.50) is the more suitable equation. However, if we wish to convert the motor to a dc voltage-controlled, position servomotor, the quantity $\theta(t)$ becomes of prime importance and equation (3.51) is thus applicable.

To convert the motor to a position servomotor we must fit a transducer to the output shaft that will convert the shaft's angular position into a voltage that can be compared with the input voltage V_a. Let the characteristics of the transducer be such that its output voltage is $V_0(t)$, where

$$V_0(t) = K_3\theta(t)$$

If $V_0(t)$ is now continuously substracted from $V_a(t)$ and the voltage difference is fed into the armature circuit, then equation (3.51) becomes

$$J\frac{d^2\theta(t)}{dt^2} + K_1\frac{d\theta(t)}{dt} = K_2\{V_a(t) - K_3\theta(t)\}$$

hence

$$J\frac{d^2\theta(t)}{dt^2} + K_1\frac{d\theta(t)}{dt} + K_2K_3\theta(t) = K_2V_a(t) \qquad (3.52)$$

Equation (3.52) is very similar to equation (3.42) describing the dynamics of the automobile suspension, and it is not surprising, therefore, that the position servomotor exhibits similar dynamic characteristics. If the motor is underdamped, e.g., if there is weak field excitation or high armature resistance, then hunting occurs such that when required to move to a new position, the motor moves quickly but oscillates about the desired point for some time before the oscillations die away. Similarly, if overdamped, e.g., if there is strong field excitation or low armature resistance the motor takes a long time to assume a new position.

Equation (3.52) is simplified by use of D transforms as follows:

$$JD^2\theta(D) + K_1D\theta(D) + K_2K_3\theta(D) = K_2V(D) \qquad (3.53)$$

which becomes

$$\left\{\frac{J}{K_2K_3}\cdot D^2 + \frac{K_1}{K_2K_3}\cdot D + 1\right\}\theta(D) = \frac{V(D)}{K_3}$$

Therefore,
$$\frac{\theta(D)}{V(D)} = \frac{1/K_3}{\dfrac{J}{K_2K_3}\cdot D^2 + \dfrac{K_1}{K_2K_3}\cdot D + 1} \qquad (3.54)$$

Let $K = 1/K_3$; also note that the coefficient of D^2 has the units of $(\text{time})^2$ and that the coefficient of D has the units of time. Thus let the coefficient of D^2 be τ_1^2 and the coefficient of D be τ_2. Equation (3.54) now becomes

$$\frac{\theta(D)}{V_a(D)} = \frac{K}{\tau_1^2D^2 + \tau_2D + 1} \qquad (3.54a)$$

An Electrical Example: The electrical circuit shown in Figure 3-19 is a typical second-order system described by a second-order differential equation. The circuit contains three components: a resistance of R ohms, a coil of inductance L henrys, and a capacitor with capacitance C farads. The input to the system is designated $e_i(t)$ (input voltage). The output can be considered to be any of the voltages across the components, but

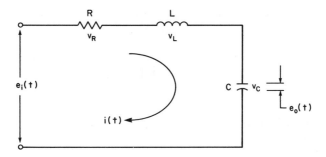

figure 3-19   a series *RLC* circuit

for this example let it be the voltage across the capacitor $e_0(t)$. The initial response to the applied voltage is the current, $i(t)$, which is forced to flow through the circuit. The output voltage (or for that matter any of the component voltages) is a secondary response induced in each component by the flow of current. However, it is usually these secondary responses that are of greater importance in such a circuit.

The system equation is derived as follows:

Applying Kirchhoff's voltage law to the single loop, we get

$$e_i(t) - v_R(t) - v_L(t) - v_C(t) = 0 \qquad (3.55)$$

but from Ohm's Law,

$$v_R(t) = Ri(t)$$

and from the theories of magnetic circuits and electrostatics we also know that:

$$v_L(t) = L\,\frac{di(t)}{dt}$$

and

$$v_C(t) = \frac{1}{C}\int^t i(t)\,dt$$

Substituting these terms into equation (3.55), we find

$$e_i(t) = Ri(t) + L\,\frac{di(t)}{dt} + \frac{1}{C}\int^t i(t)\,dt \qquad (3.56)$$

This equation is called an integro-differential equation and must be rearranged to remove the integral term, since it is very difficult to solve in this form. Another thing that must be done is to remove the term $i(t)$ and substitute $e_0(t)$, because the current is of no interest but the ouput voltage is. We will find that in substituting for $e_0(t)$ the integral term disappears, which is very convenient. Now

$$e_0(t) = \frac{1}{C} \int^t i(t)\, dt \tag{3.57}$$

Differentiating both sides of this equation gives:

$$i(t) = C \frac{de_0(t)}{dt} \tag{3.58}$$

This expression for current can now be substituted into equation (3.56) so that

$$e_i(t) = RC \frac{de_0(t)}{dt} + LC \frac{d^2 e_0(t)}{dt^2} + e_0(t) \tag{3.59}$$

rearranged in descending order of derivative gives us

$$e_i(t) = LC \frac{d^2 e_0(t)}{dt^2} e_0(t) \tag{3.59a}$$

Equation (3.59a) is a second-order differential equation that expresses the dynamic behavior of the circuit in Figure 3-19. Once again this equation can be simplified by using the D operator to become

$$e_i(D) = LCD^2 e_0(D) + RCD e_0(D) + e_0(D) \tag{3.60}$$

Therefore, $e_i(D) = (LCD^2 + RCD + 1)e_0(D)$

which gives $$\frac{e_0(D)}{e_i(D)} = \frac{1}{LCD^2 + RCD + 1} \tag{3.61}$$

From electrical theory we know that the product LC has units of (seconds)^2; also that the product RC has units of seconds. If the symbols τ_1^2 and τ_2 are, therefore, substituted for LC and RC, respectively, equation (3.61) can be rewritten as

$$\frac{e_0(D)}{e_i(D)} = \frac{1}{\tau_1^2 D^2 + \tau_2 D + 1} \tag{3.61a}$$

On comparing equations (3.44a), (3.54a), and (3.61a) we see an obvious similarity in the transfer functions for the three systems. As indicated in Section 3-5, by suitable choice of R, L, and C Figure 3-19 can be made to be an electrical analogue of either of the other two systems. By reducing R or increasing L the circuit becomes under-damped, and the output voltage oscillates about its mean value. Increasing R and reducing L can produce overdamping, which makes the output's response to a change in input very sluggish. Thus by choosing R, L, and C so that τ_1 and τ_2 for the circuit have the same values as τ_1 and τ_2 for the car chassis example in equation (3.44a), we can observe the effect of a sudden force (fast change in e_i), caused by a bump in the road, on the position and stability of the chassis by merely studying the output voltage $e_0(t)$.

In this section we have derived some second-order differential equations to describe the dynamic behavior of three different systems. From these equations we have further derived the system transfer function. None of these derivations was particularly difficult. However, as system complexity increases, we benefit from using a more visual approach for constructing the transfer function. This approach is described in Section 3-7.

3-7 BLOCK DIAGRAMS REVISITED

In Section 3-2 the very convenient and systematic method of block diagram reduction was explained and demonstrated. The mathematical statement within the final block was always found to be the transfer function of the system. In Sections 3-5 and 3-6 it was shown how, by using simple algebra, to obtain the transfer function from the system differential equation. It is reasonable to expect that the reverse procedure would produce the system differential equation from the transfer function. Hence, if we can reduce the system to one block, we can just as easily derive the transfer function and the system equation.

The advantage of this method is that it can be done by a simple step-by-step procedure that never varies, and consequently there is less chance to make a mistake. In obtaining the system equation as was shown in Sections 3-5 and 3-6 there is sometimes a confusion of "where to begin" and "what step comes next." There is very little chance of this happening when one is using block diagrams because algebraic logic makes the sequential steps fairly obvious.

So that a comparison can be made between the two methods, it would be advantageous to repeat some of the examples of the last two sections. Because we wish to use block diagrams, we must at some stage use the D transforms of the input and output variables, so that we can thus manipulate them by algebraic means.

A First-order Mechanical System *(Figure 3-13)*

Step 1.   The input and summing junction.
Two things are clear from examining the system:
a. The input is torque T and the output is angular velocity ω.
b. When the output reaches maximum, it somehow balances the input.
Question: How is the output speed, which, fed back as a torque, to be substracted from the input?

Answer:

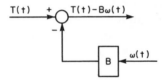

The error or actuating signal $\{T(t) - B\,\omega(t)\}$ has units of newton-meters

Step 2.   The first block.
Question: What parameter of the system is used to multiply the error to obtain a term that has something to do with speed? (For it is speed that we must ultimately end up with.)

Answer:

This operation is

$$\text{Torque} \times \frac{1}{\text{inertia}} = \text{acceleration}$$

or $(\text{Nm}) \times \left(\frac{1}{\text{kg-m}^2}\right) = \frac{1}{\text{sec}^2}$

Step 3. The next block.

Question: How is angular velocity obtained from angular acceleration?

Answer: By integration. So now we must replace all functions of time by their D transforms; thus:

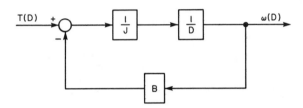

and the complete block diagram now becomes:

which reduces to

The system transfer function is therefore

$$\frac{\omega(D)}{T(D)} = \frac{1}{JD + B}$$

and may be written

$$\frac{\omega(D)}{T(D)} = \frac{\dfrac{1}{B}}{\dfrac{JD}{B} + 1}$$

which is identical to equation (3.25). The system equation is easily obtained by multiplying both sides of equation (3.25), first by T and then by $\left(\dfrac{JD}{B} + 1\right)$, which gives:

$$\frac{JD\omega(D)}{B} + \omega(D) = \frac{T(D)}{B}$$

or $$JD\omega(D) + B\omega(D) = T(D)$$

The inverse D transform of this equation is

$$J\frac{d\omega(t)}{dt} + B\omega(t) = T(t)$$

which is the system differential equation, as can be seen from comparison with equation (3.23).

A First-order Thermal System (*Figure 3-15*)

Step 1. The input and summing junction.

As before, two things are clear when we look at the system:
(a) The input is temperature T_1 and the output is temperature T_2.
(b) When the temperature of the water T_2 equals T_1, the temperature of the steam, heat will stop flowing. Thus the summing junction is as follows:

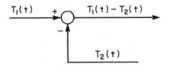

Step 2. The first block.

Question: What parameter of the system is used to multiply the error $(T_1 - T_2)$ to obtain the heat flow (q)?

Answer:

$$T_1(t) - T_2(t) \quad \boxed{\frac{1}{R}} \quad q(t)$$

This operation is

$$\text{Temperature} \times \frac{1}{\text{thermal resistance}} = \text{heat flow}$$

or $$(\,°C\,) \times \left[\frac{\text{joules}}{°C \cdot \text{sec}}\right] = \frac{\text{joules}}{\text{sec}}$$

Step 3. The next block.

Question: How is temperature obtained from heat flow?

Answer: By integrating heat flow and dividing by thermal capacitance. To do this we again introduce the D operator and substitute D transforms for the system variables; thus:

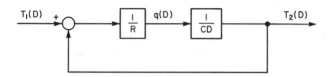

The complete block diagram now becomes

which reduces to

Hence the system transfer function is

$$\frac{T_2(D)}{T_1(D)} = \frac{1}{RCD + 1}$$

which agrees with equation (3.37).

The system differential equation is easily obtained by multiplying both sides of equation (3.37) first by $T_1(D)$ then by $(RCD + 1)$, which gives

$$RCDT_2(D) + T_2(D) = T_1(D)$$

The inverse D transform of this equation is

$$T_1(t) = RC\frac{dT_2(t)}{dt} + T_2(t)$$

which is the system differential equation, as is seen by comparison with equation (3.35).

A Second-order Electromechanical Example *(Figure 3-18)*

Step 1. The input and summing junction.

Two things are known initially about the system:

(a) The input is armature voltage V_a and the output is angular position θ.

(b) When the motor shaft reaches the required position, the output somehow balances the input.

Question: How can the output, θ, which is measured in radians, be compared with the input voltage V_a?

Answer: The output is converted to a voltage signal, by means of a transducer fitted onto the shaft, before being compared with the input.

Thus the output and input are combined at a summing junction as follows:

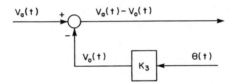

Now what do we do next? We know that the input voltage to the motor causes a current to be circulated in the armature, and that the current produces a torque that accelerates the motor. However, as the speed of the motor increases, a back emf is induced in the armature winding opposing the input voltage. Thus the summing junction is modified to become

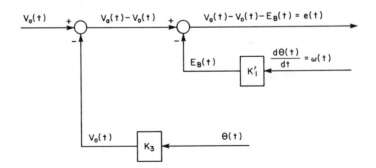

Step 2. The first block; This modified voltage causes a current I_a to be circulated in an armature whose resistance is R_a. Thus the first block is

$$\xrightarrow{\ e(t)\ }\ \boxed{\dfrac{1}{R_a}}\ \xrightarrow{\ I_a(t)\ }$$

Step 3. The second block: The current produces a torque T, and so the second block is

$$\xrightarrow{\ I_a(t)\ }\ \boxed{K'_2}\ \xrightarrow{\ T(t)\ }$$

Step 4. The remaining blocks: As stated above, the torque causes the motor to accelerate according to equation (3.49); i.e.,

$$T(t) = J\frac{d^2\theta(t)}{dt^2}$$

Thus if we double integrate T and divide the result by J, we obtain θ. To do this in the block diagram we must introduce the D operator and substitute D transforms for the system variables; that is,

We now have all the blocks defined. If we combine them all, we obtain the following block diagram.

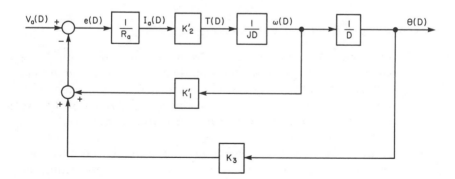

If $K_2'/R = K_2$ and $K_2' K_1'/R = K_1$, it is left to the student to show that the above block diagram can be reduced to

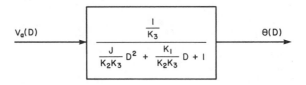

The term contained in the block is, of course, the system transfer function, as can be seen from a comparison with equation (3.54); i.e.,

$$\frac{\dfrac{1}{K_3}}{\dfrac{J}{K_2 K_3} D^2 + \dfrac{K_1}{K_2 K_3} D + 1}$$

Multiplying the left side of this equation by $V_a(D)$ and the denominator of the right-hand side, we obtain

$$JD^2\theta(D) + K_1 D\theta(D) + K_2 K_3\theta(D) = K_2 V_a(D)$$

The inverse D transform of this equation is

$$J\frac{d^2\theta(t)}{dt^2} + K_1\frac{d\theta(t)}{dt} + K_2 K_3\theta(t) = K_2 V_a(t)$$

which is identical to equation (3.52), the system differential equation.

A Second-order Electrical Example *(Figure 3-19)*

Step 1. The input summing junction: From the previous example we know that it is probable that another one of the component voltages, in addition to e_0, will be subtracted from e_i at the input summing junction to obtain the error signal. Which of the two voltages, v_R or v_L, should be subtracted from e_i in this way?

Answer: The system is of second order and will therefore contain two integrators, so let us examine the equations governing the component voltages to determine which voltages are functions of time derivatives $\left(e.g., \dfrac{di}{dt}\right)$. We can then determine which voltage is a function of the highest order of derivative and then we make that voltage the error signal thus:

The component voltages are

$$i(t) = C \frac{de_0(t)}{dt} \tag{3.62}$$

$$v_R(t) = Ri(t) \tag{3.63}$$

$$v_L(t) = L \frac{di(t)}{dt} \tag{3.64}$$

Combining equations (3.62) and (3.63) yields

$$v_R(t) = RC \frac{de_0 \, (t)}{dt^2} \tag{3.65}$$

and combining equations (3.62) and (3.64) yields

$$v_L(t) = LC \frac{d^2 e_0(t)}{dt^2} \tag{3.66}$$

and we know that

$$v_C(t) = e_0(t) \tag{3.67}$$

From an examination of equations (3.65) through (3.67) we see that $v_L(t)$ is the function with the highest order of derivative of $e_0(t)$ and when $v_L(t)$ is doubly integrated we obtain $LCe_0(t)$. Thus $v_L(t)$ will be our error signal and

$$v_L(t) = e_i(t) - v_R(t) - e_0(t)$$

Hence the summing junction is

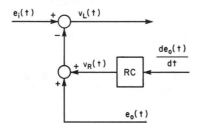

Step 2. The first block: From equation (3.66) the first block is

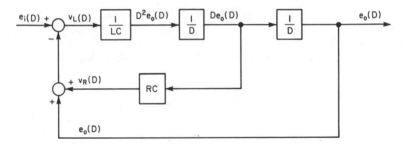

Step 3. The remaining blocks: Now to convert the output of the preceding block into $e_0(t)$ we must integrate twice. To do this we
introduce the D operator once more and transform all the system variables; thus:

Combining all the blocks, we obtain

which can be reduced to the following single block.

Therefore, the transfer function is

$$\frac{e_0(D)}{e_i(D)} = \frac{1}{LCD^2 + RCD + 1}$$

which is verified by comparison with equation (3.61). From the
transfer function we see that

$$LCD^2 e_0(D) + RCDe_0(D) + e_0(D) = e_i(D)$$

and the inverse D transform of this equation is

$$LC \frac{d^2 e_0(t)}{dt^2} + RC \frac{de_0(t)}{dt} + e_0(t) = e_i(t)$$

which is identical to equation (3.59a), the system differential equation.

3-8 SUMMARY

In this chapter we have shown how to construct a block diagram using a step-by-step approach. It was shown how the *D* operator and *D* transforms were used to advantage by allowing us to reduce the operations of calculus to simple algebraic manipulations, which in turn allowed us to use block diagrams to express differential and integral functions. A number of dynamic systems were analyzed to derive the system differential equations and demonstrate the use of the *D* operator. Finally, block diagrams were constructed for each dynamic system. We did this to demonstrate the method but also to show how even complex transfer functions, and hence differential equations, may be derived in a simple step-by-step manner.

It should now be possible to build a mathematical model of most linear control systems regardless of their apparent complexity. From the model the transfer function is developed, which is of paramount importance in the analysis of linear systems. In subsequent chapters it will be shown that the transfer function is used to gain knowledge of the characteristics of the system under investigation.

PROBLEMS

Section 3-2   1. Redraw the partial block diagrams shown in Figure 3-20 with the summing junctions moved to the right of *G*.

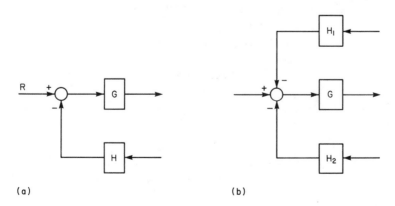

(a)   (b)

2. Redraw the partial block diagram shown in Figure 3-21 with the summing junctions combined and moved to the right of G_2.

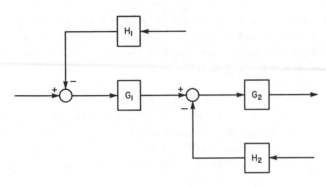

figure 3-21

3. Redraw the partial block diagram shown in Figure 3-22 with the tie point moved (a) to the right of G_3; (b) to the left of G_2.

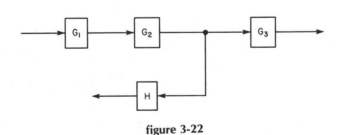

figure 3-22

4. Reduce the block diagram shown in Figure 3-23 to one block having the same system input and output.

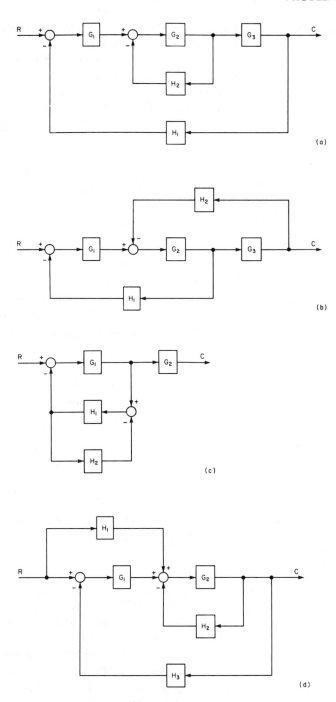

figure 3-23

Section 3-5

5. Figure 3-24 shows a diagramatic representation of the hairspring mechanism in a watch. The escape mechanism applies a torque T to the hairspring, which has a constant K. The hairspring in turn rotates the circular pendulum to an angular displacement, θ. If the pendulum has a mass M, derive the differential equation that relates θ to T.

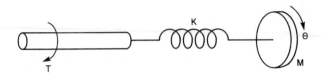

figure 3-24

6. If switch S is suddenly closed, applying an emf E to the choke in series with a resistor, as shown in Figure 3-25, derive the differential equation that describes the voltage change across R.

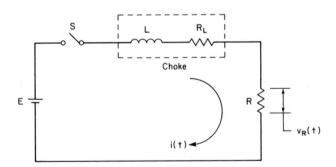

figure 3-25

7. A cake is popped into an oven to bake. The oven temperature is 190°C and the cake batter is at 20°C initially. If the cake batter has a thermal conductivity of K watts/°C and a thermal capacitance of C joules/°C, derive the differential equation that describes the rise of temperature within the cake.

8. Figure 3-26 shows a hydraulic actuator used in a power steering system. The spool valve shaft is operated by the steering mechanism, and the valve

body shaft is connected to the wheels of the vehicle. If the spool valve constant is K_v m²/sec and the main piston area is A m², derive the transfer function relating x_o to x_{in}.

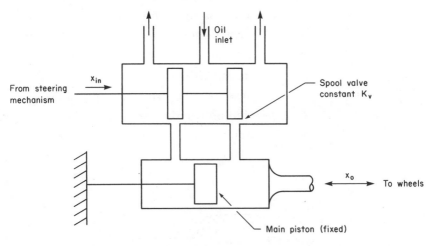

figure 3-26

Section 3-6

9. Derive the transfer functions for the circuits shown in Figure 3-27.

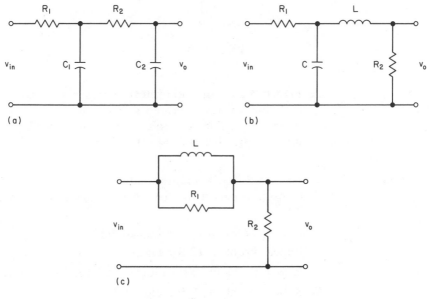

figure 3-27

10. Derive the transfer function that relates output speed to input voltage of the two-phase servo motor shown in Figure 3-28. The motor parameters are

R_C = control winding resistance   Ω

L_C = control winding inductance   H

K_m = motor torque constant   Nm/amp

F = total motor friction   Nms

J = total motor inertia   kg-m²

V_C = control winding voltage   volts

V_{ref} = reference winding voltage   volts

ω = angular velocity of the output   rad/s
shaft

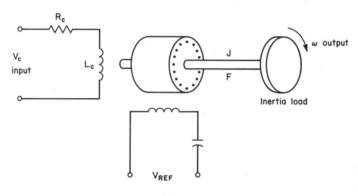

figure 3-28

11. Figure 3-29 more accurately represents an automobile than was indicated in the text. Derive the transfer function that relates the output displacement of the chassis (x_1) to the input road function (x_3).

Section 3.7   12. Draw the block diagram and hence derive the transfer function for the system shown in Figure 3-24.

13. Repeat Problem 12 for Figure 3-25.

14. Repeat Problem 12 for Figure 3-26.

15. Repeat Problem 12 for Figure 3-27 (a), (b), and (c).

16. Repeat Problem 12 for Figure 3-28.

17. Repeat Problem 12 for Figure 3-29.

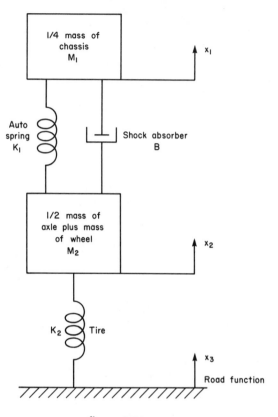

figure 3-29

system responses

"Curiouser and Curiouser," cried Alice.

Lewis Carroll, Alice in Wonderland

4-1 INTRODUCTION

The manner in which the output (or controlled variable) of a control system responds to an input demand or disturbance depends entirely upon the dynamic characteristics of the system. How these characteristics influence the response and how the response varies for different input disturbances will be examined in this chapter.

4-2 A RESPONSE REQUIRES AN INPUT

4-2.1 the input form—descriptive

Input, input variable, reference input, forcing function, input disturbance, and *driving function* are terms used to describe the signal that appears at the input to a control system. Although in a real control system the input signal may sometimes be complex and not easily described, we often make use of simple, more common functions to represent the input signal when we are attempting to analyze the control system or study its response characteristics. Some typical input functions that are often considered when one is analyzing control systems are shown in Figure 4-1.

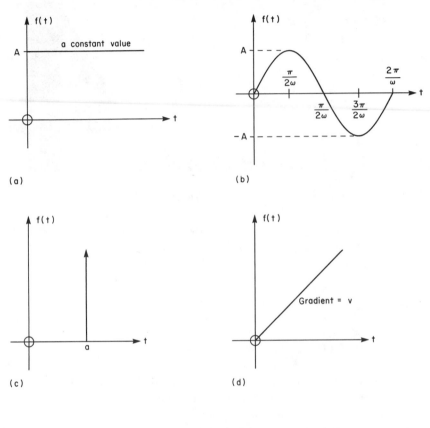

(a)

(b)

(c)

(d)

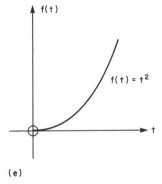

(e)

figure 4-1 inputs to control systems: (a) step function, (b) sine function, (c) unit impulse function, (d) ramp function, (e) parabolic function

Examples of such functions are abundant, except in the case of the impulse function, which can, however, be approximated with a high degree of accuracy. For example, switching a dc source into a circuit is a step function. Bursting a frangible pressure safety disc by means of sustained overpressure will also produce a step function change in fluid pressure. Sine waves are produced by all electrical generators, both ac and dc. Water flowing into a tank at a constant rate will produce a ramp function increase in the water pressure at the bottom of the tank. The distance travelled by a car moving under constant acceleration can easily be seen to be a parabolic function. True impulse functions are impossible to generate, but for practical purposes very short duration pulses are used such that the pulse which is applied to a system has so short a duration compared to the time constants of the system, that the pulse appears to be a true impulse.

4-2.2 the input form—mathematical

Step Function: The step function is described as a function of time which has a value of zero up to time $t = 0$ and for any time greater than zero has a constant value.

$$f(t) = \begin{vmatrix} 0 & t < 0 \\ A & t \geq 0 \end{vmatrix} \tag{4.1}$$

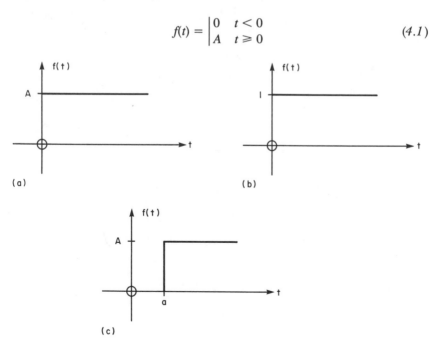

figure 4-2 step functions: (a) general step function, (b) unit step function, (c) delayed step function

A unit step function $U(t)$ is described as a function of time which has a value of zero up to time $t = 0$ and for any time greater than zero has a constant value of unity.

$$f(t) = U(t) = \begin{vmatrix} 0 & t < 0 \\ 1 & t \geq 0 \end{vmatrix} \qquad (4.1a)$$

A delayed step is described as a function of time which has a value of zero up to time $t = a$ and for any time greater than $t = a$ has a constant value, say A.

$$f(t - a) = \begin{matrix} 0 & t < a \\ A & t \geq a \end{matrix} \qquad (4.1b)$$

Sine Function: A sine function is described as a function of time which is periodically changing in sinusoidal fashion at a frequency of ω rad/second. Depending upon the magnitude of the phase angle, ϕ, at time $t = 0$ the sine function is said to have a "lagging" or a "leading" phase. In Figure 4-3, function $f_a(t)$ has a leading phase with respect to $f_b(t)$ and function $f_c(t)$ has a phase lag with respect to $f_b(t)$ such that

$$f_a(t) = A \sin(\omega t + \phi_1)$$
$$f_b(t) = A \sin(\omega t)$$
$$f_c(t) = A \sin(\omega t - \phi_2) \qquad (4.2)$$

When $\phi = \pi/2$ radians or 90 deg, we often express the function as a

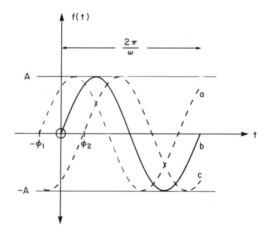

figure 4-3 sine functions

complementary sine or cosine function, but not withstanding this convenience it is still a sine function.

Impulse Function: A unit impulse function, more properly known as the Dirac-delta function (or delta function), is described as a function of time whose value is zero except at $t = a$ and where a can be any time equal to or greater than zero, such that (see Figure 4-4)

$$f(t) = \delta(t - a) = \begin{vmatrix} 0 & t \neq a \\ \infty & t = a \end{vmatrix} \qquad (4.3a)$$

and

$$\int_{-\infty}^{\infty} \delta(t - a) = 1 \qquad (4.3b)$$

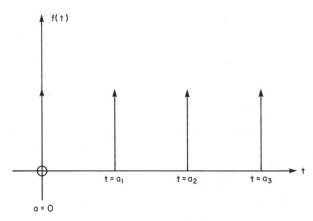

figure 4-4 impulses

A unit impulse, for example, can be derived from a rectangular pulse (Figure 4-5(a)) or from a triangular waveform (Figure 4-5(b)).

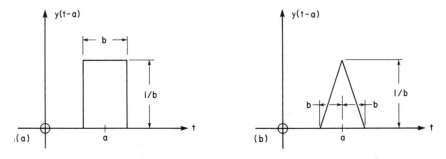

figure 4-5 (a) a rectangular wave starting at $t = a - \dfrac{b}{2}$ seconds, (b) a triangular wave starting at $t = a - b$ seconds

The area of both waveforms is unity, and this value will be maintained when either base decreases, since the amplitude is $\frac{1}{b}$. If b approaches zero length, then the amplitude will approach infinity, but the area will remain at unity. An impulse is said to have a nonzero value only at the point at which it occurs.

Ramp Function: A ramp function is a function of time which rises (or falls) in linear fashion at a constant rate. It may start at time $t = 0$, or it may be delayed so that

$$f(t) = r(t - a) = \begin{array}{ll} 0 & t < a \\ v(t - a) & t \geq a \end{array} \qquad (4.4)$$

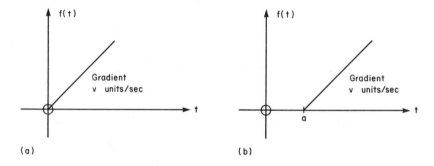

figure 4-6 ramps: (a) nondelayed ramp, (b) delayed ramp

Parabolic Function: The parabolic function follows a square law and is, therefore, nonlinear; it is described mathematically as follows:

$$f(t) = \begin{array}{ll} 0 & t < 0 \\ at^2 & t \geq 0 \end{array} \qquad (4.5)$$

4-3 THE GENERAL RESPONSE

How a system responds to an input can be found by solving the system differential equation when the independent variable in the equation is replaced by the input function. To understand what is meant by this, let us consider an easy example taken from Chapter 3. In Section 3-5

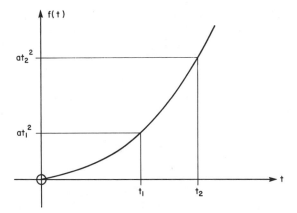

figure 4-7 parabolic function

we considered an electrical circuit (Figure 3-14) with an applied emf given as $e(t)$. We obtained an equation for this system, namely,

$$e(t) = RC \frac{dv_c(t)}{dt} + v_c(t) \qquad (3.2a)$$

where $e(t)$ is the input or independent variable and $v_c(t)$ is the dependent variable or output response. The system is shown again in Figure 4-8(a).

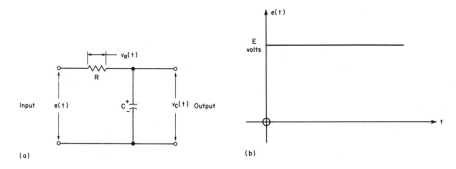

figure 4-8 (a) a simple RC circuit, (b) a step function of voltage

If the input is a step function of voltage, as shown in Figure 4-8(b),

where
$$e(t) = \begin{vmatrix} 0 \text{ volts} & t < 0 \\ E \text{ volts} & t \geqslant 0 \end{vmatrix}$$

and the equation for $e(t)$ is substituted into equation (3.29), we obtain

$$E = RC \frac{dv_c(t)}{dt} + v_c(t) \qquad (4.6)$$

The output response of this circuit, $v_c(t)$, can now be found by solving equation (4.6) in the following manner.

Since
$$E = RC \frac{dv_c(t)}{dt} + v_c(t)$$

then
$$RC \frac{dv_c(t)}{dt} = E - v_c(t)$$

Separating the variables, we obtain

$$\frac{dv_c(t)}{E - v_c(t)} = \frac{1}{RC} dt$$

Integrating both sides of the equation thus

$$\int \frac{dv_c(t)}{E - v_c(t)} = \int \frac{1}{RC} dt$$

yields the following result:

$$\text{Ln} \left[E - v_c(t) \right] = -\frac{t}{RC} + K \qquad (4.6a)$$

where K is a constant of integration.

From inspection of the circuit we know that when $t = 0$ the voltage across the capacitor $v_c(t)$ also equals zero and hence $K = \text{Ln}(E)$.

We can now substitute for K in equation (4.6a).

$$\text{Ln} \left[E - v_c(t) \right] = -\frac{t}{RC} + \text{Ln}(E)$$

or
$$\text{Ln} \left[\frac{E - v_c(t)}{E} \right] = -\frac{t}{RC} \qquad (4.6b)$$

Taking natural antilogarithms of both sides of equation (4.6b), we obtain

$$\frac{E - v_c(t)}{E} = e^{-t/RC}$$

and hence
$$v_c(t) = E - Ee^{-t/RC} \qquad (4.7)$$

or
$$v_c(t) = E(1 - e^{-t/RC}) \qquad (4.7a)$$

To verify that equation (4.7) is the correct solution, we can substitute this expression for v_c in equation (4.6) to obtain

$$(E - Ee^{-t/RC}) + RC \left(\frac{E}{RC} e^{-t/RC} \right)$$

This expression can be easily shown to equal E, which is true according to equation (4.6).

Equation (4.7) is illustrated graphically in Figure 4-9.

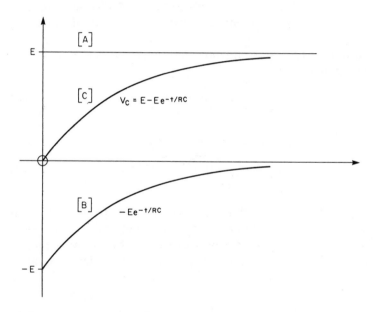

figure 4-9 the response of a first-order system to a step input

In Figure 4-9, curve [A] represents the first term on the right-hand side of equation (4.7) and curve [B] represents the second term. It can easily be seen that the sum of [A] and [B] constitutes curve [C]. It can further be seen that curve [A] starts at time $t = 0$ and rises abruptly to a magnitude of E volts, where it remains at this steady value, regardless of the increase of time t. This is the steady-state portion of the response and is always of the same form as the input—on this occasion a step function of voltage. It is also worth noting that in this case curve [A] has the same magnitude as the input, because the gain of this system is unity. Curve [B], on the other hand, has no steady value; it is constantly changing as time changes. This time-dependent term has a maximum value of $-E$ volts when $t = 0$ and decays to zero as time approaches infinity. It is, therefore, regarded as the transient portion of the response of the system and for all stable systems will die out as $t \to \infty$ (time approaches infinity). Its actual effect on the total response will be insignificant once t is such that $t >> RC$; e.g., when $t = 5RC$ the magnitude of [B] is only 0.7 percent of the magnitude of [A].

Because the output response comprises two parts, the steady state and the transient response, we can state

$$v_c(t) = v_{ss}(t) + v_T(t) \qquad (4.8)$$

where $v_{ss}(t)$ = the steady-state response as a function of time
and, $v_T(t)$ = the transient response as a function of time.

Equation (4.8) can be introduced into equation (4.6) to give a more complete equation.

$$E = RC \frac{d(v_{ss}(t) + v_T(t))}{dt} + (v_{ss}(t) + v_T(t)) \qquad (4.9)$$

therefore

$$E = RC \frac{dv_{ss}(t)}{dt} + v_{ss}(t) + RC \frac{dv_T(t)}{dt} + v_T(t) \qquad (4.9a)$$

Figure 4-9 indicates that the steady-state response at any time onward from $t = 0$ is equal to E volts. If this is so and if at the same time equation (4.9a) is true, then from equation (4.8) the transient terms of equation (4.9a) must always sum to zero, and equation (4.9) can be written as two independent equations; thus

$$RC \frac{dv_{ss}(t)}{dt} + v_{ss}(t) = E \qquad (4.9b)$$

$$RC \frac{dv_T(t)}{dt} + v_T(t) = 0 \qquad (4.9c)$$

Equations (4.9b) and (4.9c) are called the nonhomogeneous and the homogeneous equation, respectively, and in more complex systems these equations may be solved separately to obtain the steady-state response and the transient response from which the overall response can be calculated as in equation (4.8). This technique is demonstrated as follows.

4-3.1 the steady-state response

The steady-state response is found by solving equation (4.9b)

where
$$RC \frac{dv_{ss}(t)}{dt} + v_{ss}(t) = E$$

$$v_{ss}(t) = E - RC \frac{dv_{ss}(t)}{dt} \qquad (4.9d)$$

by inspection of Figure 4-9 it can be seen that

$$\frac{dv_{ss}(t)}{dt} = \frac{d(E)}{dt} = 0$$

since the time derivative of a constant is always zero. Therefore, the solution to equation (4.9d) is

$$v_{ss}(t) = E \qquad (4.10)$$

4-3.2 the transient response

The transient response is sometimes called the natural response of a system, because the behavior of the system during the transient period tells us a lot about the characteristics or the nature of the system. The transient period is a period of change when the output (or state) of the system is changing from an initial value to a new value. To produce such a change requires the input signal to have changed somehow, and the transient response of the system is characteristic of the way that it resists any attempt that is made to change its initial state. The characteristics of the system are deduced by observing the shape of the transient response. The absolute magnitude of the initial value and the final value of the output are not involved in determining this shape (i.e., solving the transient equation); only the relative difference between the two values is required.

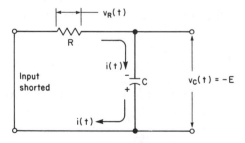

figure 4-10 a simple RC circuit with initial voltage $-E$ on the capacitor—the discharge condition

Thus if we examine the transient equation (4.9c) for the previous example, we find the solution.

$$RC \frac{dv_T(t)}{dt} + v_T(t) = 0 \qquad (4.9c)$$

Separating the variables gives

$$-\frac{dv_T(t)}{v_T(t)} = \frac{1}{RC} dt$$

Integrating both sides of this equation yields

$$Ln\,[v_T(t)] = -\frac{t}{RC} + K$$

where K is a constant of integration. This equation can be rearranged to give

$$v_T(t) = Ae^{-t/RC} \qquad (4.11)$$

where $\qquad A = e^K$

Now in our example the final value of the output is E volts higher than the initial value; alternatively we may say that the initial value of $v_T(t)$ was $-E$ volts with respect to the final value, i.e., when $t = 0$.

$$v_T(t) = -E$$

Therefore, from equation (4.11)

$$v_T(0) = A$$

hence $\qquad A = -E$

and $\qquad v_T(t) = -Ee^{-t/RC} \qquad (4.12)$

If we now add the results of equations (4.10) and (4.12) we see, once again, that

$$v_c(t) = v_{ss}(t) + v_T(t) = E - Ee^{-t/RC}$$

It is important to note that the form of first-order equations in transient analysis will always be

$$K_1 \frac{df(t)}{dt} + f(t) = 0$$

and their solutions will always be

$$f(t) = K_2 e^{-t/K_1} \qquad (4.13)$$

where K_1 and K_2 are constants. K_1 is dependent upon the system parameters, and K_2 is dependent upon the system input.

4-3.3 the total response

The analysis that we have confronted step by step in the foregoing sections is termed classical analysis. The classical method affords better insight into the interpretation of differential equations than does any of the operational methods that we shall use nearly exclusively later on. However, it presents the beginning student with some ponderous mathematical tasks to perform, and it is for this reason that we did not attempt to cover any but the simplest of systems excited by the simplest of inputs. It must be added that in terms of linear systems the elegance of the classical approach cannot be disputed. As a final example of the classical method, we shall again use a simple first-order system, but this time with a sine input. None of the following mathematics is beyond the scope of the student who has covered algebra, trigonometry, and basic calculus.

EXAMPLE 4.1

Find the total response of the circuit current $i(t)$ to an input voltage of $e(t) = E \sin \omega t$ in the circuit shown below.

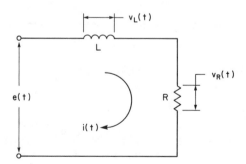

solution

This is a simple series circuit with a pure inductance L in series with a resistance R. The initial response to the applied sine input voltage is a current $i(t)$. Secondary responses (not required) are the voltage drops across the component $v_L(t)$ and $v_R(t)$.

The circuit shown forms a closed loop, and therefore Kirchhoff's voltage law may be applied so that

$$e(t) = v_L(t) + v_R(t)$$

The equation can be restated in terms of the current and circuit components thus:

$$e(t) = L\frac{di(t)}{dt} + Ri(t)$$

Since

$$e(t) = E \sin \omega t$$

then

$$E \sin \omega t = L\frac{di(t)}{dt} + Ri(t)$$

which yields

$$\frac{E}{R} \sin \omega t = \frac{L}{R}\frac{di(t)}{dt} + i(t)$$

Ohm's law states that $\dfrac{E}{R}$ is a current; therefore, let

$$I = \frac{E}{R}$$

then

$$I \sin \omega t = \frac{L}{R}\frac{di(t)}{dt} + i(t)$$

We know that the total response = steady state + transient,

or

$$i(t) = i_{ss}(t) + i_T(t)$$

and hence we can say that

$$I \sin \omega t = \frac{L}{R}\frac{di_{ss}(t)}{dt} + i_{ss}(t) + \frac{L}{R}\frac{di_T(t)}{dt} + i_T(t)$$

(i) the steady-state response

As before, we need only concern ourselves with the steady-state terms of the differential equation

$$I \sin \omega t = \frac{L}{R}\frac{di_{ss}(t)}{dt} + i_{ss}(t)$$

Since the steady-state response has the same form as the input, then $i_{ss}(t)$ will be a sine function. Because this solution must be true for all cases and not just for a specific case, we must use the generalized form for a sine function; i.e.,

$$a \cos \omega t + b \sin \omega t$$

where a and b are arbitrary constants to be determined.

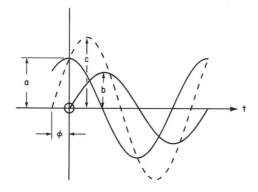

figure 4-11 the general form of the sine function a cos ωt + b sin ωt = c sin ω(t + a)

We can now say that the generalized form of the steady-state current is

$$i_{ss}(t) = a \cos \omega t + b \sin \omega t$$

and hence

$$I \sin \omega t = \frac{L}{R}\frac{d(a \cos \omega t + b \sin \omega t)}{dt} + a \cos \omega t + b \sin \omega t$$

Taking the appropriate derivatives yields

$$I \sin \omega t = -\frac{L}{R}a\omega \sin \omega t + \frac{L}{R}b\omega \cos \omega t + a \cos \omega t + b \sin \omega t$$

Collecting like terms, we obtain

$$I \sin \omega t = \left(b - \frac{L}{B} a \omega \right) \sin \omega t + \left(a + \frac{L}{R} b \omega \right) \cos \omega t$$

Equating the coefficients of similar terms on either side of the equation yields

$$b - \frac{L}{R} a \omega = I$$

and

$$a + \frac{L}{R} b \omega = 0$$

From these two equations we find that

$$a = - \frac{I \omega L / R}{1 + \dfrac{\omega^2 L^2}{R^2}}$$

and

$$b = \frac{I}{1 + \dfrac{\omega^2 L^2}{R^2}}$$

If we now substitute these two equations into the general solution, we get

$$i_{ss}(t) = \frac{I}{1 + \dfrac{\omega^2 L^2}{R^2}} \sin \omega t - \frac{I \omega L / R}{1 + \dfrac{\omega^2 L^2}{R^2}} \cos \omega t$$

The expression can be rearranged as follows:

$$i_{ss}(t) = \frac{I}{\left[1 + \dfrac{\omega^2 L^2}{R^2} \right]^{\frac{1}{2}}} \left[\frac{1}{\left[1 + \dfrac{\omega^2 L^2}{R^2} \right]^{\frac{1}{2}}} \right.$$

$$\left. \cdot \sin \omega t - \frac{\omega L / R}{\left[1 + \dfrac{\omega^2 L^2}{R^2} \right]^{\frac{1}{2}}} \cdot \cos \omega t \right]$$

Using the standard trigonometrical relationship, we obtain

$$a \sin x - b \cos x = (a^2 + b^2)^{\frac{1}{2}} \cdot \sin (x - \phi)$$

where

$$\phi = \tan^{-1} \frac{a}{b}$$

gives

$$i_{ss}(t) = \frac{I}{\left[1 + \dfrac{\omega^2 L^2}{R^2} \right]^{\frac{1}{2}}} \cdot \sin (\omega t - \phi)$$

where

$$\phi = \tan^{-1} \frac{\omega L}{R}$$

This is the steady-state current, which is a sine wave at the same frequency as the input voltage but lagging the voltage (inductive circuit) by a phase angle ϕ and having a maximum amplitude of

$$\frac{I}{\left[1 + \dfrac{\omega^2 L^2}{R^2} \right]^{\frac{1}{2}}} \text{ amps}$$

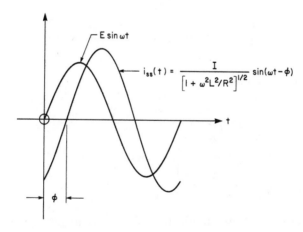

figure 4-12   the steady-state response $i_{ss}(t)$

(ii) the transient response

The transient response is obtained in precisely the same manner as before by letting the transient terms of the differential equation equal zero.

$$\frac{L}{R} \cdot \frac{di_T(t)}{dt} + i_T(t) = 0$$

The general solution is found by applying equation (4.13).

$$i_T(t) = Ke^{-tR/L}$$

where K is a constant to be determined.

The transient response is, therefore, an exponentially decaying current whose rate of decay depends upon the inductance and resistance and whose maximum value is K at time $t = 0$.

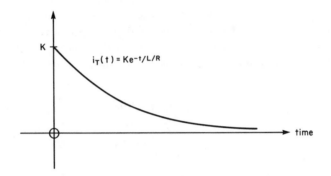

figure 4-13 the transient response $i_T(t)$

(iii) the total response

Since $i(t) = i_{ss}(t) + i_T(t)$

then $i(t) = \dfrac{I}{\left[1 + \dfrac{\omega^2 L^2}{R^2}\right]^{\frac{1}{2}}} \cdot \sin(\omega t - \phi) + Ke^{-tR/L}$

When time $t = 0$ no current will be flowing in the circuit and $i(t) = 0$; hence

$$0 = \frac{I}{\left[1 + \dfrac{\omega^2 L^2}{R^2}\right]^{\frac{1}{2}}} \cdot \sin(-\phi) + K$$

from which

$$K = \frac{I\omega L/R}{1 + \dfrac{\omega^2 L^2}{R^2}}$$

and
$$i_T(t) = \frac{I\omega L/R}{1 + \dfrac{\omega^2 L^2}{R^2}} \cdot e^{-tR/L}$$

The total response is, therefore,

$$i(t) = \frac{I}{\left[1 + \dfrac{\omega^2 L^2}{R^2} \right]^{\frac{1}{2}}} \cdot \left[\sin (\omega t - \phi) + \frac{\omega L/R}{\left[1 + \dfrac{\omega^2 L^2}{R^2} \right]^{\frac{1}{2}}} \cdot e^{-tR/L} \right]$$

where

$$\phi = \tan^{-1} \frac{\omega L}{R}$$

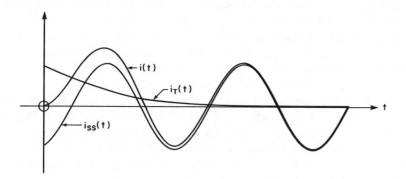

figure 4-14   the overall time response $i(t)$

It would be pointless at this juncture to demonstrate more worked examples. The solutions, although repetitious, would be just as lengthy, and there is an easier way. Summarizing the steps involved, we can say

1. Develop the system equation.
2. Equate the differential equation to the input function.
3. Substitute the steady-state and the transient terms into this equation.

4. Solve for the steady-state response by introducing the generalized form for the steady- state response, taking the derivatives where necessary and evaluating all constants.

5. Solve for the transient response when the transient terms equal zero.

6. Combine both the steady-state and transient solutions into one complete equation and, by introducing initial conditions, evaluate the constants of the transient response.

4-4 TIME CONSTANTS

The initial reaction of a system to any input disturbance is always characterized by some form of exponential change occurring in the output signal, which is described by an exponential term in the response equation. The speed with which the system assumes its steady-state condition is contingent on how fast such transients die out or decay, which is entirely dependent on the absolute magnitude of the coefficient of time in the exponent of the exponential term. In the two circuits investigated in this chapter we saw that the exponential terms were $e^{-t/RC}$ and $e^{-tR/L}$. The exponents of these two terms are $-t/RC$ and $-tR/L$, respectively. Thus as the magnitude of $1/RC$ and R/L is increased, the rate of decay of the corresponding exponential terms will increase.

When the circuit is designed, R, C, and L become fixed values, and the product RC and the quotient L/R are constant values. If R is measured in $M\Omega$ and C in μF, the product will yield units of time, i.e., seconds. Also, if L is in henrys and R in ohms, the quotient L/R will again have the units of time, i.e., seconds. RC and L/R are called the time constants of the respective exponential terms and are often denoted by the Greek letter τ (tau).

The time constants in a system completely determine the initial response of that system and how fast or how slow it will settle down to the steady-state condition. In very simple linear systems one can say that for all practical purposes the system will take five time constants to achieve steady-state conditions. For example,

when $$t = 5\tau = 5RC, \quad \text{say}$$

then $$e^{-t/RC} = e^{-5} = 0.00674$$

In equation (4.7) this would mean that

$$v_c(t) = E - 0.00674E$$

or $$v_c(t) = 0.99326E$$

i.e., the voltage across the capacitor has risen to 99.326 percent of the applied voltage, which is close enough in a real circuit to be considered fully charged.

The more complicated a system, the more time constants there are to consider. In many systems there are large time constants and small time constants. The large time constants are considered to be dominant time constants as they control the length of time that the system takes to settle down to a steady-state condition.

The following servo output device, commonly referred to as a dc torquer, is an example of a system whose dynamics include two widely differing time constants.

Device:	DC torquer size 9
Mechanical time constant:	25 ms
Electrical time constant:	300 μs
Field inductance:	7.5 mH
Field resistance at 25°C:	24.9 Ω

The dominant time constant is definitely the mechanical one of 25 ms, which is about 83 times larger than the electrical time constant (L/R), which is 300 μs. This means that to run the torquer up to full speed would take about 1/8 second, and although this is only a very short time, the maximum field current would be flowing in only 1½ milliseconds.

Apart from any particular implications of the time constants in subsequent analyses, their most important contribution to system operation is that they determine the length of the transient response.

We will be confronted by time constants many times throughout the text when such further implications will be pursued.

4-5 FROM LEIBNITZ TO LAPLACE

Gottfried Leibnitz was born in the year 1646 in Leipzig, which is now part of East Germany. Apart from the many astounding scientific and mathematical developments he made throughout his life, he developed a system of calculus at the age of 27. Eleven years later he published his work. Although prior to this event Sir Isaac Newton had also developed a similar system, which he called "fluxions," he was so secretive about his work that Leibnitz "made the scoop." Perhaps it would not have made any difference if the reverse had been true, because Leibnitz's development was better than Newton's, and it is his notation and form that is used now. In France, in the year 1749, Pierre Simon, Marquis de Laplace, was born. It is to this illustrious astronomer and mathematician that we can attribute the development of a method which greatly simplifies the solution of differential equations with constant coefficients.

One final historical word: undoubtedly it is to Oliver Heaviside, the English electrical engineer, and a nephew of Charles Wheatstone, that we owe our greatest debt of gratitude, for it is he who developed on a large scale the rules governing the "operational calculus." Heaviside was a practical man and never bothered to prove any of his mathematics from first principles. When criticized severely by the mathematicians of the day, he responded with his famous remark, "Shall I refuse my dinner because I do not fully understand the process of digestion?" It is with this philosophy in mind that we now wish to familiarize the reader with the more simple operational method of analysis.

Consider the multiplication of two numbers, such as

$$20 \times 42 = 840$$

This is a comparatively easy mental process. However, when the numbers are perhaps very large, most of us have great difficulty in accommodating the operation mentally. For instance,

$$3122 \times 157 = 490,154$$

is a difficult operation to perform without pencil and paper and gets progressively more difficult as more numbers must be multiplied or divided.

This product, as we know, can be specified in a different way, such as

$$10^{3.4944} \times 10^{2.1959} = 10^{5.6903}$$

Similarly, the quotient,

$$\frac{3122}{157} = 19.8854$$

can be written as

$$10^{3.4944} \times 10^{-2.1959} = 10^{1.2985}$$

All that we are now doing is simply adding or subtracting, where appropriate, the powers of ten to obtain our solution. If we have a comprehensive table of powers of ten, we can simply add or substract any series of multiplications or division. We are, in fact, doing logarithms.

In operational calculus we use a similar approach to solve differential equations with constant parameters in that we convert or transform derivative and integral functions of a variable (usually time) into simple algebraic expressions and then treat the complete equation as a simple piece of algebra, operating upon it with simple algebraic rules. When we finally evolve a solution, we transform the expression back into its original form. More precisely, we transform the integro-differential equations from the time domain into the operational domain, perform the appropriate algebra, and then carry out an inverse transform back to the time domain. Just as we do with logarithms, we do all our conversions by means of conversion tables.

As an example of the Laplace transform method, let us reconsider the previous problem of Example 4.1. Before we begin, it is important to understand that in the following illustrative example the precise form of the Laplace transforms are incompletely specified. Because it is our intention to demonstrate the simplicity of the transform method, we have purposely omitted the initial condition term. This and other details will be properly covered in the next chapter.

EXAMPLE 4.2

Repeat Example 4.1, using Laplace transforms.

solution

The defining equation is

$$e(t) = L\frac{di(t)}{dt} + Ri(t)$$

Substituting for $e(t)$ gives

$$E \sin \omega t = L\frac{di(t)}{dt} + Ri(t)$$

which yields

$$\frac{E}{R} \sin \omega t = \frac{L}{R}\frac{di(t)}{dt} + i(t)$$

Since

$$\frac{L}{R} = \tau$$

then

$$\frac{E}{R} \sin \omega t = \tau\frac{di(t)}{dt} + i(t)$$

We now use Laplace transforms on both sides of the equation to transform those terms which are functions of time into functions of the Laplace operator s. Tables of Laplace transforms can be found in most textbooks that deal with feedback and control theory. A table of such transforms is presented in Table 5.1 of this book. From that table we find that constants do not change and that

$$\sin \omega t \text{ transforms to } \frac{\omega}{s^2 + \omega^2}$$

$$\frac{di(t)}{dt} \text{ transforms to } sI(s)$$

$$i(t) \text{ transforms to } I(s)$$

The transformed equation now becomes

$$\frac{E}{R} \cdot \frac{\omega}{s^2 + \omega^2} = \tau s I(s) + I(s)$$

or

$$\frac{E}{R} \cdot \frac{\omega}{s^2 + \omega^2} = I(s) (\tau s + 1)$$

Solving for $I(s)$ gives

$$I(s) = \frac{E}{R} \cdot \frac{\omega}{(s^2 + \omega^2) (\tau s + 1)}$$

Referring again to the Laplace tables, we can transform this equation back into an equation where variables are functions of time; thus $I(s)$ transforms back to $i(t)$

and

$$\frac{\omega}{(s^2 + \omega^2)(\tau s + 1)}$$

transforms to

$$\frac{1}{(1 + \omega^2\tau^2)^{1/2}} \left[\sin (\omega t - \phi) + \frac{\omega\tau}{(1 + \omega^2\tau^2)^{1/2}} e^{-t/\tau} \right]$$

and hence

$$i(t) = \frac{I}{\left[1 + \dfrac{\omega^2 L^2}{R^2} \right]^{1/2}} \left[\sin (\omega t - \phi) + \frac{\omega L / R}{\left[1 + \dfrac{\omega^2 L^2}{R^2} \right]^{1/2}} \cdot e^{-Rt/L} \right]$$

where $$I = \frac{E}{R} \quad \text{and} \quad \phi = \tan^{-1}\frac{\omega L}{R}$$

EXAMPLE 4.3

The angular velocity, ω, of a fan shaft changes when a torque is applied in a manner described by equation (3.23)

$$J\frac{d\omega(t)}{dt} + B\omega(t) = T(t)$$

If applied torque is a ramp function, determine the response of the shaft in terms of its angular velocity.

solution

Let the input ramp function of torque be

$$T(t) = T_0 \cdot t$$

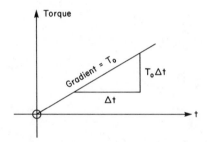

where T_0 is the linear rate of change of torque in Nm/s and t is any time in seconds. Thus the angular velocity equation becomes

$$J\frac{d\omega(t)}{dt} + B\omega(t) = T_0 t$$

Dividing both sides of the equation by B yields

$$\frac{J}{B}\frac{d\omega(t)}{dt} + \omega(t) = \frac{T_0}{B}t$$

Let $$\tau = J/B$$

then $$\tau\frac{d\omega(t)}{dt} + \omega(t) = \frac{T_0}{B}t$$

Taking Laplace transforms of all the functions of time but leaving the constants unchanged, we find from our Laplace transform tables that

$$\frac{d\omega(t)}{dt} \quad \text{transforms to} \quad s\Omega(s)$$

$$\omega(t) \quad \text{transforms to} \quad \Omega(s)$$

$$t \quad \text{transforms to} \quad \frac{1}{s^2}$$

Therefore

$$\tau s\Omega(s) + \Omega(s) = \frac{T_0}{B} \cdot \frac{1}{s^2}$$

Solving for $\Omega(s)$, we obtain

$$\Omega(s) = \frac{T_0}{B} \cdot \frac{1}{s^2(\tau s + 1)}$$

From the tables,

$$\Omega(s) \text{ transforms back to } \omega(t)$$

and $\quad \dfrac{1}{s^2(\tau s + 1)}$ transforms back to $e^{-t/\tau} + t/\tau - 1$

Hence

$$\omega(t) = \frac{T_0}{B}\left(e^{-t/\tau} + \frac{t}{\tau} - 1\right)$$

The angular velocity can easily be represented by the series of curves shown below.

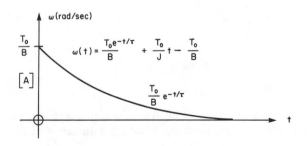

(a)

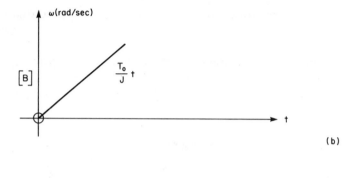

(b)

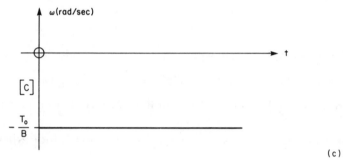

(c)

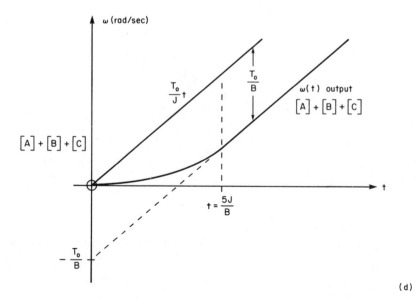

(d)

figure 4-15 fan speed response to steadily increasing torque (a), (b), and (c) Components of the response $\omega(t)$, (d) the overall response $\omega(t)$

It is easy to see from inspection of the total response that there are two parts:

1. The transient response, represented by the decaying exponent

$$\frac{T_0}{B} e^{-t/\tau}$$

2. The steady state, represented by

$$\frac{T_0}{J} \cdot t - \frac{T_0}{B}$$

The transient response will have a maximum value of $\dfrac{T_0}{B}$ radians/sec when $t = 0$ and diminish to 0.674 percent of this value in five time constants $\left(\dfrac{5J}{B}\right)$. On the other hand, the steady-state response actually has a negative value at $t = 0$ and has zero magnitude when $t = \dfrac{J}{B}$ (one time constant).

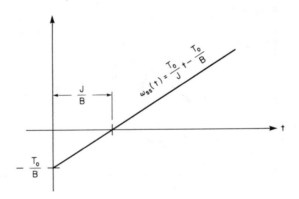

After this, the output has a positive value and increases with an acceleration of $\dfrac{T_0}{J}$ rad/s².

To conclude our demonstration of the simplicity of the Laplace transform method, let us consider a simple second-order system.

EXAMPLE 4.4

If the dc motor system shown in Figure 3-18 has a transfer function described by equation (3.54a) and specified as

$$\frac{\theta(s)}{V_a(s)} = \frac{10}{\frac{1}{4}s^2 + 1\frac{1}{4}s + 1}$$

derive an expression for the output angle $\theta(t)$ when a unit impulse is applied to the input as a voltage signal to the armature.

solution

If

$$\frac{\theta(s)}{V_a(s)} = \frac{10}{\frac{1}{4}s^2 + 1\frac{1}{4}s + 1}$$

the constants of equation (3.54a) should be identified easily by simple symmetry.

then
$$\theta(s) = V_a(s)\frac{10}{\frac{1}{4}s^2 + 1\frac{1}{4}s + 1}$$

but $V_a(s)$ is the Laplace transform of the input function. From the Laplace transform tables, the transform of the unit impulse $\delta(t)$ is unity; i.e., $\mathscr{L}[\delta(t)] = 1$.

Therefore,
$$\theta(s) = 1 \cdot \frac{10}{\frac{1}{4}s^2 + 1\frac{1}{4}s + 1}$$

or
$$\theta(s) = \frac{40}{s^2 + 5s + 4}$$

Factorizing the denominator yields

$$\theta(s) = \frac{40}{(s + 4)(s + 1)}$$

Reading directly from the Laplace tables, we find that the inverse form of the above equation is

$$\theta(t) = \frac{40}{4 - 1}(e^{-t} - e^{-4t})$$

or
$$\theta(t) = 13\frac{1}{3}e^{-t} - 13\frac{1}{3}e^{-4t} \text{ radians}$$

This output can be represented by the curves shown in Figure 4-16, where curve C represents the total output $\theta(t)$ and curves A and B represent the first and second term of the right-hand side of the output equation, respectively. The steady-state value of $\theta(t)$ in this example is zero.

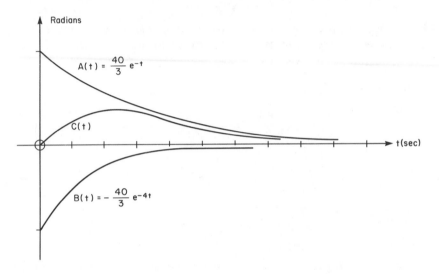

figure 4-16   the impulse response of a dc servomotor

$$C(t) = \theta(t) = \frac{40}{3} \cdot [e^{-t} - e^{-4t}]$$

4-6 SUMMARY

In this chapter we have examined the types of input signal commonly encountered in the study of control systems. We have also examined some simple control systems, and we have seen how the systems components affect the time constants of the system response and furthermore how the response changes for different input signals. Finally, we have introduced the concept of Laplace transforms as an easier means of solving control system differential equations.

PROBLEMS

Section 4.2 **1.** Specify in the time domain the waveforms shown below.

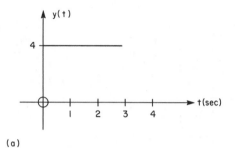

(a)

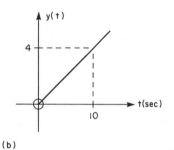

(b)

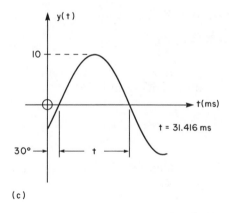

(c)

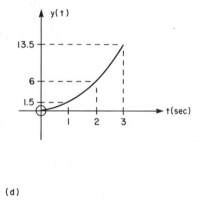

(d)

2. Specify in the time domain the waveform shown.

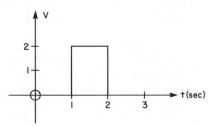

3. Sketch the waveform expressed by

$$y(t) = (8t)U(t) - 16(t - \tfrac{1}{2})U(t - \tfrac{1}{2})$$

Section 4.3 4. Derive the total response $v_R(t)$ for the circuit below, if the input is a step function of 10 volts magnitude.

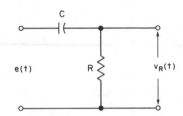

5. A certain system has an equation described by $Ky(t)$ $= \tau\dfrac{dx(t)}{dt} = x(t)$. Derive an expression for the total response $x(t)$ when $y(t) = A_o t$.

Section 4.4 6. For values of $t = 0, \tau, 2\tau, 3\tau, \ldots$, plot the following functions of time on the same graph.

$$e(t) = e^{-t/\tau}$$
$$f(t) = 1 - e^{-t/\tau}$$

Determine the error value $e(t)$ and the function value $f(t)$ at times $t = \tau, 4\tau, 5\tau$.

7. The following data are given for a servomotor:

Rotor inertia	3.82×10^{-6} kg-m^2
Stall torque	0.005 Nm
No load speed	5000 RPM
Field resistance	200 Ω
Field inductance	320 mH

(a) How long will it take the motor to attain full speed from rest?
(b) How long will it take for the field flux to reach full strength?

8. Show that the following quantities have dimensions of time: (a) RC, (b) L/R, (c) $\sqrt{LC}$.

laplace transforms—an easier way

5

. . . the Mock Turtle drew a long breath, and said,
"That's very curious!"
"It's all about as curious as it can be," said Alice.

Lewis Carroll, Alice in Wonderland

5-1 INTRODUCTION

With very little explanation the convenience and brevity of the Laplace transform method was introduced at the end of the last chapter. If, like Alice, our curiosity has been aroused, we need to know more about the transform method, not just to satisfy our curiosity, but to allow us to use this method to our best advantage when performing system analyses.

A rigorous mathematical proof concerning the validity of the Laplace transform is not presented here, as it is not necessary for our purposes. However, a very simple and easy to follow derivation can be found in *Mathematics of Engineering Systems* by F. H. Raven, published by McGraw-Hill Book Company.

This chapter starts with the definition of the transform and then goes on to explain a limited number of operational rules required when one is performing the transformation on the type of system equations found in this text.

Step by step procedures will be covered to demonstrate the tech-

131

nique of finding the inverse Laplace transform once an algebraic solution of the Laplace equations has been found. Finally, we shall return to a familiar topic—the system response—and show the ease with which a system response can be derived by using Laplace transforms and block diagrams.

5-2 SOME SIMPLE TRANSFORMATIONS

Laplace transforms may only be used to solve linear differential equations having constant parameters, where the variables are functions of time only. To obtain the transform without reference to Laplace tables, three simple steps are all that are necessary;

1. Select the function of time $f(t)$.

2. Multiply by a converging factor e^{-st}. This is a factor that will decrease to zero as time increases to infinity.

3. Integrate these terms with respect to time between the limits of zero and infinity.

These three steps can be reduced to a simple mathematical expression, namely;

$$\mathscr{L}[f(t)] \equiv \int_0^\infty f(t)\,e^{-st} \cdot dt \equiv F(s) \qquad (5.1)$$

Where $\mathscr{L}$ denotes that the time function $f(t)$ has been transformed to the Laplace function $F(s)$. Without including the mathematical process of integration, we can simply say that the Laplace transform of a function of time equals a function of s, or;

$$\mathscr{L}[f(t)] \equiv F(s) \qquad (5.1a)$$

If we want to reverse the operation and take the inverse transform back to the time domain, we write

$$\mathscr{L}^{-1}[F(s)] \equiv f(t) \qquad (5.1b)$$

The symbol $\equiv$ means "equal by definition." You may remember that

$$j \equiv \sqrt{-1}$$

In other words, j, by definition, equals the square root of minus one. There is no proof required, because it was defined this way.

Usually it will not be necessary to find the appropriate transform by performing the integral defined in equation (5.1), because there is an adequate list of transforms given in Table 5-1. However, it may be of use to illustrate the method of determining the transform from the defining equation.

EXAMPLE 5.1

Find the Laplace transform of the time-dependent function e^{-at}.

solution

Since

$$f(t) = e^{-at}$$

then

$$\mathscr{L}[f(t)] = \mathscr{L}[e^{-at}]$$

and

$$\mathscr{L}[e^{-at}] = \int_0^\infty e^{-at} \cdot e^{-st} \cdot dt$$

Simplifying this expression yields

$$\mathscr{L}[e^{-at}] = \int_0^\infty e^{-(a+s)t} \cdot dt$$

Integrating, we obtain

$$\mathscr{L}[e^{-at}] = \left[-\frac{1}{a+s} e^{-(a+s)t} \right]_0^\infty$$

Substituting the limits gives

$$\mathscr{L}[e^{-at}] = \left[-\frac{1}{a+s} e^{-(a+s)\infty} \right] - \left[-\frac{1}{a+s} e^{-(a+s)0} \right]$$

Hence

$$\mathscr{L}[e^{-at}] = \frac{1}{a+s}$$

or

$$F(s) = \frac{1}{a+s}$$

Therefore,

$$\mathscr{L}^{-1}\left[\frac{1}{s+a} \right] = e^{-at}$$

EXAMPLE 5.2

Determine $\mathscr{L}[1]$ and $\mathscr{L}[A]$, where A is any constant and hence independent of time.

solution

Because

$$\mathscr{L}[e^{-at}] = \frac{1}{s+a}$$

then if

$$a = 0$$

$$e^{-at} = 1$$

and

$$\frac{1}{s+a} = \frac{1}{s+0}$$

therefore

$$\mathscr{L}[1] = \frac{1}{s}$$

Similarly, because A does not involve time,

$$\mathscr{L}[A] = \int_0^\infty A e^{-st} \cdot dt$$

and

$$\int_0^\infty A e^{-st} \cdot dt = A \int_0^\infty 1 \cdot e^{-st} \, dt$$

Therefore,

$$\mathscr{L}[A] = A \cdot \frac{1}{s} = \frac{A}{s}$$

Conversely,

$$\mathscr{L}^{-1}\left[\frac{1}{s}\right] = 1$$

and

$$\mathscr{L}^{-1}\left[\frac{A}{s}\right] = A$$

Cyclic functions are also of primary importance in our studies; consider the next example.

EXAMPLE 5.3

Find $\mathscr{L}[\sin \omega t]$.

solution

$$\mathscr{L}[\sin \omega t] = \int_0^\infty \sin \omega \cdot t e^{-st} \cdot dt$$

This integral can be executed in either of two ways:

(i) Integration by parts.

(ii) Substituting exponential terms for the trigonometric function.

The second way is probably the simpler one in that

$$\sin \omega t = \frac{e^{j\omega t} - e^{-j\omega t}}{2j}$$

and

$$\cos \omega t = \frac{e^{j\omega t} + e^{-j\omega t}}{2}$$

therefore,

$$\mathscr{L}[\sin \omega t] = \int_0^\infty \frac{e^{j\omega t} - e^{-j\omega t}}{2j} \cdot e^{-st} \cdot dt$$

Hence

$$\mathscr{L}[\sin \omega t] = \frac{1}{2j} \int_0^\infty e^{j\omega t} e^{-st} \, dt - \frac{1}{2j} \int_0^\infty e^{-j\omega t} e^{-st} \, dt$$

The first integral above is the Laplace transform of e^{-at}, where $-a = j\omega$, and the second integral is the Laplace transform of e^{-at}, where $-a = -j\omega$.

Therefore,

$$\mathscr{L}[\sin \omega t] = \frac{1}{2j} \left[\frac{1}{s - j\omega} - \frac{1}{s + j\omega} \right]$$

$$= \frac{1}{2j} \left[\frac{(s + j\omega) - (s - j\omega)}{(s - j\omega)(s + j\omega)} \right]$$

Hence

$$\mathscr{L}[\sin \omega t] = \frac{\omega}{s^2 + \omega^2}$$

Similarly, from

$$\mathscr{L}[\cos \omega t] = \int_0^\infty \left[\frac{e^{j\omega t} + e^{-j\omega t}}{2} \right] \cdot e^{-st} \, dt$$

we obtain the following solution:

$$\mathscr{L}[\cos \omega t] = \frac{s}{s^2 + \omega^2}$$

TABLE 5-1—LAPLACE TRANSFORM TABLE*

NO.	F(S)	f(t)	COMMENTS
1.	1	$\delta(t)$	Unit impulse
2.	$\dfrac{A}{s}$	$A(t) = 0 \quad t < 0$ $ A \quad t \geq 0$	Step
3.	$\dfrac{1}{s}$	$U(t) = 0 \quad t < 0$ $ 1 \quad t \geq 0$	Unit step
4.	$\dfrac{A}{s^2}$	At	Ramp
5.	$\dfrac{2A}{s^3}$	At^2	Parabola
6.	$\dfrac{A\omega}{s^2 + \omega^2}$	$A \sin \omega t$	Sine
7.	$\dfrac{As}{s^2 + \omega^2}$	$A \cos \omega t$	Cosine
8a.	$\dfrac{A}{\tau s + 1}$	$\dfrac{A}{\tau} e^{-t/\tau}$	Free response of first-order system
8b.	$\dfrac{A}{s + a}$	Ae^{-at}	

9a.	$\dfrac{A}{(\tau_1 s + 1)(\tau_2 s + 1)}$	$\dfrac{A}{\tau_1 - \tau_2}(e^{-t/\tau_1} - e^{-t/\tau_2})$	Free response of second-order system ($\zeta > 1$)
9b.	$\dfrac{A}{(s + a)(s + b)}$	$\dfrac{A}{(b - a)}(e^{-at} - e^{-bt})$	
10a.	$\dfrac{A}{(\tau s + 1)^2}$	$\dfrac{At}{\tau^2} e^{-t/\tau}$	Free response of second-order system ($\zeta = 1$)
10b.	$\dfrac{A}{(s + a)^2}$	Ate^{-at}	
11.	$\dfrac{A\omega_n^2}{s^2 + 2\zeta\omega_n s + \omega_n^2}$	$\dfrac{A\omega_n \, e^{-\zeta\omega_n t}}{\sqrt{1 - \zeta^2}} \sin\left(\omega_n\sqrt{1 - \zeta^2}\,t\right)$	Second-order system, free response ($\zeta < 1$)
12a.	$\dfrac{A}{s(\tau s + 1)}$	$A(1 - e^{-t/\tau})$	First-order system response to a step input
12b.	$\dfrac{A}{s(s + a)}$	$\dfrac{A}{a}(1 - e^{-at})$	

TABLE 5-1—CONTINUED

NO.	F(S)	f(t)	COMMENTS
13a.	$\dfrac{A}{s^2(\tau s + 1)}$	$A\tau\left(e^{-t/\tau} + \dfrac{t}{\tau} - 1\right)$	First-order system response to a ramp input
13b.	$\dfrac{A}{s^2(s + a)}$	$\dfrac{A}{a^2}(e^{-at} + at - 1)$	
14a.	$\dfrac{A\omega}{(s^2 + \omega^2)(\tau s + 1)}$	$\dfrac{A\omega\tau}{1 + \omega^2\tau^2} \cdot e^{-t/\tau} + \dfrac{A}{\sqrt{1 + \omega^2\tau^2}}\sin(\omega t - \psi)$ where $\psi = \tan^{-1}\omega\tau$ $(0 < \psi\,\pi)$	First-order system response to a sine input
14b.	$\dfrac{A\omega}{(s^2 + \omega^2)(s + a)}$	$\dfrac{A\omega e^{-at}}{(\omega^2 + a^2)} + \dfrac{A}{\sqrt{\omega^2 + a^2}}\sin(\omega t - \psi)$ where $\psi = \tan^{-1}\omega/a$ $(0 < \psi < \pi)$	
15a.	$\dfrac{A}{s(\tau_1 s + 1)(\tau_2 s + 1)}$	$A\left[1 + \dfrac{\tau_1 e^{-t/\tau_1} - \tau_2 e^{-t/\tau_2}}{\tau_2 - \tau_1}\right]$	Second-order system response to a step input $(\zeta > 1)$
15b.	$\dfrac{A}{s(s + a)(s + b)}$	$\dfrac{A}{ab}\left[1 + \dfrac{ae^{-bt} - be^{-at}}{b - a}\right]$	

16a.	$\dfrac{A}{s(\tau s + 1)^2}$	$A\left[1 - \dfrac{(\tau + t)}{\tau}\, e^{-t/\tau}\right]$	Second-order system response to a step input ($\zeta = 1$)
16b.	$\dfrac{A}{s(s + a)^2}$	$\dfrac{A}{a^2}[1 - (1 + at)e^{-at}]$	
17.	$\dfrac{A\omega_n^2}{s(s^2 + 2\zeta\omega_n s + \omega_n^2)}$	$A\left[1 + \dfrac{e^{-\zeta\omega_n t}}{\sqrt{1 - \zeta^2}}\right.$ $\left. \cdot\, \sin\left(\omega_n\sqrt{1 - \zeta^2}\,t - \psi\right)\right]$ where $\psi = \tan^{-1}\dfrac{\sqrt{1 - \zeta^2}}{-\zeta}$ $\quad(0 < \psi < \pi)$	Second-order system response to a step input ($\zeta < 1$)
18a.	$\dfrac{A}{s^2(\tau_1 s + 1)(\tau_2 s + 1)}$	$A\left[t - \tau_1 - \tau_2 - \dfrac{\tau_2^2 e^{-t/\tau_2} - \tau_1^2 e^{-t/\tau_1}}{\tau_1 - \tau_2}\right]$	Second-order system response to a ramp input ($\zeta > 1$)

TABLE 5-1—CONTINUED

NO.	$F(S)$	$f(t)$	COMMENTS
18b.	$\dfrac{A}{s^2(s + a)(s + b)}$	$\dfrac{A}{ab}\left[t - \dfrac{a + b}{ab} - \dfrac{\dfrac{b}{a}e^{-bt} - \dfrac{a}{b}e^{-at}}{b - a} \right]$	
19a.	$\dfrac{A}{s^2(\tau s + 1)^2}$	$A[t - 2\tau + (t + 2\tau)e^{-t/\tau}]$	Second-order system response to a ramp input ($\zeta = 1$)
19b.	$\dfrac{A}{s^2(s + a)^2}$	$\dfrac{A}{a^2}\left[t - \dfrac{2}{a} + \left(t + \dfrac{2}{a} \right)e^{-at} \right]$	
20.	$\dfrac{A\omega_n^2}{s^2(s^2 + 2\zeta\omega_n s + \omega_n^2)}$	$A\left[t - \dfrac{2\zeta}{\omega_n} + \dfrac{e^{-\zeta\omega_n t}}{\omega_n\sqrt{1 - \zeta^2}} \sin\left(\omega_n\sqrt{1 - \zeta^2}\,t - \psi\right) \right]$ where $\psi = 2\tan^{-1}\dfrac{\sqrt{1 - \zeta^2}}{-\zeta}$ $\quad(0 < \psi < \pi)$	Second-order system response to a ramp input ($\zeta < 1$)

21a.
$$\frac{A\omega}{(s^2+\omega^2)(\tau_1 s+1)(\tau_2 s+1)}$$

Second-order system response to a sine input ($\zeta > 1$)

$$A\left[\frac{\tau_1^2\omega e^{-t/\tau_1}}{(\tau_1-\tau_2)(1+\omega^2\tau_1^2)}\right.$$
$$+\frac{\tau_2^2\omega e^{-t/\tau_2}}{(\tau_2-\tau_1)(1+\omega^2\tau_2^2)}$$
$$\left.+\frac{\sin(\omega t-\psi)}{[(1+\omega^2\tau_1^2)(1+\omega^2\tau_2^2)]^{1/2}}\right]$$

where $\psi = \tan^{-1}\omega\tau_1 + \tan^{-1}\omega\tau_2$

21b.
$$\frac{A\omega}{(s^2+\omega^2)(s+a)(s+b)}$$

$$A\left[\frac{\omega e^{-at}}{(b-a)(\omega^2+a^2)}+\frac{\omega e^{-bt}}{(a-b)(\omega^2+b^2)}\right.$$
$$\left.+\frac{\sin(\omega t-\psi)}{[(\omega^2+a^2)(\omega^2+b^2)]^{1/2}}\right]$$

where $\psi = \tan^{-1}\dfrac{\omega(a+b)}{ab-\omega^2}$ $(0<\psi<\pi)$

22a.
$$\frac{A\omega}{(s^2+\omega^2)(\tau s+1)^2}$$

Second-order system response to a sine input ($\zeta = 1$)

$$\frac{A}{1+\omega^2\tau^2}$$
$$\cdot\left[\frac{\omega t+2\omega\tau}{1+\omega^2\tau^2}\cdot e^{-t/\tau}+\sin(\omega t-\psi)\right]$$

where $\psi = 2\tan^{-1}\omega\tau$

TABLE 5-1—CONTINUED

NO.	F(S)	f(t)	COMMENTS
22b.	$\dfrac{A\omega}{(s^2 + \omega^2)(s + a)^2}$	$\dfrac{A}{\omega^2 + a^2}\left[\dfrac{a\omega(at + 2)e^{-at}}{\omega^2 + a^2} + \sin(\omega t - \psi)\right]$	
23.	$\dfrac{A\omega\omega_n^2}{(s^2 + \omega^2)(s^2 + 2\zeta\omega_n s + \omega_n^2)}$	$\dfrac{A\omega_n^2}{[(\omega_n^2 - \omega^2)^2 + 4\zeta^2\omega^2\omega_n^2]^{1/2}}$ $\cdot \left[\sin(\omega t - \psi_1)\right.$ $\left. + \dfrac{\omega e^{-\zeta\omega_n t}\sin(\omega_n\sqrt{1 - \zeta^2}\,t - \psi_2)}{\omega_n\sqrt{1 - \zeta^2}}\right]$ where $\psi_1 = \tan^{-1}\left(\dfrac{2\zeta\omega\omega_n}{\omega_n^2 - \omega^2}\right)$ $0 < \psi_1 < \pi$ and $\psi_2 = \tan^{-1} - \dfrac{2\zeta\omega_n^2\sqrt{1 - \zeta^2}}{\omega^2 - \omega_n^2(1 - 2\zeta^2)}$ $0 < \psi_2 < \pi$	Second-order system response to a sine input $(\zeta < 1)$

* *Reference:* Handbook of Laplace Transformation *by Floyd E. Nixon; published by* Prentice-Hall, Inc.

The order of $F(s)$ in Table 5-1 has been arranged for the convenience of the reader in that it reflects the approach to functions and systems taken in the text. For example, functions 1 through 7 consist of the five input functions described in the previous chapter. Number 8 is the transfer function of a first-order system, and numbers 9-11 denote a second-order system whose transfer function can be described in any one of three ways. In this regard consider the denominator of $F(s)$ in numbers 9-11, each denominator is a quadratic expression, namely

$$(\tau_1 s + 1) \cdot (\tau_2 s + 1) = \tau_1 \tau_2 s^2 + (\tau_1 + \tau_2)s + 1$$

$$(\tau s + 1)^2 = \tau^2 s^2 + 2\tau s + 1$$

and
$$\frac{s^2}{\omega_n^2} + \frac{2\zeta}{\omega_n} s + 1$$

The first quadratic has real but different roots; the second has real and identical roots, and it can be shown that if ζ is less than unity, the third quadratic has complex roots. In fact, all quadratic expressions fall into one of these three categories, and therefore the form of the denominator of the remainder of the functions $F(s)$ is somewhat repetitive.

Functions 12-14 show a first-order system response to step, ramp, and sinewave inputs; hence the terms $\dfrac{A}{s}, \dfrac{A}{s^2}$ and $\dfrac{A}{s^2 + \omega^2}$ are found in $F(s)$ together with the term $\dfrac{K}{\tau s + 1}$, which describes the first-order system.

Functions 15-23 can now be split into three groups of three, depicting the response to step, ramp, and sinewave inputs when ζ is greater than unity, equal to unity, or less than unity, respectively.

5-2.1 the ω_n and ζ parameters

When any control system has more than one energy-storing element, these elements will exchange energy internally once the system has been excited by the application of an input, as in a series RLC circuit where the inductance will exhibit a buildup of magnetic field, which upon collapsing will transfer energy through the resistance to build up an electric field in the capacitor. This process, followed by the reverse process, occurs repeatedly at a fixed frequency of $\omega_n = \dfrac{1}{\sqrt{LC}}$, where ω_n is considered to be the undamped natural resonant frequency of the system.

If, once the capacitor is charged, the input terminals are short-circuited, then the electric energy stored in the capacitor will oscillate back and forth between the capacitor and the coil until the energy is dissipated by the circuit resistance. The magnitude of this resistance determines the rate at which oscillations decay, and the resistance is said to provide damping into the circuit. An underdamped circuit is one where the response tends to overshoot the goal, with oscillations or "ringing" occurring very easily and decaying very slowly or not at all. An overdamped circuit has a sluggish response, and oscillations never occur. In between these two types of circuit is the circuit which has critical damping; in this circuit the response is as quick as possible without causing overshooting or oscillations to occur.

If we now consider a parallel R-L-C circuit, the resonant or natural frequency of the circuit, ω_n (with zero damping, i.e., with zero circuit resistance) is given by

$$\omega_n = \frac{1}{\sqrt{LC}}$$

as before. If R, the resistance, is non-zero, the circuit oscillates at a frequency ω_d where

$$\omega_d = \frac{1}{\sqrt{LC}} \cdot \sqrt{1 - \frac{R^2 C}{L}}$$

Let

$$\zeta^2 = \frac{R^2 C}{L}$$

then

$$\omega_d = \omega_n \cdot \sqrt{1 - \zeta^2}$$

where ζ is termed the *damping factor* or *damping ratio*.

Consider Figure 5-1; if a step input function is applied, the system can respond sluggishly in an overdamped condition as in [A], which would be the response defined by function 15 in Table 5-1, where K equals unity. Response [B] exhibits critical damping and is defined by function 16 in Table 5-1. The oscillatory response [C] shows an underdamped characteristic and is denoted by function 17 in Table 5-1. In most cases where there are oscillatory conditions, the energy-dissipating elements cause decay or damping, and the decaying sinewave shown in Figure 5-1 does not oscillate at the natural frequency ω_n, but at the damped natural frequency ω_d, which is smaller in magnitude than ω_n since

$$\omega_d = \omega_n \sqrt{1 - \zeta^2} \qquad (5.2)$$

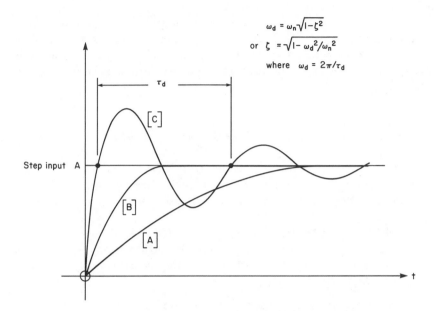

$$\omega_d = \omega_n\sqrt{1-\zeta^2}$$

$$\text{or} \quad \zeta = \sqrt{1- \omega_d^2/\omega_n^2}$$

$$\text{where} \quad \omega_d = 2\pi/\tau_d$$

figure 5-1   output responses of second-order system

Only if the oscillations are maintained at a constant amplitude as shown in Figure 5-2 are they truly at the natural frequency of the system.

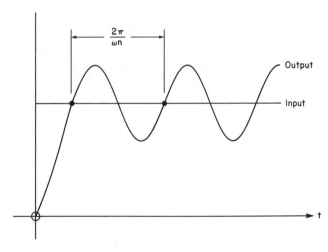

figure 5-2   undamped oscillatory response of a second-order system to a step function

It should now be apparent that any expression describing a decaying oscillation, specified either in the time domain or as a Laplace transform, will contain the parameters ω_n and ζ. To corroborate this statement the reader should refer to Table 5-1, numbers 11, 17, 20, and 23.

5-3 SOME PROPERTIES OF THE TRANSFORM

In order to make full use of Laplace transforms when solving differential equations, we must be familiar with the rules governing the manipulation of the transformed variables. These rules are explored below.

1. **Addition and Subtraction** $\mathscr{L}[\,f_1(t) + f_2(t)]$
   The Laplace transform of the sum of two functions of time, i.e., $\mathscr{L}[f_1(t) + f_2(t)]$, is equal to the sum of the transforms of each function. That is,

   $$\mathscr{L}[f_1(t) \pm f_2(t)] = \mathscr{L}[f_1(t)] \pm \mathscr{L}[f_2(t)]$$

   or $\qquad \mathscr{L}[f_1(t) \pm f_2(t)] = F_1(s) \pm F_2(s)$

EXAMPLE 5.4 $\qquad \mathscr{L}[e^{-at} + e^{-bt}] = \dfrac{1}{s+a} + \dfrac{1}{s+b}$

EXAMPLE 5.5 $\qquad \mathscr{L}[t^2 + t] = \dfrac{2}{s^3} + \dfrac{1}{s^2}$

2. **Multiplication by a Constant** $\mathscr{L}[Kf(t)]$
   The Laplace transform of the product of a constant K, and a time function $f(t)$, i.e., $\mathscr{L}[Kf(t)]$, is equal to the product of the constant and the transform of the time function. That is,

   $$\mathscr{L}[Kf(t)] = K\mathscr{L}[f(t)] = KF(s)$$

EXAMPLE 5.6 $\qquad \mathscr{L}[3 \sin wt] = 3\mathscr{L}[\sin \omega t] = 3\,\dfrac{\omega}{s^2 + \omega^2}$

EXAMPLE 5.7 $\qquad \mathscr{L}[3t^2 + 6t + 9] = \mathscr{L}[3(t^2 + 2t + 3)]$

$$= 3\mathscr{L}[t^2 + 2t + 3]$$

$$= 3\left[\frac{2}{s^3} + \frac{2}{s^2} + \frac{3}{s}\right]$$

$$= \frac{6}{s^3} + \frac{6}{s^2} + \frac{9}{s}$$

3. Differentiation

The transform of the first derivative of $f(t)$ with respect to t is s multiplied by the transform of $f(t)$ minus the value of $f(t)$ when $t = 0^+$. The value of $f(t)$ when $t = 0^+$ is the value when t is an infinitesimally small amount greater than zero and is often called the initial condition.

$$\mathcal{L}\left[\frac{df(t)}{dt}\right] = s \cdot F(s) - f(0^+)$$

The transform of the second derivative of $f(t)$ with respect to t is s multiplied by the transform of the first derivative minus the value of the first derivative when $t = 0^+$; i.e.,

$$\mathcal{L}\left[\frac{d}{dt}\left(\frac{df(t)}{dt}\right)\right] = s\mathcal{L}\left[\frac{df}{dt}\right] - \frac{df(0^+)}{dt}$$

$$= s[s \cdot F(s) - f(0^+)] - \frac{df(0^+)}{dt}$$

$$= s^2 \cdot F(s) - s \cdot f(0^+) - \frac{df(0^+)}{dt}$$

EXAMPLE 5-8

If $y(t) = 2\dfrac{dx(t)}{dt} + 3x(t)$ and $x(t) = 0$ when $t = 0$, find $Y(s)$.

solution

$$\mathcal{L}[y(t)] = \mathcal{L}\left[2\frac{dx(t)}{dt} + 3x(t)\right]$$

$$Y(s) = 2sX(s) - 2x(0^+) + 3X(s)$$

$$Y(s) = 2sX(s) - 0 + 3X(s)$$

$$Y(s) = X(s)(2s + 3)$$

EXAMPLE 5.9

If $y(t) = \dfrac{d^2a(t)}{dt^2} + 4\dfrac{da(t)}{dt} + 3a(t)$ and $\dfrac{da(t)}{dt} = 0$ and $a(t) = 0$, when $t = 0$ find $Y(s)$.

solution

$$\mathcal{L}[y(t)] = \mathcal{L}\left[\frac{d^2a(t)}{dt^2} + 4\frac{da(t)}{dt} + 3a(t)\right]$$

$$Y(s) = s^2A(s) - sa(0^+) - \frac{da(0^+)}{dt} + 4sA(s) - 4a(0^+) + 3A(s)$$

$$Y(s) = s^2A(s) - 0 - 0 + 4sA(s) - 0 + 3A(s)$$

$$Y(s) = A(s)(s^2 + 4s + 3)$$

This can also be expressed as

$$Y(s) = A(s)(s + 3)(s + 1)$$

4. Integration
The transform of an integral of a function with respect to time is the Laplace transform of the function divided by s; i.e.,

$$\mathscr{L}[\int_0^t f(t') \cdot dt'] = \frac{F(s)}{s}$$

EXAMPLE 5.10

$$\mathscr{L}[\int_0^t e^{-at'} \cdot dt'] = \frac{1}{s} \cdot \mathscr{L}[e^{-at'}]$$

$$= \frac{1}{s} \cdot \frac{1}{s+a}$$

This example can be verified as follows:

$$\int_0^t e^{-at'} dt' = \frac{-1}{a} e^{-at'} \Big|_0^t$$

$$= \frac{-1}{a} e^{-at} + \frac{1}{a}$$

$$= \frac{1}{a}(1 - e^{-at})$$

and

$$\mathscr{L}\left[\frac{1}{a}(1 - e^{-at})\right] = \frac{1}{a} \mathscr{L}[1 - e^{-at}]$$

$$= \frac{1}{a}\left[\frac{1}{s} - \frac{1}{s+a}\right]$$

$$= \frac{1}{s(s+a)}$$

5. Shift in Time
This is another way of saying that the function $f(t)$ is delayed; if,

for example, the delay is T seconds, the Laplace transform of $f(t)$ is multiplied by e^{-Ts}.

$$\mathcal{L}[f(t - T)U(t - T)] = F(s)e^{-Ts}$$

The expression in parentheses is a simple mathematical coding to denote

1. The function $f(t)$ is delayed by T seconds and is, therefore, written as $f(t - T)$.
2. It is multiplied by a unit step, $U(t - T)$, which is also delayed by T seconds. This is because $U(t - T)$ has a value of zero when t is less than T seconds and has a value of unity when t is greater than T seconds.

This might seem rather complicated, but consider the ramp functions shown in Figure 5-3.

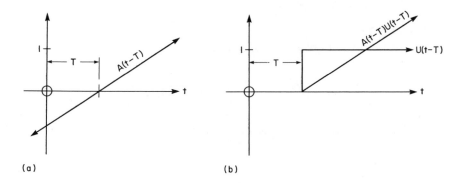

(a)   (b)

figure 5-3 ramp functions

The ramp function shown in Figure 5-3(a) has a value even when t is less than T. The ramp shown in Figure 5-3(b) is obtained by multiplying the ramp in Figure 5-3(a) by the step in Figure 5-3(b). Thus for any time $t < T$ the ramp in Figure 5-3(b) has a value of zero.

EXAMPLE 5.11

A parabola $4t^2$ starts at $t = 3$ seconds. Find its Laplace transform.

solution

$$\mathcal{L}[4(t - 3)^2 \cdot U(t - 3)] = F(s) \cdot e^{-3s}$$

Now
$$\mathscr{L}[4t^2] = 4 \cdot \frac{2}{s^3}$$

$$= \frac{8}{s^3}$$

Hence
$$\mathscr{L}[4 (t - 3)^2 \cdot U(t - 3)] = \frac{8e^{-3s}}{s^3}$$

6. Final Value Theorem

In a stable system the transients tend to decay and the system approaches its steady state as time approaches infinity. It follows that if we wish to find the steady-state value of the output signal, we need only find the limit of the time domain function when t approaches infinity. This is easily done by first finding the Laplace transform of the function, multiplying by s, and then letting s approach zero.

$$\lim_{t \to \infty} f(t) = \lim_{s \to 0} [s \cdot F(s)]$$

EXAMPLE 5.12

Suppose the Laplace transform of $f(t)$ is given by

$$\mathscr{L}[f(t)] = F(s) = \frac{10}{s(s^2 + 3s + 5)}$$

Then
$$\lim_{s \to 0} [s \cdot F(s)] = \lim_{s \to 0} \left[s \frac{10}{s(s^2 + 3s + 5)} \right]$$

$$= \lim_{s \to 0} \left[\frac{10}{s^2 + 3s + 5} \right]$$

$$= \frac{10}{5} = 2$$

Hence
$$\lim_{t \to \infty} f(t) = 2$$

That is, the final value of $f(t)$ is 2.

The final value theorem cannot be applied to every function of time. It cannot be applied to periodic functions of time or functions which do not approach a fixed value as time increases. For example, consider

$$f(t) = A \sin \omega t$$

It is obvious that when $t = 0$ the function $f(t)$ has a value of zero, but when $t = \infty$ the value of $f(t)$ is indeterminate. The application of the theorem results in an erroneous answer; thus

$$\lim_{t \to \infty} f(t) = \lim_{s \to 0} [s \cdot F(s)]$$

$$= \lim_{s \to 0} s \frac{\omega}{s^2 + \omega^2}$$

$$= 0 \text{ (which is false)}$$

This is no cause for alarm, since on the few occasions in which we shall have to apply this theorem the time functions involved will approach a steady final value as time increases.

5-4 *LAPLACE TRANSFORMS AND THE TOTAL RESPONSE*

It will be remembered that solving system differential equations using the classical method can be tedious, especially if the equations are complex enough to require splitting into homogeneous and non-homogeneous equations in order to obtain a solution. The method of solving differential equations offered to us by Laplace transforms allows us to solve even complicated equations (provided they are linear and unchanging) with only a few simple algebraic operations. This simplicity is adequately demonstrated in the following worked examples.

EXAMPLE 5.13

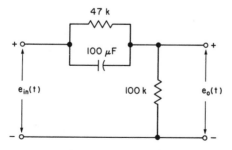

Find the output voltage $e_0(t)$ when an input of E volts dc is suddenly applied to the above circuit.

solution

It is necessary first of all to find the transfer function of the electrical system. This is probably most easily done by converting the electrical circuit to a simple block diagram and then reducing the block diagram by stages to a single block, as we shall demonstrate.

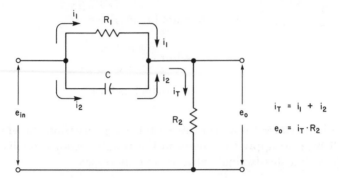

If $i_1(t)$, $i_2(t)$, and $i_T(t)$ are currents flowing in the circuit and $I_1(s)$, $I_2(s)$, and $I_T(s)$ are the respective Laplace transforms, then the circuit can be represented by the following block diagram:

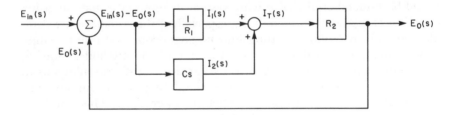

where $E_{in}(s)$ and $E_o(s)$ are the Laplace transforms of $e_{in}(t)$ and $e_o(t)$. This diagram reduces to

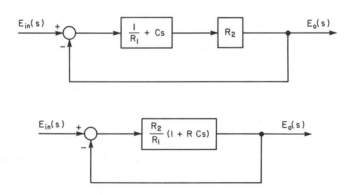

$$E_{in}(s) \xrightarrow{\hspace{1cm}} \boxed{\dfrac{R_2 + R_1 R_2 Cs}{R_1 + R_2 + R_1 R_2 Cs}} \xrightarrow{\hspace{1cm}} E_o(s)$$

whence we obtain the transfer function;

$$\frac{E_o(s)}{E_{in}(s)} = \frac{R_2}{R_1 + R_2} \cdot \frac{1 + R_1 C_2 s}{1 + \dfrac{R_1 R_2}{R_1 + R_2} \cdot Cs}$$

or

$$\frac{E_o(s)}{E_{in}(s)} = K \cdot \frac{1 + \tau_1 s}{1 + K\tau_1 s}$$

where

$$K = \frac{R_2}{R_1 + R_2} = 0.68$$

$$\tau_1 = R_1 C = 4.7 \text{ sec}$$

$$K\tau_1 = \frac{R_1 R_2 C}{R_1 + R_2} = 3.2 \text{ sec}$$

Since

$$E_{in}(s) = \frac{E}{s}$$

then

$$E_o(s) = 0.68E \cdot \frac{1 + 4.7s}{s(1 + 3.2s)}$$

Before we can use our tables of inverse Laplace transforms to determine $e(t)$, we must expand this expression by using the method of partial fractions; thus

$$\frac{1 + 4.7s}{s(1 + 3.2s)} = \frac{A}{s} + \frac{B}{1 + 3.2s}$$

where A and B are to be determined by equating coefficients of like powers of s; i.e.,

$$\frac{1 + 4.7s}{s(1 + 3.2s)} = \frac{A(1 + 3.2s) + Bs}{s(1 + 3.2s)}$$

Therefore, $\quad 1 + 4.7s = A(1 + 3.2s) + Bs = A + (B + 3.2\,A)s$

Hence $\qquad\qquad A = 1$

and $\qquad\qquad B = 4.7 - 3.2 = 1.5$

from which we obtain

$$E_o(s) = 0.68E\left[\frac{1}{s} + \frac{1.5}{1 + 3.2s}\right]$$

From Table 5-1 the inverse Laplace transform of this expression is

$$\mathscr{L}^{-1}[E_o(s)] = e(t) = 0.68E(1 + 0.47e^{-t/3.2})$$

or $\qquad e_o(t) = 0.68E + 0.32Ee^{-t/3.2}$

If we examine the expression for the output voltage, we find that when $t = 0$ the output is E volts, which is an obvious result, since the capacitor initially acts like a short circuit and all the input voltage is dropped across the 100-k resistor. However, when the capacitor has had time to charge up through R_2 it develops only 32 percent of the total circuit voltage, and the remaining 68 percent is dropped across R_2.

EXAMPLE 5.14

If a sinusoidally changing torque, $T_A = 10 \sin 2t$, is suddenly applied to the input shaft of the spring, mass, damper system shown in the diagram, determine the shaft position $\theta(t)$ as a function of time.

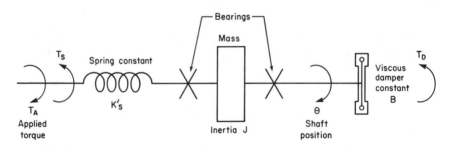

$J = 0.1$ kg-m²

$B = 0.447$ Nms/rad

$K'_s = 1$ Nm/rad

solution

Again, it is most advantageous to convert the system to a block diagram. This is best done by using Newton's second law of motion applied to rotating masses and in so doing, deriving simple equations from which the block diagram can be constructed.

$$\text{Residual torque} = \text{inertia} \times \text{angular acceleration}$$

$$T(t) = J\frac{d^2\theta(t)}{dt^2}$$

where

$$\text{Residual torque} = \text{applied torque} - \begin{pmatrix} \text{reaction} \\ \text{torque due} \\ \text{to spring} \\ \text{deformation} \end{pmatrix}$$

$$- \begin{pmatrix} \text{reaction} \\ \text{torque due to} \\ \text{movement of the} \\ \text{viscous damper} \end{pmatrix}$$

i.e., $$T(t) = T_A - T_s - T_D$$

where K'_s and B are the torque coefficients of the spring and damper, respectively.

Hence $$T_A(t) = J\frac{d^2\theta(t)}{dt^2} + B\frac{d\theta(t)}{dt} + K'_s\theta(t)$$

Because the shaft is initially stationary, the Laplace transform of this equation is

$$T_A(s) = s^2 J\theta(s) + sB\,\theta(s) + K'_s\theta(s)$$

and can be represented by the following block diagram:

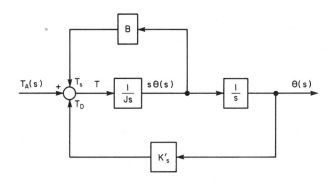

which can be reduced to

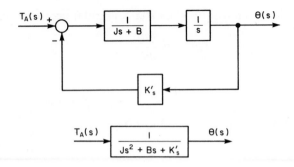

and, finally,

which yields the transfer function

$$\frac{\theta(s)}{T_A(s)} = \frac{\dfrac{1}{K'_s}}{\dfrac{J}{K'_s} \cdot s^2 + \dfrac{B}{K'_s} \cdot s + 1}$$

If numerical values for J, B, and K'_s are now substituted, we obtain

$$\frac{\theta(s)}{T_A(s)} = \frac{1}{0.1s^2 + 0.447s + 1}$$

The transfer function now has the form of

$$\frac{\theta(s)}{T_A(s)} = \frac{K}{\dfrac{s^2}{\omega_n^2} + \dfrac{2\zeta}{\omega_n}s + 1}$$

where $K = 1$

$$\omega_n = \frac{1}{\sqrt{0.1}} = 3.162 \text{ rad/s}$$

and $$\zeta = \frac{0.447\omega_n}{2} = 0.707$$

The Laplace transform of θ now becomes

$$\theta(s) = \frac{T_0\omega}{s^2 + \omega^2} \cdot \frac{K}{\dfrac{s^2}{\omega_n^2} + \dfrac{2\zeta}{\omega_n}s + 1}$$

where $\qquad T_0 = 10 \text{ Nm} \quad$ and $\quad \omega = 2 \text{ rad/s}$

The inverse Laplace transform of $\theta(s)$ is given by 23 in Table 5-1; i.e.,

$$\theta(t) = T_0 \cdot K \frac{\omega_n^2}{[(\omega_n^2 - \omega^2)^2 + 4\zeta^2\omega_n^2\omega^2]^{1/2}}$$

$$\times \left[\sin(\omega t - \psi_1) + \frac{\omega e^{-\zeta\omega_n t} \sin(\omega_n\sqrt{1 - \zeta^2} \cdot t - \psi_2)}{\omega_n\sqrt{1 - \zeta^2}} \right]$$

where

$$\psi_1 = \tan^{-1}\left[\frac{2\zeta\omega\omega_n}{\omega_n^2 - \omega^2} \right] \quad \text{and} \quad \psi_2 = \tan^{-1}\left[\frac{-2\zeta\omega_n^2\sqrt{1 - \zeta^2}}{\omega^2 - \omega_n^2(1 - 2\zeta^2)} \right]$$

Substituting appropriate values, we find that the time domain solution becomes

$$\theta(t) = 9.285 \sin(2t - 56.14°) + 8.3e^{-2.24t} \sin(2.24t + 68.2°)$$

EXAMPLE 5.15

The system shown in the block diagram is subjected to a unit step input. Find the output response $c(t)$ and sketch both the input and output signals as functions of time.

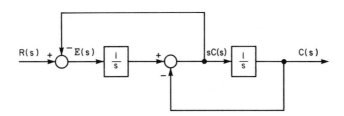

solution

To resolve the "interlocking" of the two feedback loops in the diagram, we must make one of the loops encompass the other so that we may reduce the block diagram by starting with the inner loop and working out to the outermost loop. We do this by moving the tie point at $sC(s)$ to $C(s)$ and including a gain, of value s, in the feedback loop to ensure that the correct signal is fed back to the first summing junction. Thus the block diagram becomes

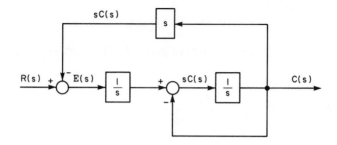

which reduces to

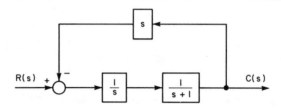

from which we can easily deduce the transfer function

$$\frac{C(s)}{R(s)} = \frac{1}{s(s + 2)}$$

Since
$$R(s) = \frac{1}{s}$$

then
$$C(s) = \frac{1}{s^2(s + 2)} = \frac{0.5}{s^2(1 + 0.5s)}$$

which from 13 in Table 5-1 yields

$$C(t) = 0.5 \times 0.5(e^{-2t} + 2t - 1)$$
$$= 0.25e^{-2t} + 0.5t - 0.25$$

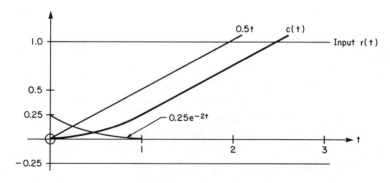

EXAMPLE 5.16

A control system has a transfer function that is described by $\dfrac{C(s)}{R(s)} = \dfrac{4s^2 + 22s + 40}{s^2 + 6s + 8}$. If the system is subjected to a unit step input, find the response $c(t)$.

solution

$$\frac{C(s)}{R(s)} = \frac{4s^2 + 22s + 40}{s^2 + 6s + 8}$$

Since $\qquad R(s) = \dfrac{1}{s}$

then $\qquad C(s) = \dfrac{4s^2 + 22s + 40}{s(s^2 + 6s + 8)} = \dfrac{4s^2 + 22s + 40}{s(s + 2)(s + 4)}$

Using partial fraction expansion, we find

$$\frac{4s^2 + 22s + 40}{s(s + 2)(s + 4)} = \frac{A}{s} + \frac{B}{s + 2} + \frac{D}{s + 4}$$

Hence

$$\frac{4s^2 + 22s + 40}{s(s + 2)(s + 4)} = \frac{A(s + 4)(s + 2) + Bs(s + 4) + Ds(s + 2)}{s(s + 4)(s + 2)}$$

Therefore,

$$4s^2 + 22s + 40 = A(s + 4)(s + 2) + Bs(s + 4) + Ds(s + 2)$$

If $s = 0$, then $40 = 8A$;   i.e., $A = 5$.

If $s = -2$, then $12 = -4B$;   i.e., $B = -3$.

If $s = -4$, then $16 = 8D$;   i.e., $D = 2$.

Therefore $\qquad C(s) = \dfrac{5}{s} - \dfrac{3}{s + 2} + \dfrac{2}{s + 4}$

and $\qquad \mathscr{L}^{-1}C(s) = c(t) = 5 - 3e^{-2t} + 2e^{-4t}$

5-5 SUMMARY

In this chapter we have investigated Laplace transforms in more detail, particularly with reference to the use of the transforms as a tool to determine how a control system responds to various disturbances, stimuli, or input signals.

PROBLEMS

Section 5-2

1. By employing the definition of the Laplace transform, find the transforms for the following functions of time.
   (a) t^3 (b) te^t (c) $t \sin \omega t$

2. Using transform 9 in Table 5-1, derive transform 11 in Table 5-1.

3. Find ζ and ω_n from the following transfer functions:

   (a) $\dfrac{8}{(s + 5)(s + 2)}$ (b) $\dfrac{6}{s^2 + 14s + 49}$

   (c) $\dfrac{0.2}{0.1s^2 + 0.6s + 1}$

4. Find ω_d from the following transfer functions

   (a) $\dfrac{0.2}{0.1s^2 + 0.3s + 1}$ (b) $\dfrac{10}{s^2 + 7s + 14.21}$

Section 5-3

5. Given that $\dfrac{d^2x(t)}{dt^2} + 5\dfrac{dx(t)}{dt} + 6x(t) = e^{2t}$, find $x(t)$ if

   $\dfrac{dx(t)}{dt} = x(t) = 0$ when $t = 0$.

6. Find the Laplace transforms that describe the following functions between the limits of $0 < t < 4$.

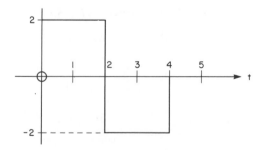

(i)

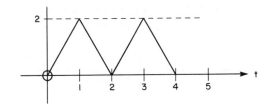

(ii)

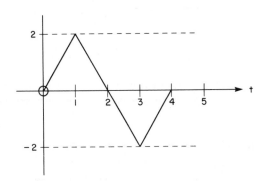

(iii)

7. Using the final value theorem, find the steady-state values of the following functions:

(a) $\dfrac{10}{(s + 10)(s + 100)}$ (b) $\dfrac{0.01}{(0.1s + 1)(0.01s + 1)}$

(c) $\dfrac{5s}{(s + 3)(s + 2)(s + 4)}$ (d) $\dfrac{20}{s^3 + 7s^2 + 10s}$

8. Show in what way the final value of

$$Q(s) = \frac{K(\tau_1 s + 1)(\tau_3 s + 1)}{s^n(\tau_2 s + 1)(\tau_4 s + 1)}$$

is determined by n. Give three examples.

Section 5-4   9. Find the output response $e_0(t)$ for the two circuits shown when the input is a 10-volt step function.

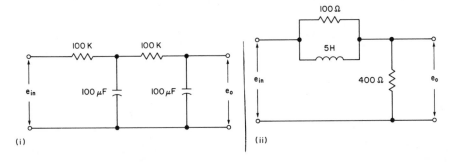

(i)   (ii)

10. Find the response $c(t)$ when a unit step is applied to the input of the system shown.

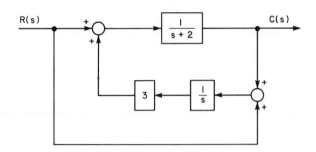

11. Find the output response $e_0(t)$ for the circuit shown when the input is a 10-volt step function.

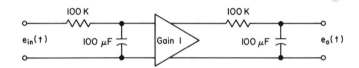

Consider the buffer amplifier to have infinite input impedance and zero output impedance.

12. A simple mercury thermometer is subjected to a rise in temperature (which is linear) of $2°K/s$. What is the thermal response of the thermometer if the thermal resistance is $10°K/watt$ and the thermal capacitance is 0.5 joules/$°K$?

part 2
the analysis of systems in time and frequency

A Problem —Instability

6

"You've no right to grow here," said the Dormouse. "Don't talk nonsense," said Alice more boldly, "You know you're growing too."
"Yes, but I grow at a reasonable pace," said the Dormouse, "not in that ridiculous fashion."

Lewis Carroll, *Alice in Wonderland*

6-1 WHAT IS INSTABILITY?

If the output of a system grows, even at a reasonable pace, when there is no input or when the input is not demanding an increase in the output, the system is said to be unstable and is of no practical use.

It is a characteristic of feedback systems that for certain values of the parameters the feedback is "positive" and the system becomes unstable; it is, therefore, very important that the system be designed not only for a speedy response but also for rigid stability within its specified operating range.

A system exhibits a tendency towards instability when it oscillates about the desired output level. However, if the oscillations die out and become zero as time approaches infinity, the system is considered absolutely stable. If the oscillations do not die out but remain at a constant level, then we have a constant amplitude, sinusoidal oscillator which has limited stability. Should the oscillations continue to increase in amplitude, then the system is absolutely unstable. To understand how instability can arise, let us consider the effects of changing certain parameters in a simple feedback system.

Figure 6-1 illustrates such a system, where H is deliberately made small.

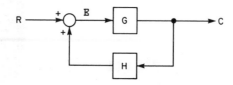

figure 6-1   a positive feedback system

The control ratio of the system is given by

$$\frac{C}{R} = \frac{G}{1 - GH} \qquad (6.1)$$

Now if G, the forward loop gain, were designed to be 5 and H were designed to be 0.1, a 1-volt input signal would initially be amplified to 5 volts at the output. Since 10 percent of the output is fed back and added to the constant 1-volt input, the new error signal (E) will become 1.5 V, and the new value of the output becomes 7.5 volts, of which 10 percent is fed back to the input, and the whole process repeats itself.

The limit, of course, is given by direct substitution of the values for G and H into equation (6.1), which yields

$$C = 1 \times \frac{5}{1 - 5 \times 0.1}$$

or $$C = 10$$

In this particular example, if the feedback signal were increased to 20 percent ($H = 0.2$) or the gain G were increased to ten (i.e., $GH = 1$) and the input were reapplied, the output would continue to grow without bounds and the limit of equation (6.1) would approach infinity. These results would still be achieved even if no input was applied to the system, because any disturbance of the input or output signals (e.g., applying power to the system) would cause the system to oscillate or "ring," and the oscillations would quickly increase in magnitude until they saturated the system amplifiers.

It can be seen from this example that by increasing the gain of the opened loop, i.e., GH to a value of $GH = 1$ causes the system to become unstable; in fact for all values of $GH \geq 1$ the system is unstable.

6-2 ABSOLUTE STABILITY—GENERAL

As we have stated earlier, the transfer function of a system contains all the information concerning the dynamic behavior of the system. From the transfer function we can determine the transient response to specific inputs and the resulting steady-state value of the output, which of course would tell us if the system is stable. However, as you might expect, the stability of a linear system can be determined directly from the transfer function without any knowledge of the input signal.

Two questions must be asked of any system:

1. Is the system absolutely stable? *After this has been determined and if the answer is yes, the second question may be asked:*
2. How stable is the system?

The answer to this question is a little more complicated, as it requires analysis of the system to determine whether it performs quickly and accurately without oscillating unduly or being too sluggish in its response. Let us deal with the first question by considering the transfer functions of first-order and second-order systems such as those listed in 8 through 11 in Table 5-1.

It can be seen that the inverse Laplace transform of the first-order system contains a decaying exponential term, $e^{-t/\tau}$; this is because the denominator includes the term $(\tau s + 1)$. Should the denominator include the term $(\tau s - 1)$, the inverse Laplace transform would contain an exponential term that became larger as time increased (namely, $e^{t/\tau}$), i.e.,

$$\mathscr{L}^{-1}\left[\frac{K}{\tau s + 1}\right] = \frac{K}{\tau}e^{-t/\tau} \qquad (6.2)$$

$$\mathscr{L}^{-1}\left[\frac{K}{\tau s - 1}\right] = \frac{K}{\tau}e^{t/\tau} \qquad (6.3)$$

If the transfer function of a first-order system happened to be that given by equation (6.3), then the response to a step input function of magnitude A would be

$$\mathscr{L}^{-1}\left[\frac{AK}{s(\tau s - 1)}\right] = AK(e^{t/\tau} - 1) \qquad (6.4)$$

If, in the time domain, we let $A = 1$ and $K = 1$, the actual response would be $(e^{-t/\tau} - 1)$, which is illustrated graphically in Figure 6-2. This diagram shows quite dramatically the continued growth of the output.

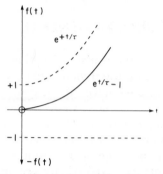

figure 6-2   unlimited growth of the response

No matter what input signal is used with this system, it will be seen that the Laplace transform of the output signal will always contain a term with $(\tau s - 1)$ in the denominator; i.e.,

$$\mathcal{L}(\text{output}) = \mathcal{L}(\text{input}) \times \frac{K}{\tau s - 1}$$

Hence, the inverse Laplace transform of the output function will always contain a term that grows exponentially.

In other words, the clue to absolute stability lies in the denominator of the transfer function; if the denominator of any transfer function is factorized so that it can be written as

$$\frac{F(s)}{(s + a_1)(s + a_2) \dots (s + a_m)}$$

where $F(s)$ is any function of s that may appear in the numerator and $a_i = \dfrac{1}{\tau_i}$, then the system is absolutely stable, provided that all the a_i values are positive. If only one of these values is negative, then the system is absolutely unstable.

Positive sign $\to$ stability $\to e^{-t}$ factor

Negative sign $\to$ instability $\to e^{t}$ factor

Precisely the same reasoning can be used in analyzing the transfer functions of second-order systems. The functions of s shown in 9, 10, and 11 of Table 5-1 are all Laplace transforms of second-order differential equations and differ only in the form of the denominator. The denominator of the transform in 9 is a quadratic expression that can be factorized into two simple factors. The same applies for 10, except that the two factors are identical. If, in either of these examples, any one of the factors in the denominator contains a negative sign, then the system described in the example is inherently unstable; i.e., if the system is excited by any type of input, then continued growth of the output would result.

In 11, however, we consider the case where the quadratic in the denominator of $F(s)$ does not have any real roots; i.e., the roots of the equation are complex numbers (involving imaginary numbers). In this example the roots are

$$[-\zeta + j\sqrt{1 - \zeta^2}]\omega_n \quad \text{and} \quad [-\zeta - j\sqrt{1 - \zeta^2}]\omega_n$$

As an example, let

$$\frac{\omega_n^2}{s^2 + 2\zeta\omega_n s + \omega_n^2} = \frac{13}{s^2 + 4s + 13} \tag{6.5}$$

If the denominator is factorized, the result is

$$\frac{13}{(s + 2 + j3)(s + 2 - j3)}$$

From here it is possible to eventually derive the expression for $f(t)$ given in 11 (Table 5-1) by proceeding in the same way as we would do with 9. The student may attempt this proof as a useful exercise. However, by using the transform given in 11 we can save ourselves a lot of work by noting that

$$\omega_n^2 = 13$$

and

$$\frac{2\zeta}{\omega_n} = \frac{4}{13}$$

Hence

$$\omega_n = 3.61 \text{ rad/s}$$

and

$$\zeta = 0.55$$

From the inverse Laplace transform we see that the transient would decay in the form of $e^{-\zeta\omega_n t}$ or e^{-2t} and the damped natural frequency of the system given by equation (5.2) is $\omega_d = \omega_n\sqrt{1 - \zeta^2}$ or $\omega_d = 3$ rad/s. As ζ gets smaller, $e^{-\zeta\omega_n t}$ gets larger, and the decay becomes more gradual until $\zeta = 0$, at which time there is no decay at all, since $e^{-\zeta\omega_n t} = 1$.

If ζ ever became negative, the decaying exponential would become a growth exponential, namely $e^{-(-\zeta)\omega_n t} = e^{\zeta\omega_n t}$, and the system would become unstable.

In conclusion, it can be said that as ζ decreases the system becomes less stable to the point where $\zeta = 0$ and marginal stability exists. Beyond this point ζ is negative, and instability results. This effect is illustrated in Figure 6-3, where the free response of a second-order system has been shown for various values of ζ.

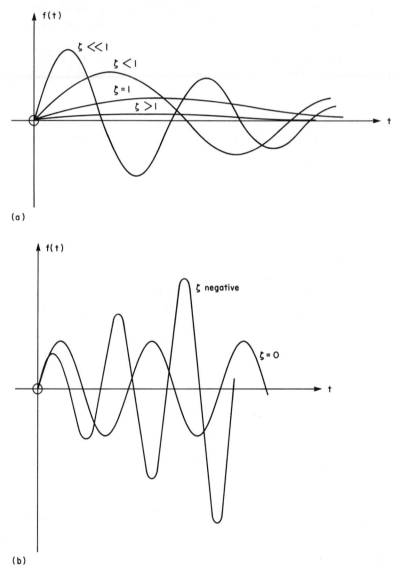

figure 6-3 (a) the stable free response of a second-order system for various values of ζ, (b) the marginally stable and unstable free response of a second-order system when ζ is zero and negative, respectively

6-3 *RELATIVE STABILITY—GENERAL*

The roots of a quadratic expression are those values of the variable that cause the value of the expression to be zero. Consider, for example, $(s + 2) (s + 3)$. This is a quadratic expression, since the product of the two terms is $s^2 + 5s + 6$. If $s = -2$ or $s = -3$, the value of the expression will be zero. Hence $s = -2$ and $s = -3$ are roots of the expression.

We have seen that all second-order systems have transfer functions with quadratic expressions in the denominator which can be factorized. (Denominators of higher-order transfer functions can also be factorized, although with greater difficulty.) We have also seen that knowledge of the value of ζ affords great insight with regard to rate of transient decay and how stable the system is.

It follows, therefore, that if we examine the roots in the denominator of the Laplace transform of a second-order system, we can easily determine the relative stability of the system.

Consider once again the general transfer function of a second-order system, namely

$$\frac{A\omega_n^2}{s^2 + 2\zeta\omega_n s + \omega_n^2}$$

The roots of the denominator are given by the expression

$$s_{1,2} = \frac{-2\zeta\omega_n \pm \sqrt{(2\zeta\omega_n)^2 - 4\omega_n^2}}{2}$$

which yields

$$s_{1,2} = -\zeta\omega_n \pm \omega_n \sqrt{\zeta^2 - 1}$$

or

$$s_{1,2} = -\zeta\omega_n \pm j\omega_n \sqrt{1 - \zeta^2} \tag{6.6}$$

Equation (6.6) reveals some interesting facts:

1. If the damping ratio ζ is greater than unity, the imaginary part of the root becomes real, making

$$s_1 = -\zeta\omega_n + \omega_n \sqrt{\zeta^2 - 1}$$

and

$$s_2 = -\zeta\omega_n - \omega_n \sqrt{\zeta^2 - 1}$$

2. If $\zeta = 1$, there is no imaginary part and both roots are real, negative, and equal.

$$s_1 = -\zeta\omega_n$$

$$s_2 = -\zeta\omega_n$$

3. If ζ is less than unity, the roots become complex conjugate pairs.

$$s_1 = -\zeta\omega_n + j\omega_n \sqrt{1 - \zeta^2}$$

$$s_2 = -\zeta\omega_n - j\omega_n \sqrt{1 - \zeta^2}$$

4. If $\zeta = 0$, the roots are entirely imaginary

$$s_1 = j\omega_n$$

$$s_2 = -j\omega_n$$

and constant amplitude oscillations can be expected to occur.

5. If ζ is negative, the real part of the roots becomes positive

$$s_1 = \zeta\omega_n + j\omega_n \sqrt{1 - \zeta^2}$$

$$s_2 = \zeta\omega_n - j\omega_n \sqrt{1 - \zeta^2}$$

and the system is absolutely unstable.

Because ζ is unitless, the roots s_1 and s_2 have the same units or dimensions as ω_n, i.e., rad/s.

It is now possible to plot these roots on the complex frequency plane more commonly called the "s" plane. A plane is simply a flat surface having an x and y coordinate system; a map or a graph is an example of a plane. Figure 6-4 shows the coordinate system of the s plane.

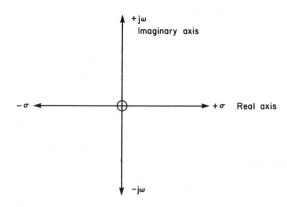

figure 6-4 the real and imaginary axes of the s plane

It is a simple task to plot values of s in the complex frequency plane. The real part $(-\zeta\omega_n)$ is measured along the real axis, and the imaginary part $(\omega_n \sqrt{1 - \zeta^2})$ is measured, both in positive and negative direction, along the $j\omega$ axis. Figure 6-5 shows the roots plotted for one condition, when ζ is less than unity.

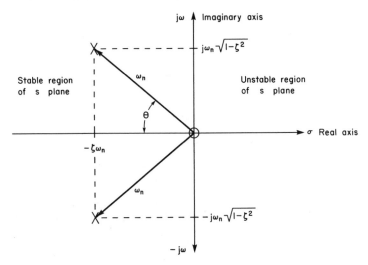

figure 6-5 the complex conjugate roots of a second-order system for $\zeta < 1$

The minimum angle of θ is zero, which denotes a damping ratio of unity; under this condition the system does not oscillate. As θ increases, the system becomes more oscillatory and hence less stable. As the roots approach the imaginary axis, the system oscillates even more and the oscillations take a long time to die out. When $\theta = 90$ deg, the system is marginally stable and the oscillations do not decay. Beyond this point ζ becomes negative and θ becomes greater than 90 deg, and the system is unstable. Thus if the roots are located in the left-hand half of the s plane, the system is stable, but if any one root is found to be in the right-hand half of the s plane, the system is completely unstable.

6-4 SUMMARY

This chapter has been a brief introduction to the concept and implications of stability. It has been shown how the transfer function can be used to determine the absolute and relative stability of a system. In the following chapter we shall pursue this theme and look at other methods of establishing the stability of a system.

how stable is the system? part i —the root locus method

Alice sighed wearily. "I think you might do something better with time," she said, "than wasting it in asking riddles that have no answers."
"If you knew time as well as I do," said the Hatter, "you wouldn't talk about wasting it. It's him."

Lewis Carroll, *Alice in Wonderland*

7-1 INTRODUCTION

We do not have to be on such intimate terms with time as the Mad Hatter to realize that time plays an immense role in our lives. The majority of control systems are designed to operate within our conception of the time domain, and in the last chapter we saw that stability was actually defined with reference to time.

However, it is not always easy to analyze control systems in the time domain, which is why historically most methods of analysis have been developed in the frequency domain. The root locus method of analysis is a convenient and powerful tool that allows speedy investigation of the relative stability of a system from a consideration of the system's dynamic behavior in the frequency domain. This at first might seem incongruous, but frequency has the dimensions of inverse time, and there exists a very close relationship between the roots of a characteristic equation (which have the units of frequency) and the time response of the relevant system.

In this chapter we shall see how plotting the roots of the characteristic equation describes a locus when the values of various system parameters are changed. We will relate the behavior (and the changes in behavior) of the system in the time domain to various positions of the roots on the locus or loci. We shall also discover that instead of plotting the roots of the closed-loop transfer function, we can usually obtain the information we want from a knowledge of the roots and zeros of the open-loop transfer function together with the application of some simple rules. Both graphical and algebraic techniques will be demonstrated to show how to extract data relating to the damping coefficient (ζ), the undamped resonant frequency (ω_n), the damped resonant frequency (ω_d), the open-loop sensitivity constant (K_0), and the closed loop gain.

7-2 ROOT LOCUS AND THE CLOSED-LOOP TRANSFER FUNCTION

In all feedback control systems there are some component parameters that can be varied easily and some which are fixed when a component is chosen. As an example, consider the simple position control system shown in Figure 7-1(a).

The system receives a reference voltage signal, which is amplified to drive the motor. The mechanical output from the motor drives the load through a reduction gear. The load angular position is relayed to a feedback potentiometer, whose slider output is fed to the input of the

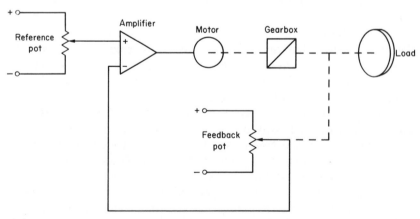

(a)

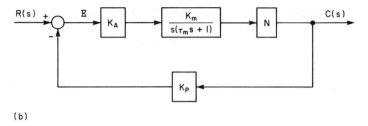

(b)

figure 7-1 (a) a simple position control system, (b) a block diagram of a simple position control system

amplifier, where it is compared with the input reference. If there is a difference between the reference and feedback signal, the motor continues to drive the load, but if they are identical in magnitude, there is no error signal and the motor stops. Figure 7-1(b) shows the simple block diagram of the system. The transfer function of each component is indicated in the appropriate block. The amplifier is considered to have a linear gain K_A with units of volts/volt. The motor is identical to that described in Chapter 3; however, the electrical time constant has been neglected and only the dominant mechanical time constant τ_m (see Chapter 4) is considered. The motor gain constant, K_m, has units of radians/volt-second. The gear ratio N is unitless. The feedback potentiometer gain K_P is obtained by considering the amount of output voltage obtained for a particular rotation of the shaft and hence has units of volts/radian.

The designer would choose the motor and gearbox to fulfill the particular load conditions; hence K_m, τ_m, and N are fixed at the outset. However, the magnitudes of K_A and K_P are adjustable, within certain limits, of course, and will be set by the designer at values calculated to give him the performance he requires.

Reduction of the block diagram in Figure 7-1(b) will yield the closed-loop transfer function of the system, which is illustrated as a single block in Figure 7-2.

$$R(s) \longrightarrow \boxed{\frac{K_A K_m N}{\tau_m s^2 + s + A K_m K_p N}} \longrightarrow C(s)$$

figure 7-2 the closed-loop transfer function of a simple position control system

The control ratio or closed loop transfer function is given by

$$\frac{C(s)}{R(s)} = \frac{K_A K_m N}{\tau_m s^2 + s + K_A K_m K_p N} \tag{7.1}$$

A more familiar form recognizable in the Laplace tables would be

$$\frac{C(s)}{R(s)} = \frac{K_A K_m N / \tau_m}{s^2 + \dfrac{1}{\tau_m} s + \dfrac{K_A K_m K_p N}{\tau_m}} \qquad (7.1a)$$

The free response or impulse response of such a system [found by setting $R(s) = 1$], as can be seen from 9, 10, and 11 in Table 5-1, depends entirely upon the roots of the quadratic expression in the denominator of the transfer function and hence upon the values of K_A and K_p. This being the case, let us consider the transient behavior of the system in relation to the roots of the system characteristic equation, namely,

$$s^2 + \frac{1}{\tau_m} + \frac{K_A K_m K_p N}{\tau_m} = 0 \qquad (7.2)$$

There are two roots to this equation, namely

$$s_1 = -\frac{1}{2\tau_m} + \sqrt{\frac{1}{4\tau_m^2} - \frac{K_A K_m K_p N}{\tau_m}} \qquad (7.2a)$$

$$s_2 = -\frac{1}{2\tau_m} - \sqrt{\frac{1}{4\tau_m^2} - \frac{K_A K_m K_p N}{\tau_m}} \qquad (7.2b)$$

Because s is a complex variable, i.e., has real and imaginary parts, we will plot the roots s_1 and s_2 in the s plane, where the imaginary portion of s is plotted against the vertical axis and the real portion of s is plotted against the horizontal axis. However, prior to doing this we must assign some values to the various parameters. Therefore, let

$$K_m = 25 \text{ rad/V se}$$

$$\tau_m = 0.05 \text{ sec}$$

$$N = \frac{1}{20}$$

We are now free to alter K_A and K_p at will. Since they are found as the product $K_A K_p$ in equations (7.2a) and (7.2b), we shall vary them as an entity. Table 7-1 lists the values of the roots for various values of the product $K_A K_p$. These roots are plotted in Figure 7-3.

Figure 7-3 shows that the locations of the roots move from 0 and -20, when $K_A K_p = 0$, to $-10 + j\infty$ and $-10 - j\infty$, when $K_A K_p = \infty$;

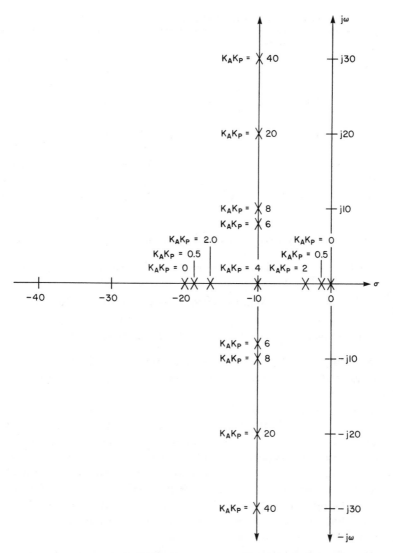

figure 7-3   the root locus of the closed-loop transfer function

also, when K_AK_p has a value of 4, the roots coincide, breaking away from the negative real axis when K_AK_p is made greater than 4. Let us now examine the relationship between the positions of the roots in the s plane and the transient behavior of the feedback position system.

When the product K_AK_p has a value between 0 and 4, the roots lie on the negative real axis and the system is stable, is nonoscillatory, and has zero overshoot.

TABLE 7-1—ROOT LOCATION AS GAIN IS VARIED

GAIN $K_A K_p$	ROOT VALUE		ω_n	ζ	$\theta = cos^{-1} \zeta$ (DEG)
	s_1	s_2			
0	0	-20			
0.5	-0.65	-19.35			
2	-2.93	-17.07			
4	-10	-10	10.00	1.0	0
6	$-10 + j7.07$	$-10 - j7.07$	12.25	0.816	35.2
8	$-10 + j10$	$-10 - j10$	14.14	0.707	45
20	$-10 + j20$	$-10 - j20$	22.36	0.447	63.4
40	$-10 + j30$	$-10 - j30$	31.62	0.316	71.5

The form of the free response in the time domain is given by 9 in Table 5-1, where τ_1 and τ_2 are the reciprocal of the roots; e.g., $\frac{1}{2.93}$ and $\frac{1}{17.07}$ when $K_A K_p = 2$. When the roots have identical values, the form of the free response is given by 10 in Table 5-1, where $\tau = \frac{1}{10}$; the system is still stable, has zero overshoot, and is nonoscillatory; however, the roots of the equation, although still real, verge on becoming complex.

If $K_A K_p$ is made only slightly greater than 4, a small amount of overshoot is noticed in the transient response of the system. As $K_A K_p$ is increased, the magnitude of the overshoot increases until slight oscillations occur in the response. Further increase in the magnitude of $K_A K_p$ causes the transient to be accompanied by an increasing amount of oscillation. However, no matter how much overshoot or oscillation occurs initially, so long as the roots of the characteristic equation remain in the left-hand half of the s plane the free response always converges to a steady value as the transient phase decays.

In terms of the system transfer function, when $K_A K_p > 4$, the roots of the characteristic equation are complex and the free response has the form of 11 in Table 5-1. In order to evaluate the expression for $f(t)$ in the table, we must first determine the values of A, ζ, and ω_n. Let us first compare the transfer function $F(s)$ from Table 5-1 with our system transfer function in equation (7.1a); i.e.,

$$\frac{K_A K_m N/\tau_m}{s^2 + s/\tau_m + K_p \cdot K_A K_m N/\tau_m} \equiv \frac{A\omega_n^2}{s^2 + 2\zeta\omega_n s + \omega_n^2}$$

From this comparison we deduce that

$$A = 1/K_p$$

$$2\zeta\omega_n = 1/\tau_m = 20 \text{ sec}^{-1}$$

$$\omega_n^2 = K_pK_AK_mN/\tau_m = 25K_AK_p \text{ sec}^{-2}$$

from which values for ζ and ω_n can be easily determined for any value of K_AK_p, as shown in Table 7-1.

Let us now turn our attention once more to the form of the transfer function in 11 of Table 5-1, and determine the general form of the roots of the characteristic equation; i.e., the roots of

$$s^2 + 2\zeta\omega_n s + \omega_n^2 = 0$$

which can be shown to be

$$s_1 = -\zeta\omega_n + j\omega_n\sqrt{1 - \zeta^2}$$

$$s_2 = -\zeta\omega_n - j\omega_n\sqrt{1 - \zeta^2}$$

In other words, if we plot the roots in the s plane, as in Figure 7-4, so that

$$s_{1,2} = \sigma \pm j\omega$$

then

$$\sigma = -\zeta\omega_n$$

and

$$\omega = \omega_n\sqrt{1 - \zeta^2}$$

From Figure 7-4 we observe two rather interesting facts. First, the distance of either root from the origin of the s plane is a direct measure of ω_n, the natural frequency of oscillation in the system.

Second,

$$\cos\theta = \frac{\zeta\omega_n}{\omega_n} = \zeta$$

Hence, when

$$\zeta = 1$$

then

$$\theta = 0°$$

and the system is fully damped with no possibility of oscillations or overshoot occurring.

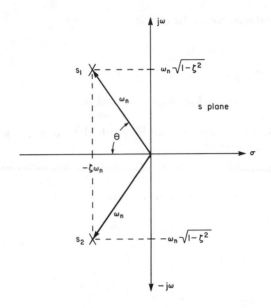

figure 7-4 complex roots of a second-order transfer function

Also, when $\qquad\qquad\qquad\qquad \zeta = 0$

then $\qquad\qquad\qquad\qquad\qquad \theta = 90°$

and the roots lie on the imaginary axis, giving completely undamped oscillations. Any increase in θ causes the roots to enter the right-hand half of the s plane, giving a completely unstable system and causing ζ to become negative.

In summary, as ζ approaches zero, the system becomes more and more underdamped. As ζ approaches unity, oscillations and overshoot disappear; $\zeta = 1$ is termed critical damping. When $\zeta = 1/\sqrt{2}$, we have the condition often known as optimum damping; i.e., the system responds quickly to sudden changes while minimizing the magnitude and duration of the output error. Of course, faster responses can be obtained with lower values of ζ if a small degree of oscillation can be tolerated, and, in fact, common practice is to use slightly underdamped systems (e.g., ζ = approximately 0.6) for this reason.

Thus from Figure 7-3 we can see that as long as

$$K_A K_p \leqslant 4$$

then $\qquad\qquad\qquad\qquad \zeta \geqslant 1 \quad \text{(overdamped)}$

when $\qquad\qquad\qquad\qquad K_A K_p = 4$

then $\qquad$ $\zeta = 1$ (critical damping)

but for $\qquad$ $4 \leqslant K_A K_p < \infty$

then as $\qquad$ $K_A K_p \to \infty$

we find $\qquad$ $\omega_n \to \infty$

and $\qquad$ $\zeta \to 0$

i.e., the system becomes more and more oscillatory as $K_A K_p$ is increased.

Another important observation that can be made in regard to the root locus in Figure 7-3 is that since the root locus never crosses the imaginary axis and hence no root ever appears in the right-hand half of the s plane the system will never be unstable for any value of $K_A K_p$. This statement is true regardless of the type of input to the system (impulse, step, sinewave, etc.), because the system stability depends only upon the location of the roots of the characteristic equation and not upon the form of the input. In other words, stability, or lack of it, is inherent in the system and does not depend upon external stimuli.

7-3 ROOT LOCUS AND THE OPEN-LOOP TRANSFER FUNCTION

7-3.1 poles and zeros

Before proceeding with this section we will introduce two terms commonly encountered when one is discussing control systems, namely, the poles and zeros of a system. These terms are perhaps a little strange but are easily explained as follows. Think of any transfer function at all, e.g.,

$$F(s) = \frac{s^2 + As + B}{s^3 + Cs^2 + Ds + E}$$

Factorize the numerator and the denominator of the transfer function; i.e., find the roots of the numerator and the roots of the denominator thus:

and $\qquad$ $s^2 + As + B \to (s + a)(s + b)$

$$s^3 + Cs^2 + Ds + E \to (s + c)(s + d)(s + e)$$

The transfer function now becomes

$$F(s) = \frac{(s + a)(s + b)}{(s + c)(s + d)(s + e)}$$

The zeros of a transfer function are any values of s that make

$$F(s) = 0$$

Thus the zeros of our transfer function are

$$s = -a; \qquad s = -b$$

The poles of a transfer function are the values of s that make

$$F(s) = \infty$$

i.e., the value of s that makes the denominator of the transfer function equal to zero. Hence the poles of $F(s)$ are

$$s = -c; \qquad s = -d; \qquad s = -e$$

The term "the poles of the system" is generally used in preference to such terms as "the roots of the system," because "the poles of a system" implies certain time response and frequency characteristics of the system, whereas the term "roots" is more properly used when one is talking about polynominal equations such as the characteristic equation. We may best be able to illustrate the difference in meaning that we attach to the words *roots* and *poles* by saying that *the poles of any system having a linear transfer function are determined by finding the roots of the system characteristic equation.*

7-3.2 the root locus

If the feedback loop of the system shown in Figure 7-1(b) is disconnected at the summing junction in the following manner:

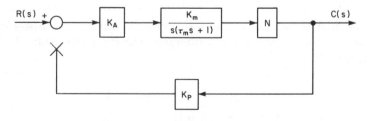

the open loop

then the open-loop transfer function can be given as

$$G(s)H(s) = \frac{K_A K_m N K_p}{s(\tau_m s + 1)} \tag{7.4}$$

Now the closed-loop transfer function is defined as

$$\frac{C(s)}{R(s)} = \frac{G(s)}{1 + G(s)H(s)} \tag{7.5}$$

and the characteristic equation is obtained by equating to zero the denominator of the closed-loop transfer function. Thus the general form of the characteristic equation is given by

$$1 + G(s)H(s) = 0 \tag{7.6}$$

As we have pointed out previously, the characteristic equation contains all the information about a system's frequency response, time response, and stability. It follows then from equation (7.6) that all this information is also contained in the equation

$$G(s)H(s) = -1 \tag{7.7a}$$

or, in polar form,

$$G(s)H(s) = 1\angle 180° \tag{7.7b}$$

where $|G(s)H(s)|$ is the open-loop gain of the system. Let us demonstrate that equation (7.7b) is true for any roots of the characteristic equation.

If, in our simple position control system, we choose a value of $K_A K_p$ such that

$$K_A K_p = 20$$

then from Table 7-1 we find that the roots of the characteristic equation of our system are a complex pair such that

$$s_1 = -10 + j20 \quad \text{and} \quad s_2 = -10 - j20$$

If we examine $G(s)H(s)$ when $s = s_1$ we find from equation (7.4) that

$$G(s)H(s) = \frac{500}{(-10 + j20)(-10 + j20 + 20)} \tag{7.8a}$$

which yields

$$G(s)H(s) = \cfrac{500}{\left[\sqrt{10^2 + 20^2} \angle \tan^{-1}\dfrac{20}{-10}\right]\left[\sqrt{10^2 + 20^2} \angle \tan^{-1}\dfrac{20}{10}\right]}$$

$$= \cfrac{500}{\left[\sqrt{500} \cdot \angle 116.6°\right]\left[\sqrt{500} \cdot \angle 63.4°\right]}$$

$$= \frac{500}{500} \angle -180 \text{ deg} \qquad\qquad (7.8b)$$

Therefore, $\qquad\qquad G(s)H(s) = 1 \angle 180 \text{ deg}$

when $s = s_1$, which is one of the closed-loop poles of the system; a similar result will be found if we choose s to be equal to s_2.

What this example illustrates is that any value of s that causes the open-loop transfer function $G(s)H(s)$ to have a gain of unity and a phase shift of 180 deg will be a pole of the closed-loop system. This fact gives us a tremendous advantage in determining how the poles of the closed-loop system behave, because the open-loop transfer function is usually much simpler and more amenable to our examination than the closed-loop transfer function.

Thus by examining $G(s)H(s)$ in the s plane and determining for what values of s equation (7.7b) is valid, we can deduce how the closed-loop poles are influenced by changes in any of the system parameters. How do we do this?

In general, when the numerator and the denominator of $G(s)H(s)$ have been factorized, $G(s)H(s)$ can be written thus.

$$GH(s) = K \cdot \frac{(s + a_1)\,(s + a_2)\,(s + a_3) \cdots}{s^n\,(s + b_1)\,(s + b_2)\,(s + b_3) \cdots} \qquad\qquad (7.9)$$

where K is the product of all the loop gains.

The terms in brackets in the numerator of equation (7.9) indicate the *open-loop zeros* (i.e., $-a_1$, $-a_2$, $-a_3$, etc.), and the terms in the denominator indicate the *open-loop poles* (i.e., $-b_1$, $-b_2$, $-b_3$, etc.), which should not be confused with the closed-loop poles, which are the roots of the closed-loop characteristic equation.

As demonstrated in our example, transfer functions such as equation (7.9) can be converted into polar notation of the form

$$G(s)H(s) = |R(s)| \angle \theta(s)$$

By using certain simple rules, which we will explain presently, we can locate the positions of the closed-loop poles in the s plane by solving graphically the equations

$$|R(s)| = 1$$

and $$\theta(s) = n\pi \qquad (n = 1, 3, 5, \text{etc.})$$

Having done this, we can also determine new locations of the poles if the gain K is changed. As K is changed from zero to infinity, the locations of the poles will trace a locus or loci within the s plane. It is this tracing of the poles or roots of the characteristic equation that gives this method of analysis its name.

Furthermore, it is often not necessary to rigorously determine the exact loci, as very often we wish only to deduce such data as the approximate value of the poles or the region where branch points exist, or maybe we wish to find at what value of K the poles enter the right-hand side of the s plane or, as is more usual, how the addition of a compensation network affects the general location and direction of the root locus. All this information can be obtained very quickly by use and understanding of the following rules.

7-4 RULES FOR PLOTTING A ROOT LOCUS

No proofs will be offered here of the mathematics involved for deriving the following rules for the construction of root loci. Such proofs can be found by consulting some of the books listed in the bibliography.[1]

To illustrate these rules, we will consider the control system denoted by the block diagram in Figure 7-5.

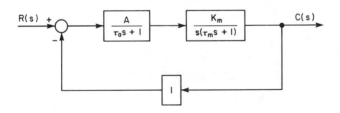

figure 7-5 a simple third-order control system

---

[1] *Feedback Control System Analysis and Synthesis,* D'Azzo and Houpis, McGraw Hill (1966).

The open-loop transfer function is given as

$$G(s)H(s) = \frac{AK_m}{s(\tau_a s + 1)(\tau_m s + 1)}$$

where $K_m = 2,$ $\tau_m = 0.05,$ $\tau_a = 0.2,$ and $0 \leqslant A < \infty$

which yields

$$G(s)H(s) = \frac{2A}{s(0.2s + 1)(0.05s + 1)}$$

Rule 1:

(a) The open-loop transfer function must be expressed with the Laplace operator, s, having unity coefficient.

$$G(s)H(s) = \frac{2A}{(0.2)(0.05)} \cdot \frac{1}{s(s + 5)(s + 20)}$$

or $G(s)H(s) = K_0 \cdot \dfrac{1}{s(s + 5)(s + 20)}$

where $K_0 = 200A$ (the open-loop sensitivity constant)

(b) The real and the imaginary axes of the s plane have identical scales. (Only the left-hand plane need be drawn and often only the left, upper quadrant is required.)

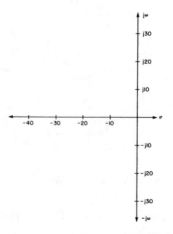

figure 7-6   uniform scaling of the s plane

(c) Locate all the zeros and poles of the open-loop transfer function $G(s)H(s)$ on the s plane. Zeros are represented by circles and poles are represented by crosses.

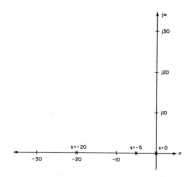

figure 7-7 the location of the poles of

$$G(s)H(s) = \frac{K_0}{s(s + 5)(s + 20)}$$

Rule 2:

Those portions of the real axis to the left of an odd number of open-loop poles and/or zeros are part of the root locus.

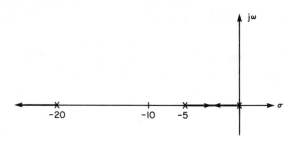

figure 7-8 negative, real axis line segments included in the root locus

For example, if in Figure 7-8 a point on the negative real axis such as -2 is selected, there is one pole to the right of this point; hence, the point, -2, is included in the locus. If a new point such as -4.9 is chosen, the same condition applies and hence the point -4.9 is included in the locus. If a third point, such as -5.2, or -12, or -18.5 is chosen, there is an even number (two) of poles (no zeros in this case) to the right of this third trial point, and therefore the point is not included in the locus segment. Any point less than -20 will be included in the locus, since any point to the left of -20 has an odd number (three) of poles plus zeros (none in this case) to the right of it.

Rule 3:

A root locus (i.e., the locus of a closed-loop pole or root) always starts at the location of one of the open-loop poles, when the gain K_0 is zero. As K_0 increases, the closed-loop roots describe loci away from the open-loop poles. A locus always terminates at an open-loop zero when K_0 approaches infinity.

Thus in Figure 7-8, the loci are shown leaving each of the open-loop poles. But what happens when two loci meet, as happens to the two loci leaving the poles at $s = 0$ and $s = -5$?

Rule 4:

If a segment of the locus on the real axis is bounded by two poles, then a breakaway point will exist between these poles, as shown in Figure 7-9.

If a segment of the locus on the real axis is bounded by two zeros, then a break-in point will exist between these zeros, as shown in Figure 7-9.

At a breakaway point the two roots, which have been approaching each other along the real axis, meet and split away from the axis in

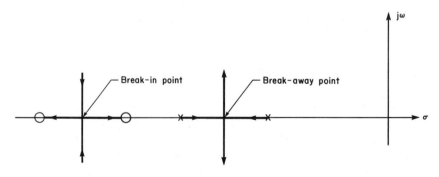

figure 7-9 break-in and breakaway points on the negative real axis

opposite directions. Thus the two closed-loop roots cease to be real and now form a complex conjugate pair.

At a break-in point the reverse process happens, a complex conjugate pair of closed-loop roots approach each other from opposite sides of the real axis, eventually meeting on the axis. On meeting they repel each other and travel in opposite directions along the axis until they reach the open-loop zeros.

The location of the branch point with respect to the two open-loop poles (or zeros) bounding the locus is affected by the proximity of the other open-loop poles and zeros. The closer these other poles and zeros are to the locus, the greater is their influence on the position of the branch point in the locus.

For example, in Figure 7-8 if the system had only two open-loop poles and they were located at $s = 0$ and $s = -5$, then the branch point would occur midway between these two points, i.e., at $s = -2.5$. The open-loop pole at $s = -20$, however, has the effect of pushing the branch point closer to the origin, and the actual position of the branch point is at $s = -2.32$. If the open-loop pole at $s = -20$ is moved closer to the origin to, say, $s = -7$, then the branch point is also "pushed" closer to the origin and would now be found to occur at $s = -1.92$.

Rule 5:

The branch locus always breaks away (or breaks in) at an initial angle of $\pm$ 90 deg to the real axis. This is illustrated in Figure 7-10.

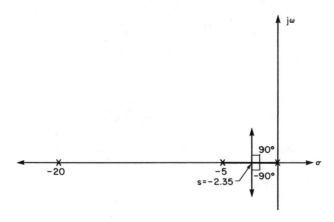

figure 7-10 root locus and location of the branch point

Rule 6:

The number of separate loci always equals the number of poles in the open-loop transfer function.

In this open-loop transfer function there are three poles and no zeros. Whenever the number of poles exceeds the zeros, at least one of the root loci ends at infinity. The number of loci ending at infinity is equal to the number of poles minus the number of zeros (i.e., three in this example). When this happens, we say that the zero or zeros are located at infinity.

It should be noted that if the open-loop transfer function had more zeros than poles, then one or more of the loci would *start* at infinity and we would say that the excess poles are located at infinity.

Rule 7:

As the gain increases and approaches infinity the loci becomes asymptotic to straight lines (or asymptotes) radiating from the centroid of the poles and zeros located on the real axis.

(a) The asymptotic angles are given by the relation

$$\psi = \frac{(2K + 1)\, 180°}{\eta_p - \eta_z}$$

where $K = 0, \pm 1, \pm 2, \ldots$ until all the angles within 360 deg are obtained, and where

$$\eta_p = \text{the number of poles}$$

$$\eta_z = \text{the number of zeros}$$

(b) The location of the centroid on the real axis is given by

$$s_c = \frac{\Sigma \text{ poles} - \Sigma \text{ zeros}}{\eta_p - \eta_z}$$

where Σ poles = the sum of the values of all the poles
and Σ zeros = the sum of the values of all the zeros

For our example the angles that the asymptotes make to the real axis are

$$\psi = \frac{(2K + 1)\, 180°}{3}$$

then for

$$K = 0 \qquad \psi = 60°$$
$$K = 1 \qquad \psi = 180°$$
$$K = 2 \qquad \psi = 300°$$
$$K = 3 \qquad \psi = 420°$$
$$\qquad\qquad = 360° + 60° \text{ (same as for } K = 0)$$

Thus there are three separate asymptotes radiating from the centroid at respective angles of 60, 120, and 300 deg to the real axis.

For the centroid of the roots, we find

$$\Sigma \text{ poles} = (0) + (-5) + (-20) = -25$$

and
$$\Sigma \text{ zeros} = 0$$

Therefore

$$s_c = \frac{(-25) - (0)}{3 - 0} = -8\tfrac{1}{3}$$

Thus the asymptotes intersect the real axis at a value of $s = -8\tfrac{1}{3}$, as illustrated in Figure 7-11.

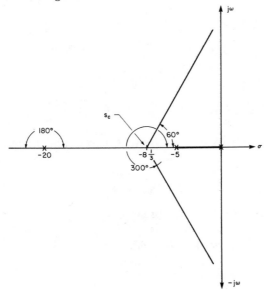

figure 7-11 location of the centroid and the asymptotes

Rule 8:

When the loci cross the imaginary axis, the system becomes unstable. The value of the open-loop sensitivity constant K_0 and the value of s at the point of crossover can be determined by substituting $j\omega$ for s in the closed-loop characteristic equation thus:

Consider our example where

$$G(s)H(s) = \frac{200A}{s(s + 5)(s + 20)}$$

The characteristic equation is

$$1 + G(s)H(s) = \frac{200A}{s(s + 5)(s + 20)} + 1 = 0$$

Substituting $j\omega$ for s yields

$$\frac{200A}{j\omega(j\omega + 5)(j\omega + 20)} + 1 = 0$$

Therefore

$$j\omega(j\omega + 5)(j\omega + 20) + 200A = 0$$

Hence

$$-j\omega^3 - 25\omega^2 + 100j\omega + 200A = 0$$

Separating the real and imaginary parts yields

$$(200A - 25\omega^2) + j(100\omega - \omega^3) = 0$$

For this equation to be true, both the real and imaginary parts must equal zero; i.e.,

$$200A = 25\omega^2 \quad \text{(real part)}$$

and
$$100\omega = \omega^3 \quad \text{(imaginary part)}$$

Solution of these simple equations yields

$$\omega = 10 \text{ rad/s}$$

and
$$A = 12.5$$

Therefore
$$K_0 = 200A = 2500$$

The complete root locus can now be plotted as shown in Figure 7-12.

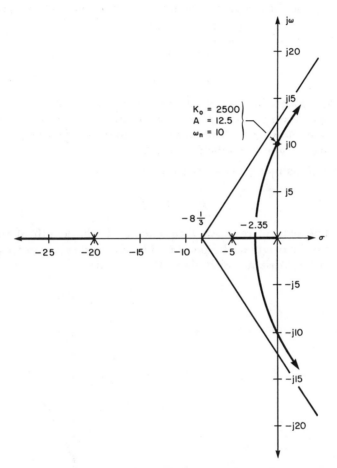

figure 7-12 the complete root locus for

$$G(s)H(s) = \frac{200A}{s(s + 5)(s + 20)}$$

Rule 9:

Direct measurement of the sensitivity constant, the natural resonant frequency, and the damping ratio can be made from the locus once it has been plotted.

Figure 7-13 illustrates the following techniques used to determine the sensitivity constant and the natural and damped resonant frequency when, for example, the system damping ratio is required to be 0.6.

(a) The sensitivity gain K_0 at any point P on the root locus is given by

$$K_0 = \frac{\Delta p_1 \cdot \Delta p_2 \cdot \Delta p_3 \cdot \ldots}{\Delta z_1 \cdot \Delta z_2 \cdot \Delta z_3 \cdot \ldots}$$

where Δp_1, Δp_2, etc. are the distances measured graphically between P and the open-loop poles p_1, p_2, etc. Similarly, Δz_1, Δz_2, etc. express the distances measured graphically between P and the open-loop zeros. If no zeros are present, then the product Δz_1, Δz_2, etc. is taken to be unity.

(b) The damping ratio ζ is given by

$$\theta = \cos^{-1} \zeta$$

(c) The natural resonant frequency ω_n is the length of the line drawn from the origin to the point P. The damped resonant frequency ω_d is the vertical distance of P above the real axis.

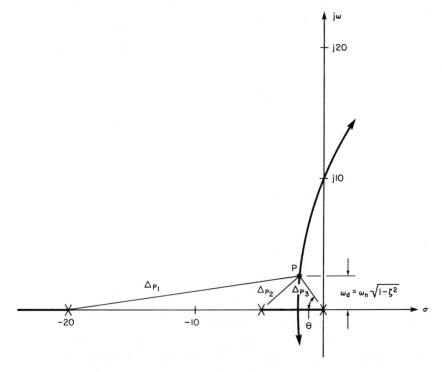

figure 7-13 evaluation of K_0, ω_n, ω_d, and $\zeta = \cos^{-1} \theta$

Evaluation of K_0: From the diagram the values of Δp are measured at:

$$\Delta p_1 = 18$$

$$\Delta p_2 = 4$$

$$\Delta p_3 = 3$$

There are no zeros, so the product of the zero distances is taken as unity.

Therefore
$$K_0 = \frac{18 \times 3 \times 4}{1} = 216$$

and hence

$$A = \frac{216}{200} = 1.08$$

Evaluation of θ:

$$\theta = \cos^{-1}(0.6)$$

hence
$$\theta = 53.13°$$

Evaluation of ω_n: The natural resonant frequency is measured as Δp_3 above and therefore equals 3 rad/s.

Evaluation of ω_d: The damped resonant frequency is measured as approximately equal to 2 rad/s but can be calculated as

$$\omega_d = \omega_n \sqrt{1 - \zeta^2} = 2.4 \text{ rad/s}$$

These rules probably seem long and involved, but they are quite simple to execute, as we shall see. There are also one or two additional rules for special cases, but we can handle those when the occasion arises.

EXAMPLE 7.1

Sketch the root locus for the system shown below as the amplifier gain is varied from zero to infinity.

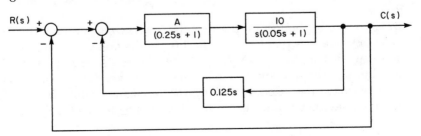

solution

The system must be simplified thus:

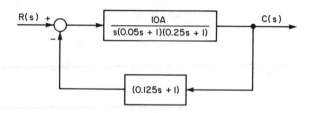

Now we may proceed using the previously given rules.

Rule 1:

Since
$$G(s)H(s) = \frac{10A(0.125s + 1)}{s(0.05s + 1)(0.25s + 1)}$$

then
$$G(s)H(s) = \frac{100A(s + 8)}{s(s + 20)(s + 4)}$$

Let
$$K_0 = 100A$$
(open-loop sensitivity constant)

There are open-loop zeros and poles at the following locations:

$$z_1 = -8 \qquad p_1 = -20$$
$$p_2 = -4$$
$$p_3 = 0$$

Rules 2 and 3:

The portions of the real axis between the origin and $s = -4$ and between $s = -8$ and $s = -20$ are root loci. Also, because the number of poles exceeds the number of zeros by two, then two of the loci will head towards infinity as K_0 increases.

Rule 4:

Because one of the loci on the negative real axis is bounded by two poles, a breakaway point exists. This breakaway point lies somewhere between $s = 0$ and $s = -4$. The position will be affected by the close proximity of the zero and to some extent by the pole at $s = -20$. The zero will have the effect of pulling the breakaway point to the left and the pole will try to push it to the right, but with less effect because

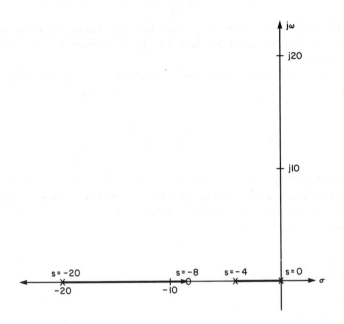

figure **7-14** The locations of the open-loop poles and zeros of

$$G(s)H(s) \frac{100A\,(s\,+\,8)}{s(s\,+\,20)\,(s\,+\,4)}$$

it is further away. A fairly simple trial and error technique to establish
this point with reasonable accuracy is as follows:

(a) Select a trial breakaway point, say $s_T = -2.24$.

(b) Measure the distance of all the poles and zeros from s_T, to
the left and to the right of s_T; let the distances to the poles be negative
and the distance to the zeros be positive, thus:

To the Left of s_T	*To the Right of s_T*
$-17.76, +5.76, -1.76$	-2.24

(c) Take the reciprocals of these distances.

Left	*Right*
$-0.06, +0.17, -0.57$	-0.45

(d) Equate the sum of the left-hand terms to the sum of the
right-hand terms. Thus:

$$-0.06 + 0.17 - 0.57 = -0.45$$

$$-0.46 = -0.45$$

If the two numbers are very close in magnitude, then the selection was very close to the real value. If the difference is large, repeat the procedure.

By following steps (a), (b), and (c) we arrive at the simple equation.

$$\frac{-1}{(20 - s_T)} + \frac{1}{(8 - s_T)} - \frac{1}{(4 - s_T)} = \frac{-1}{(s_T - 0)}$$

Of course, by finding the value of s_T that makes this equation true, we could locate the breakaway point, but because solving this equation is not always easy, we prefer to use the trial and error method described above.

Rule 5:
The branch locus breaks away at $\pm 90°$ to the real axis.

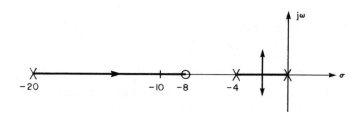

Rule 7:
Determine the angle ψ at which the asymptotes intersect the real axis.

$$\psi = \frac{(2K + 1)180°}{\eta_p - \eta_z} = \frac{(2K + 1)180°}{3 - 1}$$

Therefore $\psi = 90°$ when $K = 0, 2, 4$, etc.

and $\psi = 270°$ or $-90°$ when $K = 1, 3, 5$, etc.

The asymptotes intersect the real axis at the centroid s_c, where

$$s_c = \frac{\Sigma \text{ poles} - \Sigma \text{ zeros}}{\eta_p - \eta_z}$$

$$\Sigma \text{ poles} = (-20) + (-4) + (0) = -24$$

$$\Sigma \text{ zeros} = -8$$

Therefore $\qquad s_c = \dfrac{(-24) - (-8)}{3 - 1} = -8$

Rule 8:
This rule does not apply, because the asymptotes never cross the imaginary axis.

The root locus can now be sketched as shown.

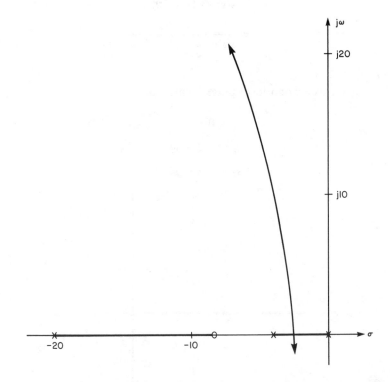

figure 7-15 The root locus of

$$G(s)H(s) = \frac{100A \, (s + 8)}{s(s + 4)(s + 20)}$$

The locus shows that the system is quite stable and can never be unstable, regardless of the value of A.

EXAMPLE 7.2

Given that the open-loop transfer function of a particular system is $\dfrac{K}{s(s^2 + 4s + 20)}$, sketch the root locus and from it determine:

(a) The system gain and damped resonant frequency when the damping ratio is 0.4.

(b) The system gain and resonant frequency at the point of marginal stability.

solution

$$G(s)H(s) = \frac{K}{s(s^2 + 4s + 20)}$$

$$= \frac{K}{s(s + 2 + j4)(s + 2 - j4)}$$

Rules 1, 2, 3, and 6:

There are three open-loop poles located at

$$s_1 = 0$$

$$s_2 = -2 + j4$$

$$s_3 = -2 - j4$$

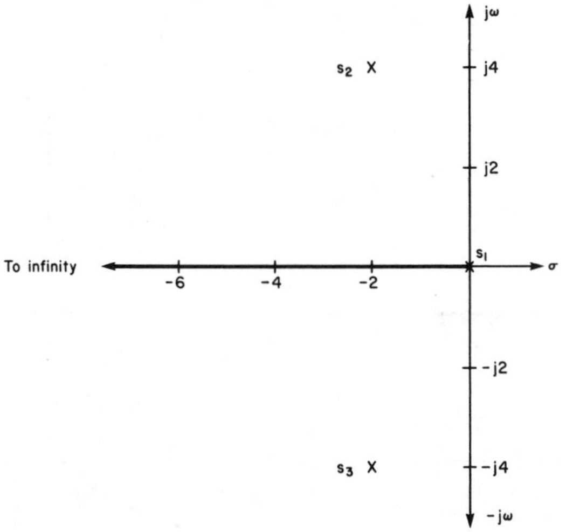

figure 7-16 The locations of the open-loop poles and zeros of

$$G(s)H(s) = \frac{K}{s(s^2 + 4s + 20)}$$

Rules 4 and 5:
These rules do not apply since all the loci end at infinity and only one open-loop pole appears on the negative real axis.

Rule 7:
$$\psi = \frac{(2K + 1)180°}{\eta_p - \eta_z} = \frac{(2K + 1)180°}{3 - 0}$$

Therefore $\psi = 60°, 180°, 300°$

and
$$s_c = \frac{\Sigma \text{ poles} - \Sigma \text{ zeros}}{\eta_p - \eta_z}$$

Therefore
$$s_c = \frac{(0) + (-2 + j4) + (-2 - j4)}{3 - 0} = -\frac{4}{3}$$

Rule 8:
The intercept of the locus with the imaginary axis is found by solving for ω when $j\omega$ is substituted for s in the characteristic equation, namely,

$$\frac{K}{s(s^2 + 4s + 20)} + 1 = 0$$

or
$$K + s^3 + 4s^2 + 20s = 0$$

Substituting $j\omega$ for s, we obtain:

$$K + (j\omega)^3 + 4(j\omega)^2 + 20j\omega = 0$$

which yields

$$(K - 4\omega^2) + j(20\omega - \omega^3) = 0$$

For this equation to be true both the real and imaginary parts must equal zero; hence

$$K - 4\omega^2 = 0$$

and
$$20\omega - \omega^3 = 0$$

which yields
$$\omega = \sqrt{20} = \pm 4.47$$

and hence
$$K = 80$$

Rule 9:
This rule can be applied after the root locus is constructed.

In order to determine the angle of departure of the root locus from a complex pole we need to introduce another rule.

Rule 10:

The angle of departure from a complex pole or the angle of arrival at a complex zero can be determined by summing all the angles made when lines are drawn from the complex pole (*or zero*) to all the other poles and zeros and then adding 180 deg to this sum. The angles to other poles are added negatively, and the angles to other zeros are added positively. Thus:

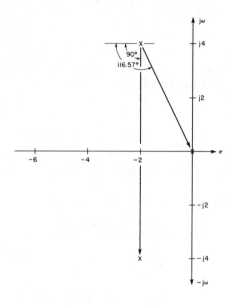

The angle of departure from the upper complex pole is given by

$$-(116.57° + 90°) + 180° = -26.57°$$

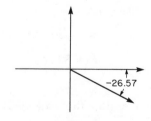

The root locus can now be constructed. Because two conjugate complex poles exist, both the second and the third quadrants must be shown.

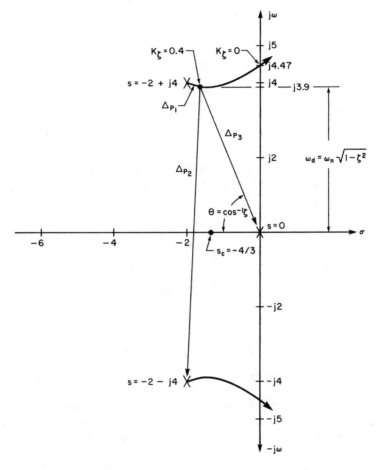

figure 7-17 The root locus of $G(s)H(s) = \dfrac{K}{s(s^2 + 4s + 20)}$

(a) When $\zeta = 0.4$, $\qquad \cos^{-1} \zeta = 66.4°$

Once the loci have been plotted, a line, which subtends an angle of 66.4° to the negative real axis, can be drawn from the origin to intercept the upper root locus. Measuring the length of this line and its vertical component, we find

$$\omega_n \approx 4.25 \text{ rad/s}$$

and
$$\omega_d \approx 3.9 \text{ rad/s}$$

Now
$$K = K_0 = \frac{\Delta p_1 \cdot \Delta p_2 \cdot \Delta p_3}{1}$$

where measurements reveal that

$$\Delta p_1 \approx 0.25$$

$$\Delta p_2 \approx 7.95$$

$$\Delta p_3 \approx 4.25$$

Hence
$$K = K_0 = 8.45$$

(b) At the point of marginal stability the locus crosses the imaginary axis. Thus Rule 8 yields the system resonant frequency at this point; i.e.,

$$\omega_n = \omega_d = 4.47 \text{ rad/s}$$

and
$$K = 80$$

The same values could be determined from direct measurements taken from the diagram.

EXAMPLE 7.3

(a) If the open-loop transfer function of a system is given as $G(s)H(s) = \frac{K(0.25s + 1)}{s(0.4s + 1)}$, sketch the locus as K varies from zero to infinity.

(b) From the locus determine the smallest value of damping ratio that the system could have and the damped resonant frequency at this value. What is the value of K for this condition?

solution

(a) *Rule 1:*

$$G(s)H(s) = \frac{K(0.25s + 1)}{s(0.4s + 1)}$$

or
$$G(s)H(s) = \frac{0.25K}{0.4} \cdot \frac{(s + 4)}{s(s + 2.5)}$$

Hence
$$G(s)H(s) = \frac{K_0(s + 4)}{s(s + 2.5)}$$

where
$$K_0 = 0.625K$$

The open-loop poles and zero are located as shown in Figure 7-18.

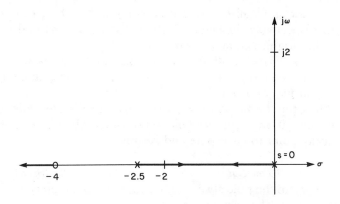

figure 7-18 The locations of the open-loop poles and zeros of

$$G(s)H(s) = \frac{0.625K(s + 4)}{s(s + 2.5)}$$

Rules 2 and 3:
The portions of the real axis that form part of the root locus are shown in the diagram, where it is seen that all the negative real axis to the left of $s = -4$ is part of the locus.

Rule 6:
From this rule we know that because the number of open-loop poles exceeds the number of open-loop zeros by one, one locus must go to infinity as K_0 increases. Because only one locus goes to infinity, it must do so along the negative real axis. (An even number of loci must be involved for them to go to infinity other than along the negative real axis, as they do in Example 7.2.)

If we now consider our diagram again, we see that a zero is at $s = -4$ and hence a locus is seen approaching this point from the left. But we have just said that one locus goes to infinity along the real axis, and from rule 2 above all the real axis left of $s = -4$ is part of the locus. Thus we have loci heading away from each other at some point to the left of $s = -4$. For this condition to occur, a break-in point must exist somewhere between $s = -4$ and $s = -\infty$.

Where did the two loci that meet at the break-in point originate? The answer, of course, is given to us by Rule 4.

Rule 4:
A breakaway point exists between $s = 0$ and $s = -2.5$. Thus the two loci breaking away at this point eventually return to the negative real

axis at the break-in point to the left of $s = -4$, whereupon one of the loci goes to infinity and the other locus goes to the zero at $s = -4$.

The locations of the breakaway point and the break-in point are shown, respectively, in Figure 7-19 at distances of σ_1 and σ_2 to the left of the origin, i.e., at $s = -\sigma_1$ and $s = -\sigma_2$.

Because of the small number of poles and zeros in this example, we can determine the values of σ_1 and σ_2 by using the simple formula outlined in Example 7.1.

Thus for the breakaway point at σ_1 there is one pole and one zero to the left of it and one pole to the right. The distance from the breakaway point to each pole and zero is

To the Left	*To the Right*
$(2.5 - \sigma_1)$ to the pole at $s = -2.5$	$(\sigma_1 - 0)$ to the pole at the origin
$(4 - \sigma_1)$ to the zero at $s = -4$	

Now we take the reciprocal of each of these quantities and sum them together, letting the pole distances be negative and the zero distances be positive. Finally, we equate the sum of the left-hand terms to the sum of the right-hand terms, as follows:

$$\frac{-1}{(2.5 - \sigma_1)} + \frac{1}{(4 - \sigma_1)} = \frac{-1}{(\sigma_1 - 0)} = \frac{-1}{\sigma_1}$$

By finding the common denominator of these three terms, we can rewrite the equation as

$$-\sigma_1(4 - \sigma_1) + \sigma_1(2.5 - \sigma_1) = -(2.5 - \sigma_1)(4 - \sigma_1)$$

Therefore

$$-1.5\sigma_1 = -10 + 6.5\sigma_1 - \sigma_1^2$$

and

$$\sigma_1^2 + 8\sigma_1 + 10 = 0$$

which has solutions

$$\sigma_1 = 1.55$$

and

$$\sigma_2 = 6.45$$

As can be seen, because we have solved a quadratic equation (which has two solutions), we have not only determined the breakaway point, but we have also automatically determined the break-in point; i.e.,

The breakaway point is at

$$s = -\sigma_1 = -1.55 \text{ rad/s}$$

and break-in point is at

$$s = -\sigma_2 = -6.45 \text{ rad/s}$$

Rule 7:

Because the locus returns to the axis, the interception point and the angle of the asymptotes have no real significance, for, of course, the only pole that can go to infinity is constrained to doing so along the negative real axis (i.e., ϕ = 180 deg).

The root locus can now be sketched. So that the scale can be as large as possible, only the second quadrant need be used.

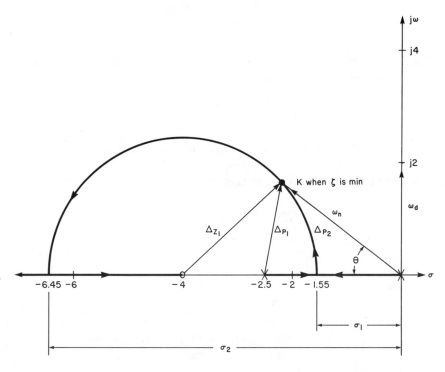

figure 7-19 The root locus of $G(s)H(s) = \dfrac{0.625K(s + 4)}{s(s + 2.5)}$

(b) The smallest value of ζ is given when the line ω_n is a tangent to the locus. Measurements taken from the locus indicate that

$$\theta \simeq 38°$$

and hence $\quad \zeta = \cos\theta \simeq 0.79$

Also $\quad\quad \omega_n \simeq 3$ rad/s

and $\quad\quad\quad \omega_d \simeq 1.9$ rad/s (measured)

or $\quad\quad\quad\quad$ 1.85 rad/s (calculated from $\omega_n\sqrt{1 - \zeta^2}$)

Now
$$K_o = \frac{\Delta p_1 \cdot \Delta p_2}{\Delta z_1}$$

where
$$\Delta z_1 = 2.45$$
$$\Delta p_1 = 1.9$$
$$\Delta p_2 = \omega_n = 3$$

Hence
$$K_o = 2.33 = 0.625K$$

and therefore
$$K = 3.72$$

If we wish to cross-check these values, we can substitute K_o into the characteristic equation, thus:

$$G(s)H(s) = \frac{2.33(s + 4)}{s(s + 2.5)}$$

Now $\quad G(s)H(s) + 1 = 0$ (the characteristic equation)

Rewriting the equation, we obtain

$$2.33(s + 4) + s(s + 2.5) = 0$$

Hence
$$s^2 + 4.83s + 9.32 = 0$$

This equation has the form

$$s^2 + 2\zeta\omega_n s + \omega_n^2 = 0$$

Hence $\quad \omega_n = 3.05$ rad/s (close enough to measured value of $\omega_n = 3$ rad/s)

and $\quad \zeta = 0.79$ (which is in agreement with the measured value)

If this quadratic equation is factorized, the roots of the equation will give the location of the intercept of the line ω_n with the locus; i.e., the roots of the equation

$$s^2 + 4.83s + 9.31 = 0$$

are
$$s_{1,2} = -2.42 \pm j1.87$$

or
$$s_{1,2} = 3.05 \angle \pm 37.7°$$

which compares well with the measured values of

$$\omega_n = 3 \text{ rad/s}$$

and $$\theta = 38°$$

7-5 SUMMARY

With the explanation of root locus given in Section 7-2 and the ten rules of construction explained and demonstrated in Section 7-4, the reader should be able to construct a root locus from the open-loop transfer function with a reasonable amount of confidence and accuracy.

These techniques are intended to allow rapid construction and quick analysis, and the reader should practice these techniques by sketching as many root loci as possible. Such practice will provide greater insight and understanding when other pole-zero topics are discussed later in the text.

PROBLEMS

Section 7-2   1. Plot the root locus of the closed-loop poles as K varies from zero to infinity in the following.

(a)

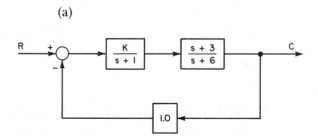

(b)

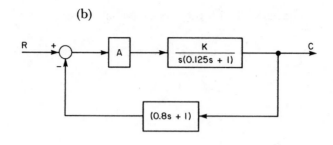

(c)

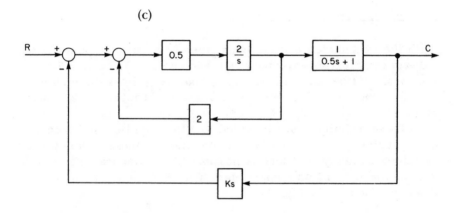

Section 7-4   2. Sketch the root locus of the following open-loop transfer functions as K varies from zero to infinity.

(a) $G(s)H(s) = \dfrac{K}{s(0.5s + 1)}$ (b) $G(s)H(s) = \dfrac{K(s+1)}{s(0.5s+1)}$

(c) $G(s)H(s) =$
$$\dfrac{K}{s^2 + 15s + 50}$$

(d) $G(s)H(s) =$
$$\dfrac{K}{s(s^2 + 15s + 50)}$$

(e) $\dfrac{K(s+1)}{s(s^2 + 7s + 10)}$

3. The following open-loop transfer functions represent various systems. Find the value of gain constant K when the damping ratio = 0.5.

(a) $\dfrac{K}{s(s + 1)(0.2857s + 1)}$

(b) $\dfrac{K(0.2s + 1)}{s(s + 1)(0.4s + 1)(0.01s + 1)}$

(c) $\dfrac{K}{s(0.1s + 1)(s^2 + 6s + 13)}$

(d) $\dfrac{K(0.2s + 1)}{s^2(s + 1)(0.4s + 1)(0.01s + 1)}$

4. If

$$G(s)H(s) = \frac{K(s + a)}{s(s + 4)(s + 8)(s + 16)}$$

sketch the root loci when $a = 2, 5, 8,$ and 12. Sketch all the loci on the same graph. Comment upon the various positions of the zero.

5.

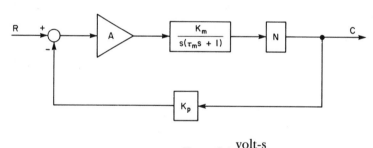

$$K_m = 5.1 \frac{\text{volt-s}}{\text{rad}}$$

$$\tau_m = 0.3125 \text{ s}$$

$$N = 0.25$$

$$K_p = 4.83 \text{ volts/rad}$$

(a) In the system shown above, for what value of A does the system become unstable?
(b) What value of A will give a damping ratio of 0.347?
(c) What is the damped resonant frequency at the value of gain determined in (b)?

how stable is the system? part ii —frequency response methods

8

The next moment soldiers came running through the wood, at first in twos and threes, then ten or twenty together, and at last in such crowds that they seemed to fill the whole forest, Alice got behind a tree, for fear of being run over, and watched them go by.

Lewis Carroll, *Through the Looking Glass*

8-1 INTRODUCTION

The root locus method of analysis dealt with the transient response and stability of linear systems by considering what happened to the poles and zeros of the system *when the system parameters were changed.*

A number of other methods exist for evaluating system performance which consider the behavior of the open-loop gain and phase of the system *when the frequency of an imaginary sinusoidal input is varied.* These methods are usually termed *frequency response methods of analysis.*

8-2 FREQUENCY RESPONSE—GENERAL

If the input to a negative feedback control system is varied in a periodic manner, the output response will either lag or lead the input, depending upon the type of components in the system. If the input changes sinusoidally, the lag or lead time of the output is expressed as an angular difference from the input. When the angular difference between the input signal and the feedback signal approaches 180 deg, the negative feedback loop starts to become a positive feedback loop and the system may become unstable. The possibility of instability is raised to a certainty if the open-loop gain of the system also happens to be greater than unity.

The open-loop gain and phase of a linear system at a particular frequency are independent of the magnitude of the input signal and can be determined solely from a knowledge of the system and its components. For example, consider the linear circuit shown in Figure 8-1. We may assume for our purposes that this circuit represents the open-loop arrangement of components in a control system such that the input signal is E and the feedback signal is V_c.

figure 8-1 simple *rc* circuit with variable input frequency

At any frequency, the voltage across the capacitor is given by

$$V_c = I \cdot X_c \qquad (8.1)$$

where
$$I = \frac{E}{Z}$$

$$X_c = 1/\omega C$$

$$Z = \sqrt{R^2 + X_c^2}$$

If the phase angles associated with X_c and Z are included, the voltage across the capacitor can be determined from

$$V_c = \frac{E \cdot X_c \, \angle -90°}{\sqrt{R^2 + X_c^2} \, \angle -\tan^{-1}\dfrac{X_c}{R}} \qquad (8.2)$$

from which

$$V_c(\omega) = \frac{E(\omega)\,\dfrac{1}{\omega C}}{\sqrt{R^2 + \left(\dfrac{1}{\omega C}\right)^2}} \angle\left(-90° + \tan^{-1}\frac{X_c}{R}\right)$$

(8.3)

or

$$V_c(\omega) = \frac{E(\omega)}{\sqrt{(\omega RC)^2 + 1}} \angle\left(-90° + \tan^{-1}\frac{1}{\omega RC}\right)$$

(8.4)

When the frequency is 723.4 Hz, equation (8.4) yields a value of

$$V_c = 7.07 \angle -45° \text{ volts}$$

(8.5)

If the frequency is doubled to 1446.8 Hz, then

$$V_c = 4.47 \angle -63.4° \text{ volts}$$

(8.6)

In fact, as the frequency of the circuit is varied from zero to infinity, the magnitude and phase shift of V_c are plotted in Figure 8-2 from the data in the accompanying table.

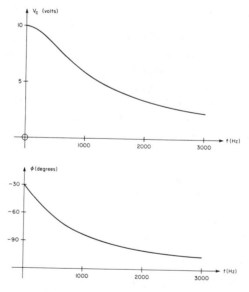

f(Hz)	1	20	100	200	400	723.4	1446.8	2000	4000	8000	10000
Vc(volts)	10	10	9.96	9.64	8.75	7.07	4.47	3.4	1.78	0.9	0.72
φ(degrees)	−1.0	−1.6	−7.9	−15.45	−29	−45	−63.4	−70	−79.8	−84.8	−85.9

figure 8-2 output voltage and phase versus frequency for the circuit shown in figure 8-1

8-3 RESONANCE

A phenomenon often encountered when frequency response techniques are used and one which can be a problem to the designer, if not allowed for, is resonance.

Resonance can occur in a system if it is both lightly damped and contains two or more complementary energy-storing elements, e.g., capacitance and inductance in an electrical circuit or a spring connected to an inertial load in a mechanical device. In a completely undamped system, resonance describes the condition when any disturbance to the system causes a sustained oscillation of energy between the energy-storing elements at a natural frequency, ω_n, determined by the values of these elements.

In all practical systems, in the absence of an external energy supply, the oscillations will decay as some energy is always dissipated at each oscillation. In fact, the amount of damping in a system is increased by increasing the amount of energy dissipated within the system, e.g., by adding resistors to the electrical circuit or a viscous damper to the mechanical device. If, however, this dissipated energy is replaced from an external power source, at a similar frequency to ω_n, resonant oscillations will be maintained. In this type of application the frequency of the external source is termed the forcing frequency, ω. If ω and ω_n differ, the effort of maintaining the system oscillation increases dramatically as the gap between ω and ω_n widens.

Thus in all practical systems the resonant frequency of the system can be defined as that frequency which causes minimum impedance to be offered to the external forcing function.

To illustrate these ideas, let us consider the electric circuit shown in Figure 8-3.

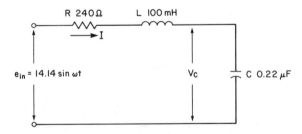

figure 8-3　a simple RLC series circuit

If resonance is the condition of minimum impedance, then

$$X_c - X_L = 0 \qquad (8.7)$$

or
$$X_c = X_L \tag{8.8}$$

Hence
$$\omega_n = \sqrt{\frac{1}{LC}} \tag{8.9}$$

i.e.,
$$\omega_n = 6742 \text{ rad/s (or 1073 Hz)} \tag{8.10}$$

The characteristic equation of this circuit is

$$LCs^2 + RCs + 1 = 0 \tag{8.11}$$

which has the same form as

$$\frac{1}{\omega_n^2} s^2 + \frac{2\zeta}{\omega_n} s + 1 = 0 \tag{8.12}$$

Hence, for the damping ratio,

$$\zeta = \frac{R}{2}\sqrt{\frac{C}{L}}$$

which yields
$$\zeta = 0.18$$

Since ω_d, the damped natural frequency, is given by

$$\omega_d = \omega_n\sqrt{1 - \zeta^2}$$

then
$$\omega_d = 6528.4 \text{ rad/s or 1039 Hz}$$

The steady-state amplitude of V_c can be obtained from 23 of Table 5-1 as

$$V_{c_{ss}}(\omega) = \frac{A\omega_n^2}{[(\omega_n^2 - \omega^2)^2 + (2\zeta\omega\omega_n)^2]^{1/2}} \angle -\tan^{-1}\left[\frac{2\zeta\omega\omega_n}{\omega_n^2 - \omega^2}\right] \tag{8.13}$$

where
$$A = 14.14 \text{ volts}$$

Equation (8.13) is the phasor form of the steady-state solution.
    Let us now examine the value of V_c at three values of ω_n.

1.
    Let
$$\omega = \omega_n/2$$

    i.e.,
$$\omega = 3371 \text{ rad/s}$$

    Hence
$$V_{c_{ss}}(\omega) = 12.97\angle -13.5° \text{ volts}$$

2.

Let $\qquad\qquad \omega = \omega_n$

i.e., $\qquad\qquad \omega = 6742$ rad/s

Hence $\qquad V_{c_{ss}}(\omega) = 27.78\angle -90°$ volts

3.

Let $\qquad\qquad \omega = 2\omega_n$

i.e., $\qquad\qquad \omega = 13484$ rad/s

Hence $\qquad V_{c_{ss}}(\omega) = 3.24\angle -166.5°$ volts

Figure 8-4 shows a sketch of the voltage and phase of $V_{c_{ss}}$ as the frequency is varied.

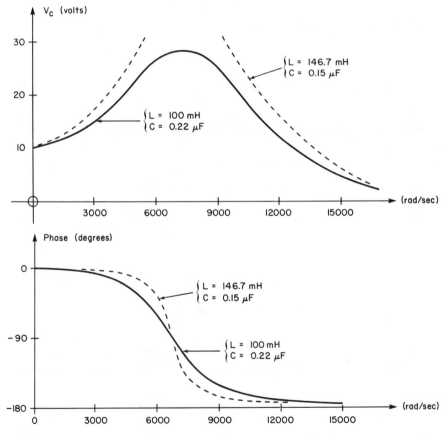

figure 8-4   voltage and phase of $V_{c_{ss}}$ versus frequency for the *RLC* circuit of figure 8-3

If now the capacitor is changed to a value of 0.15 μF and the inductance to 146.7 mH, the resonant frequency remains unchanged at 6742 rad/s, but the damping ratio is reduced by 33 percent to 0.12. The effect of the reduction in the damping ratio on $V_{c_{ss}}$ is to increase the amplitude of the oscillations, as can be seen from the following.

1. When
$$\omega = \omega_n/2$$
$$V_{c_{ss}}(\omega) = 13.17\angle -9.1° \text{ volts}$$

2. When
$$\omega = \omega_n$$
$$V_{c_{ss}}(\omega) = 41.67\angle -90° \text{ volts}$$

3. When
$$\omega = 2\omega_n$$
$$V_{c_{ss}}(\omega) = 3.29\angle -170.9° \text{ volts}$$

Curves for this condition are plottd in Figure 8-4, where the effects of reducing ζ are clearly demonstrated.

Our discussion of resonance can be summarized as follows:

1. Resonance requires at least two energy-storing elements in the system so that a reciprocating energy transfer can be maintained between them.

2. The resonant frequency depends entirely upon the values of the energy-storing elements.

3. Resonance occurs only when the damping ratio is less than unity ($\zeta < 1$).

4. The smaller the damping ratio, the greater the resonant peak.

5. The phase shift is 0 deg when the input frequency is zero and has maximum value of $\pm N \times 90$ deg (-180 deg for our example) w. en the input frequency approaches infinity, where N is the number of energy storing components in the system.

6. The phase shift at resonance for our second-order example is -90 deg.

8-4 FREQUENCY RESPONSE AND STABILITY

In Section 6-1 we saw that a system can have positive feedback and remain marginally stable provided the magnitude of the feedback signal is not too large. The system considered in Section 6-1 is shown in Figure 8-5 for convenience, where G and H can be assumed to be constant gains, unaffected by the frequency content of the input R.

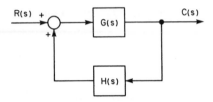

figure 8-5 block diagram showing positive feedback

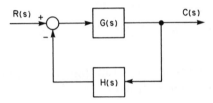

figure 8-6 block diagram showing negative feedback

The overall transfer function of the positive feedback system in Figure 8-5 is given by

$$\frac{C(s)}{R(s)} = \frac{G}{1 - GH} \qquad (8.14)$$

If $H = 1$, then G has to be less than unity to ensure any degree of stability. As G approaches unity, the overall gain of the system approaches infinity, a condition which denotes instability. In fact, positive feedback has little or no practical application and considerable effort is usually expended when control systems are designed to avoid its occurrence.

Consider now Figure 8-6, which is a similar system but which has negative feedback loop and frequency-dependent, complex gains, $G(s)$ and $H(s)$. The transfer function for this system is

$$\frac{C(s)}{R(s)} = \frac{G(s)}{1 + G(s)H(s)} \qquad (8.15)$$

The obvious difference between the two transfer functions is the change of the sign in the denominator caused by changing from positive feedback to negative feedback.

However, in equation (8.15), because $G(s)$ and $H(s)$ are frequency-dependent complex numbers, the frequency of R can be adjusted until $G(s) H(s)$ has a real and negative value, say,

$$G(s)H(s) = -|GH|$$

then equation (8.15) becomes

$$\frac{C(s)}{R(s)} = \frac{G(s)}{1 - |GH|} \tag{8.16}$$

which has a denominator identical to that of equation (8.14). Therefore, we can say that at this frequency, positive feedback has occurred in this system. Also if $|GH|$ is greater than or equal to unity, the positive feedback will cause instability; i.e., positive feedback occurs and the system becomes unstable when

$$G(s)H(s) = -1 \tag{8.17}$$

But what has actually happened to $G(s)H(s)$ when it changes sign? In other words, what is meant when a frequency-dependent term such as $G(s)H(s)$ is multiplied by -1? Suppose

$$A = -1$$

Then A can be written in complex notation as

$$A = -1 + j0$$

which can be transformed to the following polar form:

$$A = 1\angle 180° \text{ or } 1\angle -180°$$

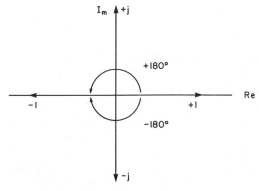

figure 8-7 argand diagram showing equivalence of -1 and $1\angle 180$ deg

from which it is seen that multiplying a term such as $G(s)H(s)$ by -1 is equivalent to a phase change in $G(s)H(s)$ of ± 180 deg.

In summary, therefore, whereas at low frequencies the value of $G(s)H(s)$ may be real and positive, giving only positive terms in the denominator of equation (8.15), as the frequency of the input signal R increases, the phase of $G(s)H(s)$ changes. If the system contains enough energy storage components, the phase change will eventually exceed 180 deg, which is equivalent to a change of the sign in the denominator of equation (8.15).

In addition to a change in the phase of $G(s)H(s)$, the magnitude of this term will also be changed and is usually attenuated (unless resonance occurs) as frequency increases. As a general rule (when $G(s)H(s)$ contains no positive poles), if the magnitude of $G(s)H(s)$ is less than unity, when the phase change is 180 deg the system can be considered to be stable.

It is reasonable, therefore, to assume that by plotting the magnitude and phase of $G(s)H(s)$ for a given system as a function of frequency, we could determine at what frequency the phase change exceeds 180 deg and the magnitude of $G(s)H(s)$ at this frequency. In other words, we could determine if the system is stable. However, before we can examine a system's stability, we must decide on a method of plotting the gain and phase of $G(s)H(s)$.

There are several methods of performing frequency plots, the most frequently encountered being Nichols charts, polar plots, and cartesian plots (Bode diagram). In this book we will confine ourselves to using only the Bode diagram; we will discuss the use of Bode diagrams in some depth for the remainder of this chapter.

8-5 BODE DIAGRAMS

Bode diagrams are the most commonly used and easiest to understand of the various frequency response techniques of analysis. Construction of a Bode diagram requires plotting the gain and phase of the system open-loop transfer function, $G(s)H(s)$, as the input frequency is changed. Linear-logarithmic graph paper is used for the plots so that the gain is expressed in decibels (db) and plotted on the vertical linear axis while frequency, ω, is plotted the horizontal logarithmic axis. Figures 8-9 and 8-10 are examples of a Bode diagram drawn for a phase-lag circuit.

If the use of decibels is new to the reader, this is no cause for concern. Decibels are only another way of expressing the gain of the system, and the rule for conversion is very simple. If the magnitude or gain of $G(s)H(s)$ is given by A, then the gain in decibels (db) is given by the expression

$$20 \log_{10} (A)$$

The first question we ask ourselves when we want to construct a Bode diagram is: How do we calculate the gain and phase of $G(s)H(s)$ as a function of frequency? Well, given that we already have the transfer function of $G(s)H(s)$, we use a simple rule that lets us substitute $j\omega$ wherever s appears in the transfer function. For example, the transfer function for the system shown in Figure 8-3 is

$$T(s) = \frac{V_c(s)}{E_{in}(s)} = \frac{1}{LCs^2 + RCs + 1} \qquad (8.17)$$

Hence

$$V_c(s) = \frac{E_{in}(s)}{LCs^2 + RCs + 1} \qquad (8.18)$$

If we had used Ohm's Law to calculate the value of V_c as a function of frequency ω, i.e., $V_c(\omega)$, we would have eventually derived the following expression.

$$V_c(\omega) = \frac{E_{in}(\omega)}{R + j\omega L + 1/j\omega C} \cdot \frac{1}{j\omega C} \qquad (8.19)$$

Multiplying by $j\omega C$ and dividing by $E_{in}(\omega)$ yields

$$T(\omega) = \frac{V_c(\omega)}{E_{in}(\omega)} = \frac{1}{LC(j\omega)^2 + RC(j\omega) + 1} \qquad (8.20)$$

Comparing equations (8.17) and (8.20), we see that the gain and phase of $T(s)$ can be calculated directly from the transfer function by making the substitution $j\omega$ for s.

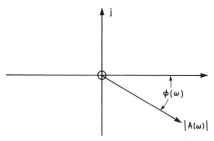

figure 8-8   the polar form of $T(\omega)$

Equations (8.17) and (8.20) show that the transfer function $T(s)$ can be converted into $T(\omega)$, which is an expression containing only complex numbers and ω. This expression is then easily transformed into polar form so that

$$T(\omega) = A(\omega)\angle\phi(\omega)$$

where A is the gain and ϕ is the phase of the transfer function. To construct a Bode diagram for this system, we now simply calculate A and ϕ for various values of ω and plot the two curves on log-linear graph paper. Let us look at an example of how to construct a Bode diagram for a first-order transfer function.

EXAMPLE 8.1

A first-order system has an open-loop transfer function

$$G(s)H(s) = \frac{K}{1 + \tau s} \qquad (8.21)$$

where $\qquad K = 2 \qquad$ and $\qquad \tau = 0.1$ sec

Thus in the frequency domain, by substituting $j\omega$ for s, the open-loop transfer function can be expressed as

$$G(\omega)H(\omega) = \frac{K}{1 + j\tau\omega} \qquad (8.22)$$

In polar form this becomes

$$G(\omega)H(\omega) = \frac{K}{(1 + \tau^2\omega^2)^{1/2}} \angle -\tan \omega\tau \qquad (8.23)$$

i.e., of the form

$$G(\omega)H(\omega) = |\text{gain}|\angle\text{Phase}$$

Expressing the gain term in decibels yields

$$\text{Gain} = 20 \log_{10} \frac{K}{(1 + \tau^2\omega^2)^{1/2}} \text{ db} \qquad (8.25)$$

from which can be stated

$$\text{Gain} = 20 \log_{10} K + 20 \log_{10} \frac{1}{(1 + \tau^2\omega^2)^{1/2}} \text{ db} \qquad (8.26a)$$

or $\qquad$ Gain $= 20 \log_{10} K + 20 \log_{10} (1 + \tau^2\omega^2)^{-1/2}$ db $\qquad$ (8.26b)

Hence $\qquad$ Gain $= 20 \log_{10} K - 10 \log_{10} (1 + \tau^2\omega^2)$ db $\qquad$ (8.26c)

Substituting for K and τ and evaluating the terms of equation (8.26c) yields the following:

$$20 \log_{10} K = 6 \text{ db}$$

which is constant for all values of ω, since ω does not appear in the expression. The latter term of expression (8.26c) is tabulated here for various values of ω.

ω (rad/sec)	$-10 \log_{10} (1 + \tau^2\omega^2)$ (db)
0	0
5	$- 0.97$
10	$- 3.01$
20	$- 6.99$
40	-12.3

The db gain/log frequency curve for $G(\omega)H(\omega)$ is plotted in Figure 8-9.

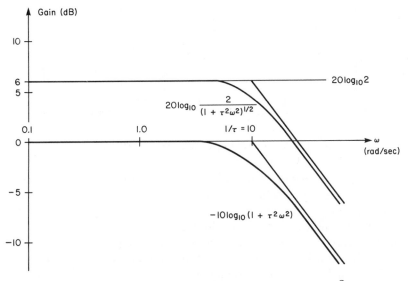

figure 8-9 gain versus frequency for $G(\omega)H(\omega) = \dfrac{2}{1 + j0.1\omega}$

The phase of $G(\omega)H(\omega)$ is given by the expression

$$\phi = \angle -\tan^{-1}(\tau\omega)$$

or

$$\phi = \angle -\tan^{-1}(0.1\omega)$$

which for various values of ω is tabulated below.

ω (rad/sec)	$\phi = \angle -tan^{-1}(\tau\omega)$ (degrees)
0	0
5	-26.5
10	-45
20	-63.5
40	-76

The phase/log frequency curve for $G(\omega)H(\omega)$ is plotted in Figure 8-10. A Bode diagram always comprises both a gain and a phase plot as in Figures 8-9 and 8-10.

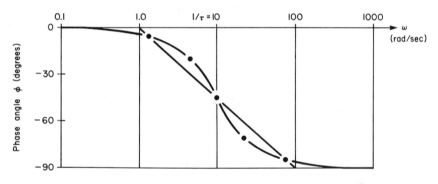

figure 8-10   phase versus frequency for $G(\omega)H(\omega) = \dfrac{2}{1 + j0.1\omega}$

Why is the use of Bode diagrams so popular? The major reason is that they can be sketched very quickly and yet still retain reasonably good accuracy. We will talk about the rules that enable us to sketch Bode diagrams in a moment. For now, look at Figures 8-9 and 8-10 and note that the curves in both figures can be approximated by the use of two or three straight lines. In Figure 8-9 the gain curve is approximated by two lines, each of which is asymptotic to the curve; furthermore, the intersection of these two lines occurs at a frequency of $\omega = 1/\tau$ rad/s. (We will show later that the angle of intersection is also easily determined from the number of poles and zeros.) In Figure 8-10 the phase curve is approximated by three lines: $\phi = 0$ deg, $\phi = 90$ deg, and a line that spans two decades of ω running from 0 deg to -90 deg with its midpoint at $\omega = 1/\tau$.

From these notes it would appear that the straight line approximations to the gain and phase curves can be sketched by using only the values of the poles, the zeros, and the steady-state gain of $G(s)H(s)$. The following rules explain how these three quantities influence the angles and points of intersection of the straight lines in a Bode diagram sketch.

8-6 SIMPLE RULES FOR PLOTTING BODE DIAGRAMS

8-6.1 introduction

The open-loop transfer function for a linear system can always be written in the form

$$G(s)H(s) = \frac{K(1 + \tau_1 s)(1 + \tau_3 s) \ldots}{s^n(1 + \tau_2 s)(1 + \tau_4 s) \ldots} \qquad n \geq 0 \qquad (8.27)$$

Open-loop zeros = $1/\tau_1$, $1/\tau_3$, etc.; open-loop poles = $1/\tau_0$, $1/\tau_2$, $1/\tau_4$, etc. Also, as we shall see, a Bode diagram can be sketched for any general system by simply adding the effects of each pole and each zero in order to determine the angles and intersection points of the asymptotes. Therefore, in order to be able to sketch a Bode diagram for any transfer function, we must establish a set of rules that tells us, first, how a single pole (or zero) affects the asymptote angle and intersection point, and second, how these angles are affected by the number of poles and zeros in the open-loop transfer function.

8-6.2 the dc gain

If in equation (8.27), $n = 0$, then

$$G(\omega)H(\omega) = \frac{K(1 + j\omega\tau_1)(1 + j\omega\tau_3) \ldots}{(1 + j\omega\tau_2)(1 + j\omega\tau_4) \ldots} \qquad (8.28)$$

Also when the frequency ω is zero

$$G(O)H(O) = K \qquad (8.29)$$

For a transfer function with no pure s terms in the denominator, as in equation (8.28), the value of K is known as the dc gain (i.e., zero frequency) or steady-state gain. Figure 8-11 shows a Bode diagram of three different constants, i.e.,

$$K = \tfrac{1}{2}$$

$$K = 2$$

$$K = 5$$

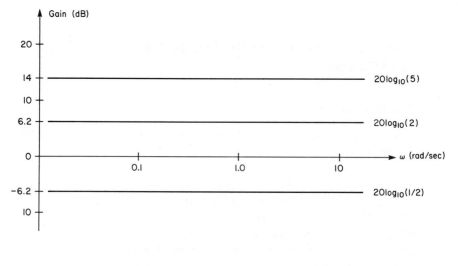

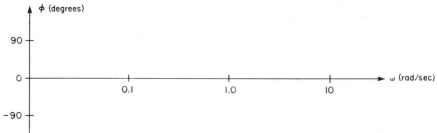

figure 8-11 a bode diagram showing the gain and phase of a constant

The Bode diagram of a constant comprises a horizontal line for the gain curve, with a value of $20 \log_{10}(K)$; the phase curve is also a horizontal line with a value $\phi = 0$ deg. It will be noticed that, while for the gain plot the horizontal line rises as K is increased, the horizontal line in the phase plot is unchanged. In other words, the magnitude of the constant K in no way affects the phase of $G(\omega)H(\omega)$, and when ω is very small so that

$$G(\omega)H(\omega) \simeq K$$

the phase of $G(\omega)H(\omega)$ will be zero.

8-6.3 pure differentiators $[G(\omega)H(\omega) = (j\omega)^m]$ (multiple zeros at the origin of the s-plane)

If $m = 1$, then

$$G(\omega)H(\omega) = j\omega \qquad (8.30)$$

thus $\qquad\qquad G(\omega)H(\omega) = |\omega|\angle 90° \qquad (8.31)$

i.e., the db gain is $20 \log_{10}(\omega)$ and the phase of $G(\omega)H(\omega)$ is fixed at 90 deg. If ω is doubled, then

$$\text{db gain} = 20 \log_{10}(2\omega)$$
$$= 20 \log_{10}(\omega) + 20 \log_{10}(2)$$
$$\text{db gain} = 20 \log_{10}(\omega) + 6 \text{ db} \qquad (8.31)$$

by which we see that by doubling the frequency the gain of $G(\omega)H(\omega)$ is increased by 6 db. A doubling of frequency is called an *octave*, and the rate of increase of the gain of a pure integrator $(j\omega)$ is expressed as 6 db per octave (also given in some texts as 20 db per decade, i.e., per tenfold increase of frequency).

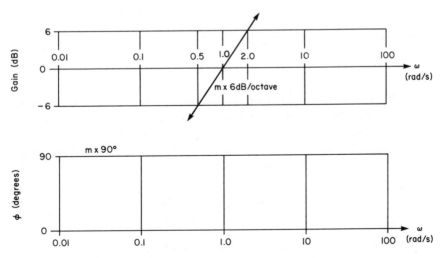

figure 8-12 bode diagram of $G(\omega)H(\omega) = j\omega$

The Bode plot of a pure differentiator ($m = 1$) is shown in Figure 8-12. It will be seen that the phase is fixed at $+90$ deg and the gain curve is a straight line with a slope of 6 db/octave, which intersects the 0 db line at $\omega = 1$ rad/s; i.e., $G(\omega)H(\omega) = 1$ when $\omega = 1$.

If now we assume that m has a value other than 0 or 1, then

or
$$G(\omega)H(\omega) = (j\omega)^m \qquad (8.32)$$

i.e.,
$$G(\omega)H(\omega) = |\omega^m| \angle m \cdot 90°$$

Also,
$$\text{db gain} = 20m \log_{10}(\omega) \qquad (8.33)$$

$$\phi = m \cdot 90° \qquad (8.34)$$

and doubling ω would now increase the db gain by $m \times 6$ db. Thus, as shown in Figure 8-12, the gain curve increases at $6m$ db/octave, while the phase of $G(\omega)H(\omega)$ is fixed at $m \times 90$ deg.

8-6.4 pure integrators $[G(\omega)H(\omega) = 1/(j\omega)^n]$ (multiple poles at the origin of the s-plane)

For a pure integrator ($n = 1$)

$$G(\omega)H(\omega) = \frac{1}{j\omega} \qquad (8.35)$$

or
$$G(\omega)H(\omega) = \left|\frac{1}{\omega}\right| \angle -90° \qquad (8.36)$$

Therefore,
$$\text{db gain} = -20 \log_{10}(\omega) \qquad (8.37)$$

and
$$\phi = -90° \qquad (8.38)$$

By reasoning similar to that used in the section on pure differentiators, we can say the Bode plot of the gain curve for a pure integrator is a straight line intersecting the 0 db axis at $\omega = 1$ rad/s and having a negative slope of -6 db per octave while the phase of an integrator is fixed at -90 deg. If now

$$G(\omega)H(\omega) = \frac{1}{(j\omega)^n} \qquad (8.39)$$

then, as shown in Figure 8-13, the slope of the gain curve is $-n \times 6$ db/octave and the phase of $G(\omega)H(\omega)$ is $-n \times 90$ deg.

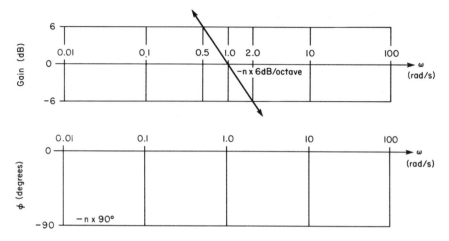

figure 8-13   bode diagram of $G(\omega)H(\omega) = (1/j\omega)^n$

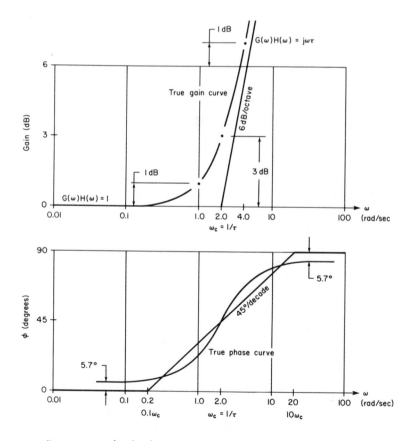

figure 8-14   bode diagram of $G(\omega)H(\omega) = (1 + j0.5\omega)$

8-6.5 simple zero [$G(\omega)H(\omega) = 1 + j\omega\tau$]

The Bode diagrams that we have looked at so far (for a constant gain, a differentiator, and an integrator) have consisted entirely of straight line plots. The transfer functions considered in the remainder of this chapter have Bode diagrams that can only be approximated by the use of straight lines. The approximations, however, are accurate enough for them to be used by control engineers as a practical tool when designing control systems.

The first such transfer function that we will consider is that of a simple zero, i.e.,

$$G(\omega)H(\omega) = 1 + j\omega\tau \qquad (8.40)$$

The Bode diagram of this system is shown in Figure 8-14 for the case when $\tau = 0.5$ s.

Gain Curve: Let us examine the gain curve first. This curve has two asymptotes, and for most of the frequency range the true curve runs very close to one or the other of these two lines. The gain curve is, therefore, approximated by these two straight lines. But how do we know where to draw these lines? Where do they intersect and at what angle? These questions are easily answered. For very small frequencies the asymptotic value of $G(\omega)H(\omega)$ is given by

$$G(0)H(0) = 1 + j\tau \times 0 = 1 \qquad (8.41)$$

Thus one of the asymptotes is the horizontal line corresponding to the dc gain, which for this example is given by equation (8.41), i.e.,

$$\text{db gain} = 0 \text{ db} \qquad (8.42)$$

The second asymptote corresponds to the values of $G(\omega)H(\omega)$ when ω is large, so that when

$$\omega \gg 1 \qquad (8.43)$$

we find that equation (8.40) tends to

$$G(\omega)H(\omega) \simeq j\omega\tau \qquad (8.44)$$

The equation of the second asymptote is, therefore,

$$G(\omega)H(\omega) = j\omega\tau \qquad (8.45)$$

and $$|G(\omega)H(\omega)| = \omega\tau \qquad\qquad (8.45a)$$

which will be recognized as the transfer function of a pure differentia-tor. Hence, as we saw in Section 8-6.3, the slope of this asymptote is 6 db/octave.

We can now draw one asymptote completely (the dc gain in deci-bels) and we know the slope of the second asymptote, but where do the two asymptotes intersect? Answer: at the frequency that causes the equa-tions of the two lines to be equal; thus, equating equations (8.41) and (8.45a) yields

$$\omega = \omega_c = \frac{1}{\tau} \qquad\qquad (8.46)$$

where ω_c = Corner Frequency

Consequently, to sketch the gain curve on a Bode plot for a simple zero we need only draw two straight lines; the first is horizontal and drawn at the value of the dc gain in db. The second line intersects the first at $\omega_c = 1/\tau$ and has a slope of 6 db/octave.

Phase Curve: The phase of equation (8.40) is given by

$$\phi = \tan^{-1}(\omega\tau) \qquad\qquad (8.47)$$

at the corner frequency ϕ is 45 deg and as ω becomes very small the phase curve approaches the 0 deg axis in an asymptotic manner. Simi-larly, as ω becomes very large, the phase curve approaches the asymptote $\phi = 90$ deg. A good approximation to how the phase curve changes between 0 deg and 90 deg is given, as shown in Figure 8-14, by a line with a slope of 45 deg per decade passing through the point $\phi = 45$ deg when $\omega_c = 1/\tau$. The maximum phase error when this approximation is used is 5.7 deg; it occurs where two lines intersect, i.e., at $\omega = 1/10\tau$ and $\omega = 10/\tau$, as shown in Figure 8-14.

8-6.6 simple pole [$G(\omega)H(\omega) = 1/(1 + j\omega\tau)$]

Figure 8-15 is the Bode diagram for the transfer function

$$G(\omega)H(\omega) = \frac{1}{1 + j\omega\tau} \qquad\qquad (8.48)$$

when $$\tau = 0.5 \text{ s}$$

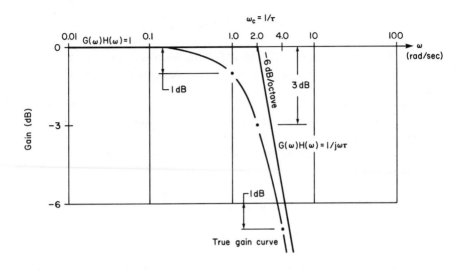

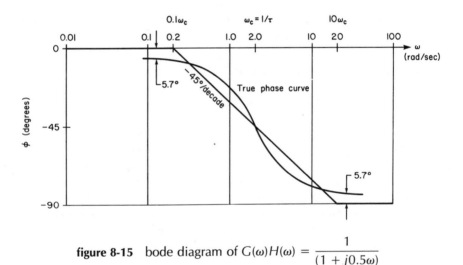

figure 8-15 bode diagram of $G(\omega)H(\omega) = \dfrac{1}{(1 + j0.5\omega)}$

This diagram is very similar to Figure 8-14, the differences being that, as ω increases, the gain curve now decays at -6 db per octave instead of increasing at $+6$ db per octave, and the phase of $G(\omega)H(\omega)$ now goes from 0 deg to -90 deg instead of from 0 deg to $+90$ deg. The dc gain, the corner frequency, and the points of intersection of the asymptotes in both sketches are determined in exactly the same way as in the previous section.

It is left as an exercise for the reader to work through the arithmetic for deriving the sketches shown in Figure 8-15.

8-6.7 complex poles $[G(\omega)H(\omega) = \omega_n^2/((j\omega)^2 + 2\zeta\omega_n(j\omega) + \omega_n^2)]$

If the open-loop transfer function is such that

$$G(\omega)H(\omega) = \frac{\omega_n^2}{(j\omega)^2 + 2\zeta\omega_n(j\omega) + \omega_n^2} \qquad (8.49)$$

and $$\zeta < 1$$

the two roots of the denominator of $G(\omega)H(\omega)$ are complex and can not be plotted in the same way as the real pole dealt with in Section 8-6.6.

The Bode diagram for equation (8.49) is illustrated in Figure 8-16 when

$$\zeta = 0.4 \qquad (8.50)$$

and $$\omega_n = 2 \text{ rad/s} \qquad (8.51)$$

It will be seen that the gain curve can once again be approximated by two asymptotes, as was the case for the simple pole; however, this time, the high frequency asymptote decays at -12 db/octave rather than -6 db/octave.

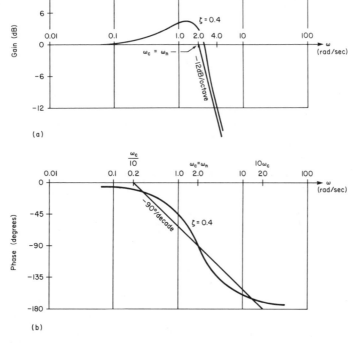

figure 8-16 bode diagram of $G(\omega)H(\omega) = \dfrac{4}{(j\omega)^2 + 1.6(j\omega) + 4}$

Why is the slope of the high frequency asymptote different for complex poles, and how do we determine where the asymptotes intersect? Once again, the asymptotes represent the behavior of the gain at very low and very high frequencies. That is, when the frequency is small, equation (8.49) becomes

$$G(O)H(O) = 1 \qquad (8.52)$$

thus
$$\text{db gain} = 0 \text{ db} \qquad (8.53)$$

and
$$\phi = 0° \qquad (8.54)$$

When the frequency is very high such that

$$|(j\omega^2)| \geq |2\zeta\omega_n(j\omega) + \omega_n^2| \qquad (8.55)$$

then equation (8.49) becomes

$$G(\omega)H(\omega) = \frac{\omega_n^2}{(j\omega)^2} = \left|\frac{\omega_n^2}{\omega^2}\right| \angle -180° \qquad (8.56)$$

i.e.,
$$|G(\omega)H(\omega)| = \left(\frac{\omega_n^2}{\omega^2}\right) \qquad (8.57)$$

or
$$\text{db gain} = 40 \log_{10} \omega_n - 40 \log_{10} (\omega) \qquad (8.58)$$

and
$$\phi = -180° \qquad (8.59)$$

From equation (8.58) it is easy to see that if ω is multiplied by 2 the db gain is reduced by 12 db; hence the slope of the high frequency asymptote is -12 db/octave, i.e., -6 db/octave for each of the two roots. Also from equation (8.59) the phase of $G(\omega)H(\omega)$ at high frequency is approximately -180 deg.

Where on the dc gain line (horizontal asymptote) do the two asymptotes intersect? As before, the answer is when equation (8.57) equals the dc gain, i.e., unity in this case. Thus the intersection point of the two gain asymptotes for a quadratic denominator with complex poles occurs when

$$\omega = \omega_c = \omega_n \qquad (8.60)$$

The actual shapes of the gain and phase curves vary a great deal with the value of ζ. Consequently, when dealing with complex poles, the control engineer will use a standard Bode diagram that has previously been drawn for various values of ζ. Such a Bode diagram is shown in

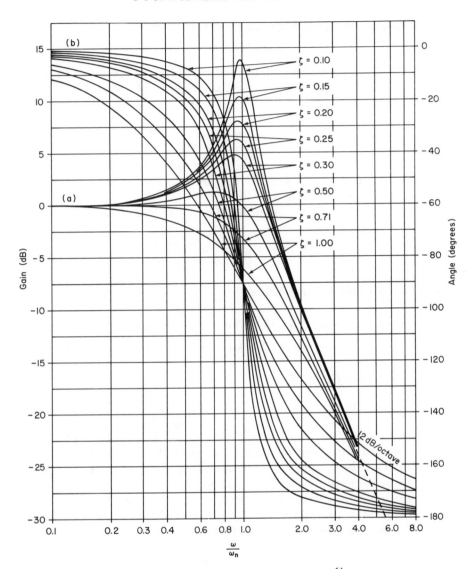

figure 8-17 bode diagram of $G(\omega)H(\omega) = \dfrac{\omega_n}{(j\omega)^2 + 2\zeta\omega_n(j\omega) + \omega_n^2}$

(a) gain versus frequency poles, (b) phase versus frequency poles

Figure 8-17 where the results for any value of ζ between 0.1 and 1.0 can be obtained quite accurately by simple interpolation. To use this diagram for any value of dc gain and ω_n, simply add the decibel value of dc gain to the numbers along the vertical axis and multiply the numbers along the horizontal axis by ω_n.

8-6.8 how to sketch a bode diagram for a general system

Generally the open-loop transfer function will contain a number of poles and zeros, both simple and complex, i.e., in general

$$G(\omega)H(\omega) = K \frac{(1 + j\omega\tau_1)(1 + j\omega\tau_3) \cdots}{(1 + j\omega\tau_2)(1 + j\omega\tau_3) \cdots} \qquad (8.61)$$

How do we sketch the straight line approximation to the Bode diagram for this general system? First let us see what the dc gain and the phase of such a system look like.

We could rewrite equation (8.61) in the following polar form.

$$G(\omega)H(\omega) = K \frac{|A_1| \angle\phi_1 |A_3| \angle\phi_3 \cdots}{|A_2| \angle\phi_2 |A_4| \angle\phi_4 \cdots} \qquad (8.62)$$

Thus the db gain of $G(\omega)H(\omega)$ is

$$\text{db gain} = 20 \, (\log_{10}(K) + \log_{10}(A_1) + \log_{10}(A_3) + \cdots \qquad (8.63)$$
$$- \log_{10}(A_2) - \log_{10}(A_4) - \cdots)$$

and the phase of $G(\omega)H(\omega)$ is

$$\phi = (\phi_1 + \phi_3 + \phi_5 + \cdots - \phi_2 - \phi_4 - \cdots) \qquad (8.64)$$

Now A_1, A_2, etc. and ϕ_1, ϕ_2, etc. are all frequency dependent. Therefore for any given frequency the value of the db gain and the value of ϕ can be obtained by adding, algebraically, the logs of A_i and all the ϕ_is, respectively. What this means is that if we were to sketch Bode diagrams for each pole and zero, we could then sketch the overall Bode diagram by simply adding together the individual graphs. We will demonstrate this practice with a number of examples.

EXAMPLE 8.2
Suppose

$$G(\omega)H(\omega) = 2 \frac{(1 + j0.4\omega)}{(1 + j0.1\omega)(1 + j\omega)} \qquad (8.65)$$

Thus the dc gain is

$$G(O)H(O) = 2 \ (\equiv 6\text{db})$$

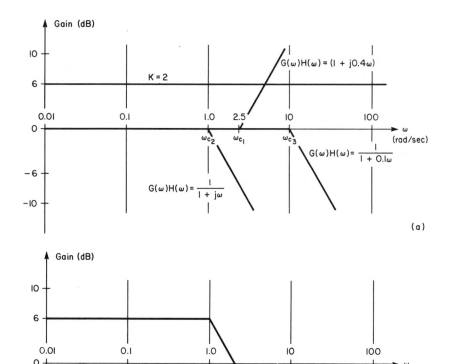

figure 8-18 sketch of the gain of $G(\omega)H(\omega) = \dfrac{2(1 + j0.4\omega)}{(1 + j0.1\omega)(1 + j\omega)}$
(a) for individual poles and zeros, (b) overall gain

and $$\phi = 0°$$

$G(\omega)H(\omega)$ has one zero at

$$\omega_{c_1} = 2.5 \text{ rad/s}$$

and two poles at $$\omega_{c_2} = 1 \text{ rad/s}$$

and $$\omega_{c_3} = 10 \text{ rad/s}$$

where ω_{c_1}, ω_{c_2}, and ω_{c_3} are the respective corner frequencies.

Figure 8-18(a) shows sketches of the gain for the individual poles and zeros, while Figure 8-18(b) shows how these sketches are summed together to produce the total gain curve. At very low frequencies the gains of all the poles and zeros are unity and the overall gain is given by K, i.e., 6 db. Therefore, for the total curve at low frequencies, we draw a line along 6 db in Figure 8-18(b). At $\omega_{c_2} = 1$ rad/s the gain from one of the poles starts to decrease at -6 db/octave; thus the overall gain must also start to decrease from 6 db at a rate of -6 db/octave. The overall gain continues to decrease at this rate until $\omega_{c_1} = 2.5$ rad/s, when the zero term causes the gain to start increasing at 6 db/octave. However, as the gain was previously decreasing at -6 db/octave, the net effect on the gain is zero and the gain remains constant at -1.6 db as the frequency is increased. When $\omega_{c_3} = 10$ rad/s the second pole also starts to reduce in gain at -6 db/octave, producing a corresponding decay in the overall gain.

Thus the pattern for sketching the overall gain is quite simple.

1. Determine the low frequency value of dc gain.
2. Determine the values of all the poles and zeros (ω_{c_i}).
3. At every corner frequency increase the gain by 6 db/octave for a simple zero or decrease the gain by -6 db/octave for a simple pole.

Also

4. If the transfer function contains complex poles (or zeros), decrease the gain, at ω_n, by -12 db/octave (increase by $+12$ db/octave for complex zeros).

As can be seen from Figures 8-19(a) and 8-19(b), the phase curve can be sketched just as easily by following some similar rules; i.e., for the phase curve,

1. Determine the phase of $G(\omega)H(\omega)$ at low frequencies.
2. A pure gain causes zero phase shift.
3. At one decade below every corner frequency, increase the phase for two decades at the rate of 45 deg per decade for a simple zero or decrease the phase at the rate of -45 deg per decade for a simple pole.

Also

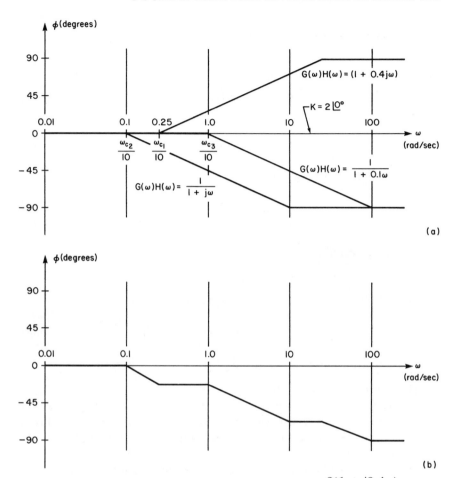

figure 8-19 sketch of the phase of $G(\omega)H(\omega) = \dfrac{2(1 + j0.4\omega)}{(1 + j0.1\omega)(1 + j\omega)}$

(a) for the individual poles and zeros, (b) overall phase

4. If the transfer function contains complex poles (or zeros) use a standard Bode diagram, such as Figure 8-17, to determine the effect on the phase shift.

EXAMPLE 8.3

Sketch the Bode diagrams for the following open-loop transfer functions.

1. $G(s)H(s) = \dfrac{20}{s(1 + 0.1s)}$

2. $G(s)H(s) = 20 \cdot \dfrac{100}{s^2 + 8s + 100}$

3. $G(s)H(s) = \dfrac{20}{s(1 + 0.2s)(1 + 0.1s)}$

solution 1.

Substituting $j\omega$ for s gives

$$G(\omega)H(\omega) = \frac{20}{j\omega(1 + j0.1\omega)}$$

$G(\omega)H(\omega)$ contains three terms:

[A] *A Constant: 20*

The value of this constant expressed in decibels is

$$20 \log_{10}(20) = 26 \text{ db} \qquad (8.66)$$

and the phase contribution of the constant is zero.

[B] *A Pure Integration: $1/j\omega$*

When ω is zero, this term has an infinite value. However, for nonzero values of ω we can sketch the gain of this term. From Section 8-6.4 we know that the gain curve decreases at -6 db/octave and crosses the 0 db axis at $\omega = 1$ rad/s.

The phase sketch for a pure integrator is shown in Figure 8-13, i.e., the phase is constant at -90 deg.

[C] *A Simple Pole: $1/(1 + j0.1\omega)$*

Gain and phase plots for this term are shown in Figure 8-15. The corner frequency for this pole is

$$\omega_c = 10 \text{ rad/s} \qquad (8.67)$$

The gains of the three terms are first plotted individually in Figure 8-20(a). The three curves are then summed together, in the same way as in the previous example, to produce the overall gain curve [D].

In Figure 8-20(b) the phase plots for the pole and the integrator are plotted individually and combined in the usual way to produce the overall phase plot for $G(\omega)H(\omega)$.

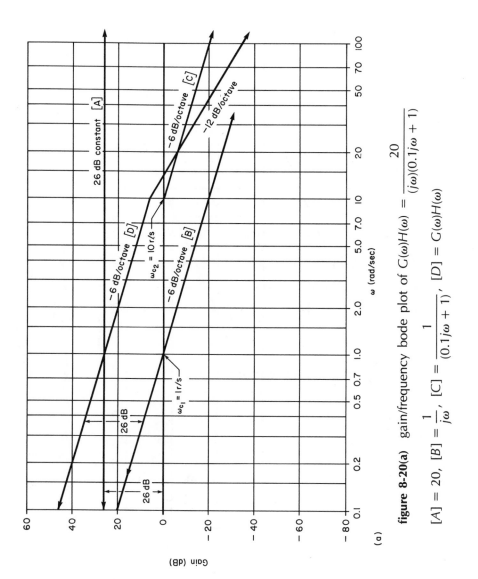

figure 8-20(a) gain/frequency bode plot of $G(\omega)H(\omega) = \dfrac{20}{(j\omega)(0.1j\omega + 1)}$

$[A] = 20$, $[B] = \dfrac{1}{j\omega}$, $[C] = \dfrac{1}{(0.1j\omega + 1)}$, $[D] = G(\omega)H(\omega)$

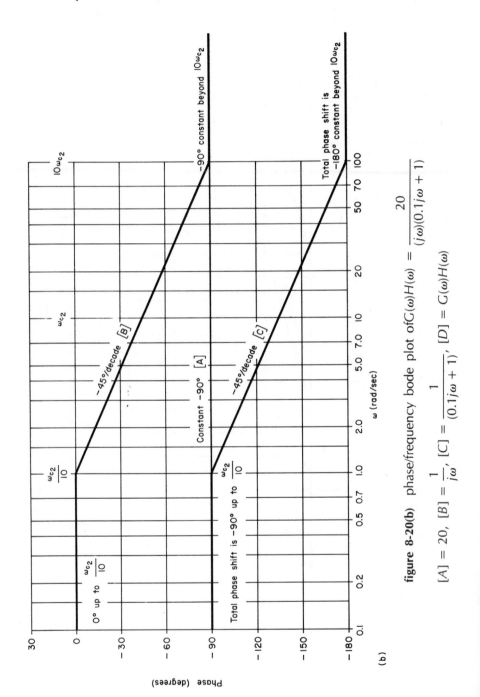

figure 8-20(b) phase/frequency bode plot of $G(\omega)H(\omega) = \dfrac{20}{(j\omega)(0.1j\omega + 1)}$

$[A] = 20$, $[B] = \dfrac{1}{j\omega}$, $[C] = \dfrac{1}{(0.1j\omega + 1)}$, $[D] = G(\omega)H(\omega)$

(b)

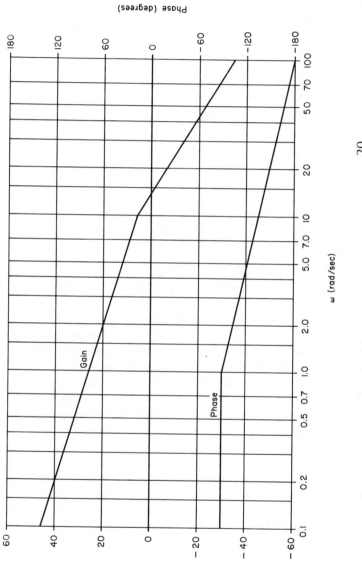

figure 8-21 complete bode diagram for $G(\omega)H(\omega) = \dfrac{20}{(j\omega)(0.1j\omega + 1)}$

The complete Bode diagram for this transfer function is presented in Figure 8-21.

solution 2

The open-loop transfer function is

$$G(\omega)H(\omega) = 20 \cdot \frac{100}{(j\omega)^2 + 8(j\omega) + 100}$$

The function has the form

$$G(\omega)H(\omega) = K \cdot \frac{\omega_n^2}{(j\omega)^2 + 2\zeta\omega_n(j\omega) + \omega_n^2}$$

Hence

$$K = 20$$

$$\omega_n = 10 \text{ rad/s}$$

and

$$\zeta = 0.4$$

Gain Curve:

First we draw a line representing the dc gain (i.e., $20 \log_{10} K = 26$ db) on the gain plot of Figure 8-22. Next the high frequency asymptote is drawn, having a slope of -12 db/octave and crossing the 26 db line at $\omega_n = 10$ rad/s. Reference is then made to Figure 8-17 to assess how the gain curve for $\zeta = 0.4$ deviates from the two asymptotes. Because ζ is fairly small, the system is underdamped and a small resonant peak can be seen to occur in the gain at about 8.2 rad/s ($\omega_n\sqrt{1 - 2\zeta^2}$). The gain curve for $\zeta = 0.4$ can be transferred from Figure 8-17 by adding 26 db to the numbers on the left-hand vertical axis of Figure 8-17 and by multiplying the horizontal axis by ω_n (i.e., multiply by 10).

Phase Curve:

The phase curve is best constructed by direct reference to Figure 8-17 for the case $\zeta = 0.4$. Values can then be transferred directly to the phase plot in Figure 8-22 after multiplying the horizontal axis of Figure 8-17 by 10.

solution 3

The open-loop transfer function is

$$G(\omega)H(\omega) = \frac{20}{j\omega(1 + j0.2\omega)(1 + j0.1\omega)}$$

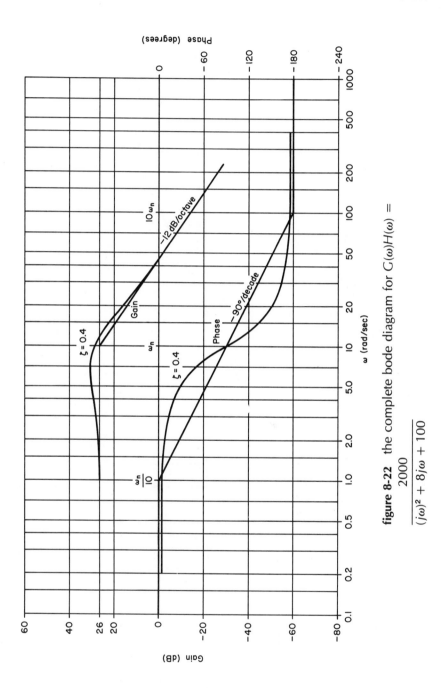

figure 8-22 the complete bode diagram for $G(\omega)H(\omega) =$
$$\frac{2000}{(j\omega)^2 + 8j\omega + 100}$$

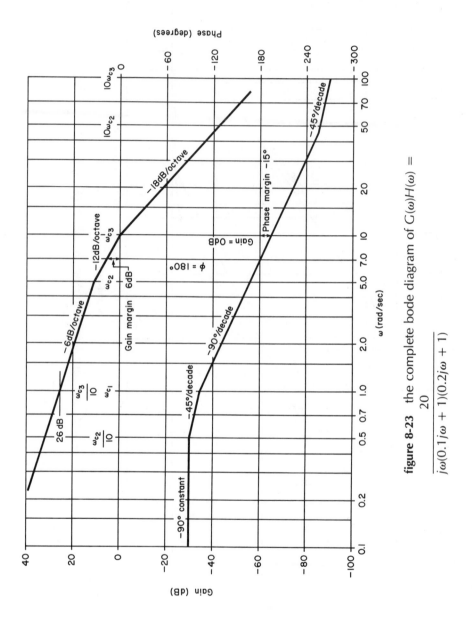

figure 8-23 the complete bode diagram of $G(\omega)H(\omega) =$
$$\frac{20}{j\omega(0.1j\omega + 1)(0.2j\omega + 1)}$$

This transfer function has three poles and hence three corner frequencies, i.e., at

$$\omega_{c_1} = 0$$

$$\omega_{c_2} = 1/0.2 = 5 \text{ rad/s}$$

and $\qquad\qquad \omega_{c_3} = 1/0.1 = 10 \text{ rad/s}$

Thus the low frequency asymptote has a slope of -6 db/octave and crosses the 26 db line [$20 \log_{10}(20)$] at $\omega = 1$ rad/s. Also, because $G(\omega)H(\omega)$ contains no zeros, the slope is altered by a further -6 db/octave at each successive corner frequency. The Bode diagram of $G(\omega)H(\omega)$ is sketched in Figure 8-23.

The phase angle of $G(\omega)H(\omega)$ at low frequency is -90 deg (i.e., $1/j\omega$) and the poles at ω_{c_2} and ω_{c_3} cause changes in the phase angle such that:

Between $\omega_{c_2}/10$ and $\omega_{c_3}/10$ the phase decreases at $-45°$/decade.

Between $\omega_{c_3}/10$ and $10\omega_{c_2}$ the phase decreases at $-90°$/decade.

Between $10\omega_{c_2}$ and $10\omega_{c_3}$ the phase decreases at $-45°$/decade.

Above $10\omega_{c_3}$ the phase is constant (i.e., $-270°$).

The phase sketch is plotted in Figure 8-23.

8-7 THE GAIN MARGIN AND PHASE MARGIN

In order to assess system stability we measure two quantities from our frequency plot, namely the gain margin and the phase margin of the system.

As discussed earlier in this chapter, if

$$G(\omega)H(\omega) = 1\angle 180° \qquad\qquad (8.68)$$

then the system is unstable. It is, therefore, desirable to keep the magnitude of $G(\omega)H(\omega)$ less than unity (0 db) when the phase change is 180 deg; conversely, we also want to keep the phase change less than 180 deg when the magnitude of $G(\omega)H(\omega)$ is unity (0 db).

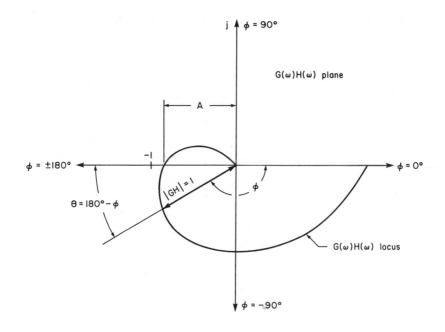

figure 8-24   polar plot of $G(\omega)H(\omega)$,

Phase margin = θ

Gain margin = $20 \log_{10}(A)$ db

The gain margin measures how much less than unity is the gain of $G(\omega)H(\omega)$ when the phase change is 180 deg, and the phase margin is the angle by which the phase change is less than 180 deg when the magnitude of $G(\omega)H(\omega)$ is unity. Figure 8-24 shows how the gain margin and phase margin can be measured by using a polar plot.

With reference to Figure 8-23, it can be seen that the gain margin and phase margin can be read directly from the Bode diagram for the two conditions $\phi = 180$ deg and unity gain, respectively. From this diagram, the gain margin is +6 db, which means that the gain of $G(\omega)H(\omega)$ is greater than unity when $\phi = 180$ deg and hence the system examined in Example 8.3(3) is unstable. As a rule of thumb, a desirable value for the gain margin is -12 db ($|GH| = 0.25$) and for the phase margin we try to achieve a value for ϕ of approximately 60 deg.

To stabilize this system and achieve the desired gain and phase margins we must alter $G(\omega)H(\omega)$. A simple way of changing $G(\omega)H(\omega)$

is to reduce its gain. (e.g., by reducing the amplification in an electronic circuit).
What we have is

$$\text{Gain margin} = +6 \text{ db}$$

$$\text{Phase margin} = -15°$$

What we want is

$$\text{Gain margin} = -12\text{db approximately}$$

$$\text{Phase margin} = 60° \text{ approximately}$$

Because we are only reducing the gain of $G(\omega)H(\omega)$ we are unlikely to achieve both desired conditions simultaneously. For example, if we reduce the gain of $G(\omega)H(\omega)$ by 18 db we would achieve the desired gain margin, but the phase margin would be only $+38$ deg ($\phi = 142$ deg), which may be acceptable if an underdamped system can be tolerated. On the other hand, to achieve a phase margin of 60 deg (i.e., $\phi = 120$ deg) the gain has to be reduced by 22.5 db, which leaves a gain margin of -16.5 db ($|GH| = 0.15$). Although the system is now stable and has no tendency to oscillate the gain may now be too low and the closed-loop response will be sluggish.

To achieve a larger gain while retaining the desired phase margin and gain margin, we must add extra frequency components into $G(\omega)H(\omega)$. These components are referred to as compensation networks; we will examine them further in Chapter 10.

8-8 SUMMARY

We have discussed the use of frequency analysis for establishing the stability of systems having linear, constant coefficient, transfer functions. In particular we have examined the Bode diagram approach to stability analysis and have described how Bode diagrams can be quickly sketched for any general transfer function. Finally we have talked about how to establish the stability of a system, using Bode diagrams, and a simple method of stabilizing the system.

PROBLEMS

Section 8-2
1. Plot a graph of the output voltage magnitude and phase versus frequency for the following circuits.

(a)

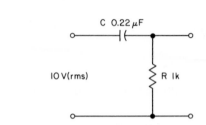

(b)

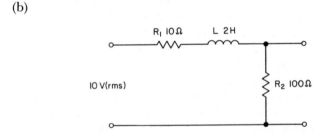

(c)

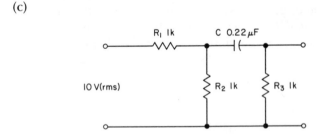

2. A band pass circuit has the following open-loop transfer function

$$G(\omega)H(\omega) = \frac{0.25j\omega}{(0.25j\omega + 1)(0.1j\omega + 1)}$$

Sketch a graph of the open loop gain and phase versus frequency.

Section 8-3   3. A system has the following closed-loop transfer function

$$\frac{C(s)}{R(s)} = \frac{1}{0.01s^2 + 0.07s + 1}$$

If the input to the system is $5 \sin \omega t$ use transform pair 23, Table 5-1, and sketch the output as a function of input frequency. Determine from the graph that the damped resonant frequency $\omega_d = \omega_n \sqrt{1 - \zeta^2}$.

Section 8-6   4. Sketch the Bode diagrams for the following open-loop transfer functions.

(a) $\dfrac{5}{0.2j\omega + 1}$ 　　　　　　　 (b) $\dfrac{5}{j\omega\,(0.2j\omega + 1)}$

(c) $\dfrac{0.25(0.2j\omega + 1)}{j\omega(0.4j\omega + 1)}$ 　　　 (d) $\dfrac{0.25\,(0.2j\omega + 1)}{(j\omega)^2(0.4j\omega + 1)}$

(e) $\dfrac{10}{j\omega(0.01(j\omega)^2 + 0.07j\omega + 1)}$

part
design and modification— real life feedback systems

3

steady-state errors

The Hatter was the first to break the silence, "What day of the month is it?" he said, turning to Alice: he had taken his watch out of his pocket, and was looking at it uneasily, shaking it every now and then, and holding it to his ear.
Alice considered a little, and then said "The fourth."
"Two days wrong!" sighed the Hatter.

Lewis Carroll, *Alice in Wonderland*

9-1 INTRODUCTION

The steady-state error of the Mad Hatter's pocket watch is enormous and would certainly not be acceptable in a modern control system. His watch, however, is an open-loop system, and it demonstrates a disadvantage of such a system. Feedback can sometimes obviate the problem of steady-state errors, but not always, because system error depends upon two things: the type of input and the type number of the control system.

A steady-state error is the difference between the output response and the input command, when all the transients have died away.

To determine this error it is necessary to know the nature of the input signal, but in many practical systems the input is not known ahead

259

of time. In fact, the inputs to many control systems may vary in a very random fashion with respect to time. For example, for the control system of an air-to-air guided missile the speed and direction of the target are unpredictable, particularly if the pilot of the target aircraft is maneuvering to evade the attacking missile.

In such an example no single mathematical expression will suit all the different ways by which the input to the missile control system can change. Designers try to get around this difficulty when testing their control system designs, by studying how the control system behaves when its input signal is changed in a particular manner. The input signal can, of course, have any units; e.g., the input may be demanding changes in temperature, direction, or pressure, depending on what the control system is designed for. Irrespective of the units of the input signal, we generally vary the input in three standard ways in order to study the steady-state error characteristics of the system. The three ways in which the input is changed, shown in Figure 9-1, are known as a step change, a ramp change, and a parabolic change.

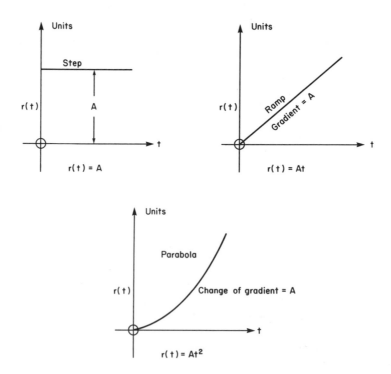

figure 9-1 three simple functions often considered as input signals when steady-state errors are studied

Information gained from studying a system's response to these input signals can be used to predict system performance when more erratic inputs are encountered. Such predictions are, however, outside the scope of this text, and we will confine ourselves here to a discussion of what steady-state errors are, how they arise, and how they are influenced by the type number of the system.

9-2 WHAT IS THE STEADY-STATE ERROR?

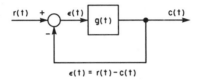

figure 9-2 a simple unity feedback control system: the error $\epsilon(t)$ is the difference between the input and output signal.

Consider the unity feedback control system in Figure 9-2. If the input demand is $r(t)$ and the output is $c(t)$, then, provided a steady-state condition exists, $c(t)$ will eventually settle to behaving in a steady fashion. If at this stage $c(t)$ is not equal to $r(t)$, i.e.,

$$c(t) \neq r(t)$$

then the resulting error, ϵ_{ss}, is termed the steady-state error, where

$$\epsilon_{ss} = \epsilon(t) = r(t) - c(t) \qquad (9.1)$$
$$\scriptstyle t \to \infty$$

System steady-state errors can be grouped into three categories:

1. Zero error. Output follows input without error.
2. Finite and constant error. Output follows input with some fixed magnitude of error.
3. Infinite error. Output diverges from the input with ever-increasing magnitude of error. This really means that the system cannot follow the input at all.

Figure 9-3 illustrates these three types of error as they may occur for each of the three types of input change.

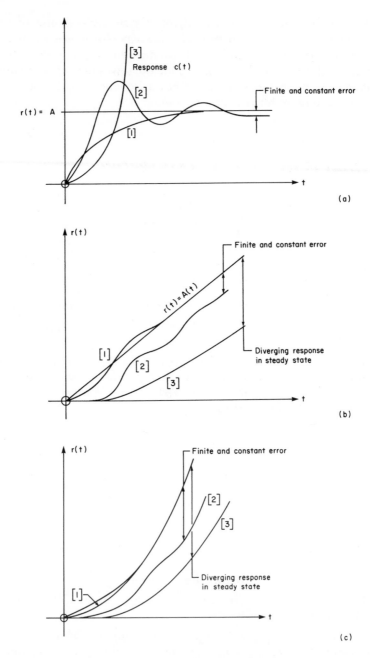

figure 9-3 typical steady-state errors and responses for the three types of input change: [1] zero error; [2] finite and constant steady-state error; [3] increasing error with time. (a) step change of input $r(t) = A(t > 0)$. (b) ramp change of input $r(t) = At$. (c) parabolic change of input $r(t) = At^2$.

9-3 WHEN INPUT AND OUTPUT SIGNALS HAVE DIFFERENT UNITS

In many feedback control systems, it would appear that the output signal cannot be substracted from the input signal because the input and output signals have different units. The units differ, usually, because it is more convenient to have the input signal in a form other than the output; e.g., consider the servo-position control system in Figure 9-4. The output of the servo is θ, the angular position of the load (expressed in degrees), whereas the input signal is $V_{\theta D}$, a voltage whose magnitude varies as θ_D, the demanded value for θ. Figure 9-5 is the block diagram for this control system.

$G(s)$ represents the dynamics of the ac motor and $H(s)$ is the gain of the feedback transducer. This transducer is a circular potentiometer

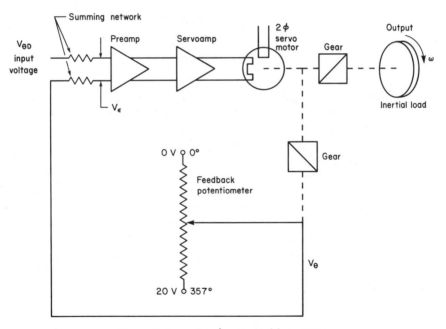

figure 9-4 a simple ac position servo

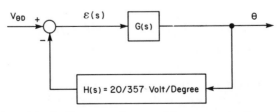

figure 9-5 block diagram of ac position servo in figure 9-4

which has a range of 0 to 357 deg and in this example, 20 volts has been applied across the full range terminals. Thus a 357 deg movement of the load, from rest, causes the potentiometer to rotate 357 deg, which causes V_θ to increase from 0 V to 20 V. Of course, if the load only rotates θ deg, the corresponding change in V_θ (assuming that θ does not exceed 357 deg) is

$$V_\theta = H \cdot \theta \text{ volts} \tag{9.2}$$

i.e.,

$$V_\theta = \frac{20}{357} \cdot \theta \text{ volts}$$

Now, if a 357 deg rotation of the load causes V_θ to change by 20 volts, then $V_{\theta D}$ must have a range of 20 V in order to be able to demand the full 357 deg rotation of the load; i.e.,

$$\theta_D = \frac{357}{20} \cdot V_{\theta D} \text{ degrees}$$

or

$$\theta_D = \frac{1}{H} V_{\theta D} \text{ degrees} \tag{9.3}$$

Therefore, although $V_{\theta D}$ (the input signal) and θ (the output signal) have different units, it is still possible to calculate the error or error signal, by converting one of the signals into the other set of units. When talking of the *error signal* in any control system, we mean that signal leaving the input summing junction, designated ε in Figure 9-5. In the steady-state condition this signal is also referred to as the *steady-state actuating signal*. Thus in our example ϵ has the same units as $V_{\theta D}$; hence θ must be converted to V_θ in order to calculate ε. We can, of course, calculate the *system error* or *output error* directly in degrees, by converting $V_{\theta D}$ into θ_D. However, as we stated earlier, the input units are usually more convenient for observing and manipulating; when we want to know the system error (as expressed in the output units), we have only to divide ε by H; i.e., for our example,

$$\text{Error in degrees} = \varepsilon \cdot \frac{360}{20} = \frac{\varepsilon}{H} \tag{9.4}$$

9-4 STEADY-STATE ERROR AND THE FINAL VALUE THEOREM

The steady-state error is the value of $\epsilon(t)$ when t is large enough for all the system transients to have decayed. We usually define the steady-state error by the expression

$$\epsilon_{ss} = \lim_{t \to \infty} [\epsilon(t)] \tag{9.5}$$

which says that ϵ_{ss}, the steady-state error, is the limiting value of $\epsilon(t)$ as t gets very large (i.e., approaches infinity). Trying to evaluate ϵ_{ss} by evaluating the system response and calculating $\epsilon(t)$ when all the transients have decayed is a lengthy and laborious process. However, a simpler method is at hand.

In Section 5.3(6) we saw how the final value theorem was used to solve equations such as equation (9.5); i.e.,

$$\epsilon_{ss} = \lim_{t \to \infty} [\epsilon(t)] = \lim_{s \to 0} [s \cdot \mathcal{E}(s)] \tag{9.6}$$

where $\mathcal{E}(s)$ is the Laplace transform of $\mathcal{E}(t)$. Thus if we can find an expression for $\mathcal{E}(s)$ we can easily determine ϵ_{ss} without having to evaluate the system response.

Figure 9-6 is a block diagram of a general control system, where $\mathcal{E}(s)$ is given by

$$\mathcal{E}(s) = R(s) - H(s)C(s)$$

$$= R(s) - H(s)G(s)\mathcal{E}(s)$$

Therefore, $\mathcal{E}(s)[1 + H(s)G(s)] = R(s)$

and

$$\mathcal{E}(s) = \frac{R(s)}{1 + G(s)H(s)} \tag{9.7}$$

Hence from equation (9.6) the steady-state error for any control system can be calculated directly by evaluating

$$\epsilon_{ss} = \lim_{s \to 0} \left[\frac{s \cdot R(s)}{1 + G(s)H(s)} \right] \tag{9.8}$$

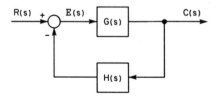

figure 9-6   block diagram of a general control system

EXAMPLE 9.1

If the open-loop transfer function of a control system is

$$G(s)H(s) = \frac{90}{s + 10}$$

what is the system steady-state error when the input signal is a step function of unity magnitude?

From equation (9.8)

$$\epsilon_{ss} = \lim_{s \to 0} \left[s \cdot \frac{R(s)}{1 + 90/(s + 10)} \right]$$

$$\epsilon_{ss} = \lim_{s \to 0} \left[s \cdot \frac{R(s)\,(s + 10)}{s + 10 + 90} \right]$$

$$= \lim_{s \to 0} \left[s \cdot \frac{1(s + 10)}{s(s + 100)} \right]$$

$$= \lim_{s \to 0} \left[\frac{s + 10}{s + 100} \right]$$

$$\epsilon_{ss} = 0.1$$

Thus the steady-state error is 0.1, or 10 percent.

9-5 THE TYPE NUMBER OF A CONTROL SYSTEM

We said in the introduction to this chapter that the steady-state error was influenced by the combination of the input signal and the type number of the control system. We will now explain what we mean when we talk of the control system type number.

We stated in Section 8-6.1 that the open-loop transfer function of a linear control system can always be written in the form:

$$G(s)H(s) = K \cdot \frac{(1 + s\tau_1)\,(1 + s\tau_3) \ldots}{s^n(1 + s\tau_2)\,(1 + s\tau_4)} \qquad (n \geqslant 0) \qquad (9.9)$$

where K is a constant. Thus $G(s)H(s)$ has zeros located at $s = 1/\tau_1$, $1/\tau_3$, etc., while the poles of $G(s)H(s)$ are such that n poles are located at $s = 0$ and single poles are located at $s = 1/\tau_2$, $1/\tau_4$, etc.

With $G(s)H(s)$ written in this form, the type number of a control system is defined by the value of the index n in equation (9.9); e.g.,

Example 9.2 $\qquad G(s)H(s) = \dfrac{20(1 + 3s)}{(1 + 4s)\,(1 + s)}$

is said to be a control system of type 0, because the denominator does not contain a pure s term.

Example 9.3 $\qquad G(s)H(s) = \dfrac{20(s + 6)}{s(s + 2)\,(s + 5)}$

is a control system of type 1, because the denominator contains a single pure s term, i.e., s^1; hence $n = 1$.

Example 9.4 $\qquad G(s)H(s) = \dfrac{20(s^2 + 3s + 2)}{s^2(s^2 + 4s + 5)}$

is a control system of type 2, because the denominator contains an s^2 term, i.e., $n = 2$.

It will be noted from Example (9.4) that it is not necessary to completely factorize the denominator and numerator of $G(s)H(s)$ in order to determine the type number. Neither is it necessary, as we see in Example (9.3), for the terms of the numerator and denominator to be in the form of $(1 + s\tau)$ in order to determine n. All that is necessary is to determine the degree of the pure s term in the denominator. Let us illustrate these points with one final example.

Example 9.5 $\quad G(s)H(s) = \dfrac{30s(1 + 5s)}{(20s^2 + 4s)\,(s^2 + 3s + 5)s^2}$

Extracting s where possible from the numerator and denominator gives

$$G(s)H(s) = \frac{30s(1 + 5s)}{s(20s + 4)\,(s^2 + 3s + 5)s^2}$$

Grouping pure s terms, we obtain

$$G(s)H(s) = \frac{30s(1 + 5s)}{s^3(20s + 4)\,(s^2 + 3s + 5)}$$

$$= \frac{30(1 + 5s)}{s^2(20s + 4)\,(s^2 + 3s + 5)}$$

Thus after cancelling the pure s term in the numerator with an s term from the denominator, we are left with a s^2 term in the denominator; i.e., the control system is of type 2.

Let us now examine what kind of steady-state errors we get when we combine our three types of input signal with type 0, type 1, and type 2 control systems. (Control systems of higher type number are not considered in the text and are not often encountered in practice.)

9-6 STEADY-STATE ERRORS FOR A TYPE 0 SYSTEM

Such a system is specified by

$$G(s)H(s) = \frac{K(1 + sT_1)\,(1 + sT_3)\, . \, . \, .}{s^0(1 + sT_2)\,(1 + sT_4)\, . \, . \, .}$$

which is the same as

$$G(s)H(s) = \frac{K(1 + sT_1)\,(1 + sT_3)\, . \, . \, .}{(1 + sT_2)\,(1 + sT_4)\, . \, . \, .}$$

A Step Input Function: Consider a step input where

$$r(t) = A$$

Hence
$$R(s) = \frac{A}{s}$$

Applying equation (9.7) yields

$$\mathcal{E}(s) = \frac{\dfrac{A}{s}}{1 + \dfrac{K(1 + sT_1)(1 + sT_3)\, . \, . \, .}{(1 + sT_2)(1 + sT_4)\, . \, . \, .}}$$

Also from equation (9.8) the steady-state error is given by

$$\epsilon_{ss} = \lim_{t \to \infty}\,[\epsilon(t)] = \lim_{s \to 0}\left[s \cdot \frac{\dfrac{A}{s}}{1 + \dfrac{K(1 + sT_1)(1 + sT_3). \, . \, .}{(1 + sT_2)(1 + sT_4). \, . \, .}} \right]$$

Therefore, $\quad \epsilon_{ss} = \lim_{s \to 0} \left[\dfrac{A}{1 + \dfrac{K(1 + sT_1)(1 + sT_3). \ldots}{(1 + sT_2)(1 + sT_4). \ldots}} \right]$

If zero is now substituted for s, then

$$\epsilon_{ss} = \frac{A}{1 + K} \tag{9.10}$$

Equation (9.10) shows us that a type 0 system does have a steady-state error when the input is a step function. It also shows us that the error can be made smaller by increasing K. Although, in practical terms, the error can never be removed, since that would imply that K had an infinite value, which is not possible. The loop gain constant K, appearing in equation (9.10), is usually designated as K_p and given the name *position constant*, so that equation (9.10) becomes

$$\epsilon_{ss} = \frac{A}{1 + K_p} \tag{9.10a}$$

A Ramp Input Function: Consider now a ramp input where

$$r(t) = At$$

Hence $\qquad\qquad R(s) = \dfrac{A}{s^2}$

Applying equation (9.7) yields

$$\mathcal{E}(s) = \frac{\dfrac{A}{s^2}}{1 + \dfrac{K(1 + sT_1)(1 + sT_3). \ldots}{(1 + sT_2)(1 + sT_4) \ldots}} \tag{9.11}$$

From equation (9.8) the steady-state error is

$$\epsilon_{ss} = \lim_{s \to 0} \left[s \cdot \frac{\dfrac{A}{s^2}}{1 + \dfrac{K(1 + sT_1)(1 + sT_3) \ldots}{(1 + sT_2)(1 + sT_4) \ldots}} \right] \tag{9.12}$$

Therefore, $\epsilon_{ss} = \lim_{s \to 0} \left[\dfrac{A}{s \left[1 + \dfrac{K(1 + sT_1)(1 + sT_3) \ldots}{(1 + sT_2)(1 + sT_4) \ldots} \right]} \right]$ 　　　(9.12a)

or 　　$\epsilon_{ss} = \lim_{s \to 0} \left[\dfrac{A}{s + \dfrac{sK(1 + sT_1)(1 + sT_3) \ldots}{(1 + sT_2)(1 + sT_4) \ldots}} \right]$ 　　　(9.12b)

If zero is now substituted for s, then

$$\epsilon_{ss} = \frac{A}{0} \qquad\qquad (9.12c)$$

and the steady-state error will be infinity. In other words, a type 0 system cannot follow a ramp input.

A Parabolic Input Function: If we consider a parabolic input where

$$r(t) = \frac{1}{2}At^2$$

then 　　　　　　　　　　$R(s) = \dfrac{A}{s^3}$

By applying the same analytical procedure that we used for the ramp input function, we would also find that

$$\epsilon_{ss} = \frac{A}{0} \qquad\qquad (9.13)$$

i.e., the steady-state error again will be infinite, because the only function that a type 0 control system can follow, with a finite error, is a step function.

EXAMPLE 9.6

Find the steady-state error in a control system where

$$G(s)H(s) = 10 \frac{(1 + 0.2s)\,(1 + 0.1s)}{(1 + 0.15s)\,(1 + 0.3s)}$$

for

1. A step function of magnitude 10; i.e., $R(s) = 10/s$.

2. A ramp function $r(t) = 10t$; i.e., $R(s) = 10/s^2$.
3. A parabolic function $r(t) = 10t^2$; i.e., $R(s) = 20/s^3$.

solution 1.

$$R(s) = \frac{10}{s}$$

Therefore, $\quad \epsilon_{ss} = \lim_{s \to 0} \left[s \cdot \dfrac{\dfrac{10}{s}}{1 + \dfrac{10(1 + 0.2s)\,(1 + 0.1s)}{(1 + 0.15s)\,(1 + 0.3s)}} \right]$

i.e., $\qquad\qquad \epsilon_{ss} = \dfrac{10}{1 + 10} = 0.909$

solution 2.
From table 9-1 and from previous examples it can be seen that

when $\qquad\qquad R(s) = \dfrac{10}{s^2}$

then $\qquad\qquad \epsilon_{ss} = \infty$

solution 3
Again, from Table 9-1 and from previous examples it can be seen that when

$$R(s) = \frac{20}{s^3}$$

then $\qquad\qquad \epsilon_{ss} = \infty$

9-7 STEADY-STATE ERRORS FOR A TYPE I SYSTEM

Such a system is specified by

$$G(s)H(s) = \frac{K(1 + sT_1)\,(1 + sT_3)\,\ldots}{s^1(1 + sT_2)\,(1 + sT_4)\,\ldots} \qquad (9.14)$$

which is the same as

$$G(s)H(s) = \frac{K(1 + s\tau_1)(1 + s\tau_3) \ldots}{s(1 + s\tau_2)(1 + s\tau_4) \ldots} \qquad (9.14a)$$

A Step Input Function: Consider a step input where

$$R(s) = \frac{A}{s}$$

Applying equations (9.7) and (9.8) successively yields

$$\epsilon_{ss} = \lim_{s \to 0} \left[s \cdot \frac{\dfrac{A}{s}}{1 + \dfrac{K(1 + s\tau_1)(1 + s\tau_3) \ldots}{s(1 + s\tau_2)(1 + s\tau_4) \ldots}} \right] \qquad (9.15)$$

which reduces to

$$\epsilon_{ss} = \lim_{s \to 0} \left[\frac{A}{1 + \dfrac{K(1 + s\tau_1)(1 + s\tau_3) \ldots}{s(1 + s\tau_2)(1 + s\tau_4) \ldots}} \right] \qquad (9.15a)$$

If zero is now substituted for s, then

$$\epsilon_{ss} = \frac{A}{1 + \infty} = \frac{A}{\infty} \qquad (9.16)$$

It is obvious that equation (9.16) yields zero for the magnitide of the steady-state error. In other words, a type 1 system follows a step input perfectly (without error), once the transient terms have had time to decay.

A Ramp Input Function: Now consider a ramp input where

$$R(s) = \frac{A}{s^2}$$

Applying equation (9.7) and (9.8) successively yields

$$\epsilon_{ss} = \lim_{s \to 0} \left[s \cdot \frac{\dfrac{A}{s^2}}{1 + \dfrac{K(1 + sT_1)(1 + sT_3) \ldots}{s(1 + sT_2)(1 + sT_4) \ldots}} \right] \qquad (9.17)$$

which in turn yields

$$\epsilon_{ss} = \lim_{s \to 0} \left[\frac{A}{s + \dfrac{K(1 + sT_1)(1 + sT_3) \ldots}{(1 + sT_2)(1 + sT_4) \ldots}} \right] \qquad (9.17a)$$

If zero is now substituted for s, then

$$\epsilon_{ss} = \frac{A}{K} \qquad (9.18)$$

Equation (9.18) shows us that a type 1 system does have a steady-state error when the input is a ramp function. Also, as in the case of a type 0 system with a step input function, the error can be made smaller (but never zero) if K is increased in magnitude. The loop gain constant K which appears in equation (9.18) is usually designated as K_v and given the name *velocity constant*, so that equation (9.18) becomes

$$\epsilon_{ss} = \frac{A}{K_v} \qquad (9.18a)$$

A Parabolic Input Function: If we consider a parabolic input where

$$R(s) = \frac{A}{s^3}$$

then applying the same analytical procedure as above will yield

$$\epsilon_{ss} = \frac{A}{0} \qquad (9.19)$$

The steady-state error will be infinite, because a type 1 system cannot follow a parabolic input function.

EXAMPLE 9.7

If
$$G(s)H(s) = \frac{10}{s(1 + 0.25s)(1 + 0.3s)}$$

find the steady-state error for the following input functions:

1. $r(t) = 10, \quad t > 0$
$\qquad\qquad 0, \quad t \leqslant 0$

2. $r(t) = 10t$

3. $r(t) = 10t^2$

solution 1.

$$R(s) = \frac{10}{s}$$

Therefore, $\epsilon_{ss} = \lim_{s \to 0} \left[s \cdot \dfrac{\dfrac{10}{s}}{1 + \dfrac{10}{s(1 + 0.25s)(1 + 0.3s)}} \right]$

i.e.,
$$\epsilon_{ss} = \frac{10}{1 + \infty} = 0$$

solution 2.

$$R(s) = \frac{10}{s^2}$$

Therefore, $\epsilon_{ss} = \lim_{s \to 0} \left[s \cdot \dfrac{\dfrac{10}{s^2}}{1 + \dfrac{10}{s(1 + 0.25s)(1 + 0.3s)}} \right]$

Therefore, $\epsilon_{ss} = \dfrac{10}{0 + 10} = 1$

solution 3.

$$R(s) = \frac{20}{s^3}$$

Therefore, $\quad \epsilon_{ss} = \lim_{s \to 0} \left[s \cdot \dfrac{\dfrac{20}{s^3}}{1 + \dfrac{10}{s(1 + 0.25s)(1 + 0.3s)}} \right]$

Therefore, $\quad \epsilon_{ss} = \infty$

9-8 STEADY-STATE ERROR FOR A TYPE 2 SYSTEM

Such a system is specified by

$$G(s)H(s) = \frac{K(1 + s\tau_1)(1 + s\tau_3) \dots}{s^2(1 + s\tau_2)(1 + s\tau_4) \dots} \qquad (9.20)$$

A Step Input Function: Consider a step input where

$$R(s) = \frac{A}{s}$$

Once again, applying equation (9.7) and (9.8) successively and taking the limit as $s \to 0$ yields

$$\epsilon_{ss} = \frac{A}{\infty} = 0 \qquad (9.21)$$

The result indicates that a type 2 system follows a step input with zero steady-state error.

A Ramp Input Function: Consider a ramp input where

$$R(s) = \frac{A}{s^2}$$

The result can once again be found to be

$$\epsilon_{ss} = \frac{A}{\infty} = 0 \qquad (9.22)$$

indicating that a type 2 system will also follow a ramp input with zero steady-state error.

A Parabolic Input Function: If the input is now parabolic, where

$$R(s) = \frac{A}{s^3}$$

the result is no longer a zero error, since

$$\epsilon_{ss} = \lim_{s \to 0} \left[s \, \frac{\dfrac{A}{s^3}}{1 + \dfrac{K(1 + s\tau_1) \, (1 + s\tau_3) \ldots}{s^2(1 + s\tau_2) \, (1 + s\tau_4) \ldots}} \right] \qquad (9.23)$$

which can easily be shown to yield

$$\epsilon_{ss} = \frac{A}{K} \qquad (9.24)$$

Equation (9.24) says that a type 2 system does have a finite steady-state error when the input is a parabolic function and, of course, also shows that the error can be made smaller, but not zero, by increasing K. The loop gain constant K in equation (9.24) is usually designated as K_α and given the name *acceleration constant*, so that the steady-state error for a parabolic input is

$$\epsilon_{ss}(t) = \frac{A}{K_\alpha} \qquad (9.24a)$$

EXAMPLE 9.8

$$G(s)H(s) = \frac{20(1 + 0.5s) \, (1 + 0.2s)}{s^2(1 + 0.25s) \, (1 + 0.3s)}$$

Find the steady-state error for the following input functions:

1. $R(s) = \dfrac{10}{s}$

2. $R(s) = \dfrac{10}{s^2}$

3. $R(s) = \dfrac{20}{s^3}$

solution 1.

$$R(s) = \frac{10}{s}$$

Therefore, $\epsilon_{ss} = \lim_{s \to 0} \left[s \cdot \dfrac{\dfrac{10}{s}}{1 + \dfrac{20(1 + 0.5s)\,(1 + 0.2s)}{s^2(1 + 0.25s)\,(1 + 0.3s)}} \right]$

i.e., $\epsilon_{ss} = 0$

solution 2.

$$R(s) = \frac{10}{s^2}$$

Therefore, $\epsilon_{ss} = \lim_{s \to 0} \left[s \cdot \dfrac{\dfrac{10}{s^2}}{1 + \dfrac{20(1 + 0.5s)\,(1 + 0.2s)}{s^2(1 + 0.25s)\,(1 + 0.3s)}} \right]$

i.e., $\epsilon_{ss} = 0$

solution 3.

$$R(s) = \frac{20}{s^3}$$

$$\text{Therefore,} \quad \epsilon_{ss} = \lim_{s \to 0} \left[s \cdot \dfrac{\dfrac{20}{s^3}}{1 + \dfrac{20(1 + 0.5s)(1 + 0.2s)}{s^2(1 + 0.25s)(1 + 0.3s)}} \right]$$

i.e., $\quad \epsilon_{ss} = 1$

A summary of the error data derived in Sections 9-6, 9-7, and 9-8 is presented in Table 9-1.

TABLE 9-1—SYSTEM TYPE AND STEADY-STATE ERRORS

SYSTEM TYPE NUMBER, n	POSITION ERROR DUE TO STEP INPUT	POSITION ERROR DUE TO RAMP INPUT	POSITION ERROR DUE TO PARABOLIC INPUT
0	$\dfrac{A}{1 + K_p}$	∞	∞
1	0	$\dfrac{A}{K_v}$	∞
2	0	0	$\dfrac{A}{K_\alpha}$
3	0	0	0

9-9 THE EFFECT OF FEEDBACK ON SYSTEM DISTURBANCES

Apart from the steady-state type of error that we have considered so far in this chapter, system errors can also be caused when some parameter or condition, other than the input signal, is changed or disturbed. The disturbance may take the form of a fixed change in the system (e.g., a change in load requirements), or the change may be of a random or fluctuating nature (e.g., system noise producing perturbation of the output signal).

Whatever their nature, these disturbances result in a change (steady or fluctuating) in the output signal. With reference to Figure 9-7, the disturbance to the system is represented by $y(t)$ and the resulting output is $c'(t)$, where $c(t)$ was the output prior to the appearance of $y(t)$. Thus

$$C(s) = G(s)R(s) \qquad\qquad (9.25)$$

and
$$C'(s) = C(s) + Y(s) \qquad\qquad (9.26)$$

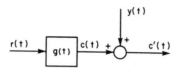

figure 9-7   the effect of system disturbances in an open-loop control system

The system shown in Figure 9-7 is an open-loop control system. Let us now add a feedback loop to this system, as shown in Figure 9-8, and try to find the transfer function between $Y(s)$ and $C'(s)$.

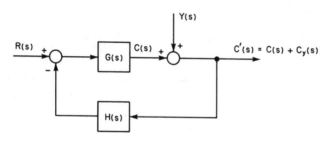

figure 9-8   system disturbances in a closed-loop control system

$C'(s)$ now comprises two components, namely $C(s)$ and $C_y(s)$. To find the transfer function, let us redraw the block diagram, as shown in Figure 9-9, and for the moment ignore the effect of $R(s)$, so that

$$C'(s) = C_y (s)$$

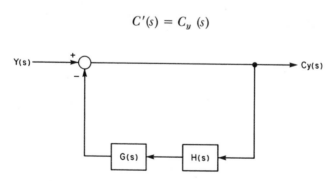

figure 9-9   closed-loop control system with unity forward gain and feedback $G(s)H(s)$

The control system shown in Figure 9-9 now has unity forward gain, and the feedback loop gain is $G(s)H(s)$; hence the transfer function between $Y(s)$ and $C_y(s)$ is

$$\frac{C_y(s)}{Y(s)} = \frac{1}{1 + G(s)H(s)} \tag{9.27}$$

Thus, although the disturbance to the system is $Y(s)$, the feedback loop will try to counter the action of the disturbance and will reduce its effect on the output to an amount represented by $C_y(s)$, where

$$C_y(s) = \frac{Y(s)}{1 + G(s)H(s)} \tag{9.28}$$

i.e., when both $R(s)$ and $Y(s)$ are present,

$$C'(s) = C(s) + \frac{Y(s)}{1 + G(s)H(s)} \tag{9.29}$$

To summarize, by comparing equation (9.26) and (9.29) it can be seen that the effect of feedback on system disturbances is to attenuate their influence on the output signal by a factor of $1/[1 + G(s)H(s)]$. We will illustrate this principle with an example.

EXAMPLE 9.9

A 200 V, dc motor running at a steady speed of 900 rpm suffers a 10 percent reduction of speed when the load torque is increased by 100 Nm. This system is illustrated in the diagram.

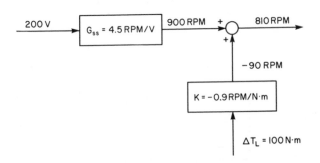

Design a simple feedback loop that will limit the effect of the load increase to a 4 percent reduction of the motor speed.

solution

The steady-state gain of the motor is

$$G_{ss} = 4.5 \text{ rpm/V}$$

The block diagram for the closed-loop control system will be of the form shown below. Note that in addition to a tachogenerator (feedback gain H), a power amplifier will be required in the forward gain to increase the dc drive voltage as the motor attempts to slow down. Let A be the gain of this amplifier.

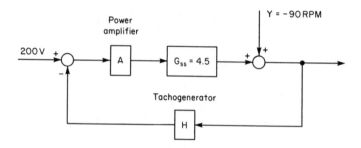

For our control system we still wish the 200 V input voltage to be equivalent to a demand for a motor speed of 900 rpm. Thus the tachogenerator gain H must have a value of

$$H = \frac{200}{900} = \frac{1}{4.5} \text{ V/rpm}$$

The loop gain of the control system $G(s)H(s)$ is $AG_{ss}H$; hence

$$G(s)H(s) = A$$

Now the disturbance to the system without feedback is a step change in speed; i.e., $y(t) = -90$ rpm. Thus

$$Y(s) = \frac{-90}{s}$$

What we want to do is reduce the steady-state change in the output speed (C_{yss}) to -4 percent of 900 rpm; i.e.,

$$C_{yss} = -36 \text{ rpm}$$

From equation (9.28)

$$C_y(s) = \frac{Y(s)}{1 + G(s)H(s)}$$

Also, from the final value theorem,

$$C_{yss} = \lim_{s \to 0} [s \cdot C_y(s)]$$

$$= \lim_{s \to 0} \left[s \cdot \frac{Y(s)}{1 + G(s)H(s)} \right]$$

$$= \lim_{s \to 0} \left[s \cdot \frac{-90}{s} \cdot \frac{1}{1 + AG_{ss}H} \right]$$

Therefore, from the above,

$$-36 = \frac{-90}{1 + A}$$

Hence the gain of the power amplifier is

$$A = 1.5$$

Thus the effect of the load disturbance can be reduced from a 10 percent reduction of speed to a 4 percent reduction by use of a power amplifier and a tachogenerator in a straightforward feedback circuit. Further reductions, of course, are possible, although the penalty of increasing instability must be paid for such reductions.

9-10 SUMMARY

In this chapter we have reviewed what is meant by steady-state errors and have examined how these errors are affected by the combination of input function and system type. Also, in the latter part of the chapter, we have considered how a simple feedback loop can be used to reduce errors resulting from system disturbances. In the next chapter we will look at compensation networks, which can be used not only to reduce errors but also to improve system stability.

PROBLEMS

Section 9-4   1. For the speed control system shown below, find the steady-state actuating signal which results when the input potentiometer is sharply rotated through 30 deg. The system components have the following values:

K_I = 5.06 volts/rad

A = 4

K_m = 5 rad/volt-s

K_g = 3 volts/1000 rpm

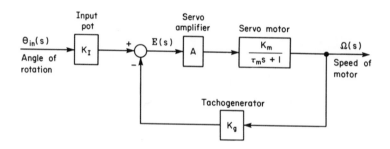

Sections 9-6, 9-7, and 9-8

2. Find the steady-state error resulting from each of the three input functions $\dfrac{10}{s}, \dfrac{10}{s^2}, \dfrac{10}{s^3}$ for the control systems with the following open-loop transfer functions.

(a) $\dfrac{4}{(0.1s + 1)\,(0.2s + 1)}$

(b) $\dfrac{4}{s(0.1s + 1)\,(0.2s + 1)}$

(c) $\dfrac{4s}{(0.1s + 1)\,(0.2s + 1)}$

(d) $\dfrac{4}{(0.01s^2 + 0.14s + 1)}$

(e) $\dfrac{4(s + 2)\,(0.4s + 1)}{s^2(0.01s^2 + 0.14s + 1)}$

(f) $\dfrac{800(0.5s + 1)\,(0.4s + 1)}{s^2(s^2 + 14s + 100)}$

3. Find K_p, K_v, and K_α for the system with the following open-loop transfer function:

$$G(s)H(s) = \frac{4(s + 2)\,(0.4s + 1)}{(0.1s + 1)\,(s + 5)}$$

Section 9-9   4. For the system described in Problem 1 above, show that by increasing the amplifier gain there will be a reduction in error due to load torque disturbances. By selection of various values of gain A, establish a percentage error reduction due to a fixed load torque disturbance T_L. Assume that the load disturbance T_L creates a change of 10 % full load speed.

improving the system performance

10

"The time has come," the Walrus said, "To talk of many things,—of shoes—and ships —and sealing wax,—of cabbages—and kings——."

Lewis Carroll, *Through the Looking Glass*

10-1 INTRODUCTION

The design for a control system will usually be required to fulfill a number of specifications, e.g., accuracy, speed of response, allowable overshoot, maximum duration of the settling time, and, of course, stability (which can be specified in terms of the gain margin and phase).

When designing a control system, however, we usually find that our first attempts are directed towards defining the structure of the control system correctly. As a case in point, consider the speed control system in the example of Section 9-9. The first criterion in this system is to control the speed of the motor, with a certain accuracy, in the presence of fluctuating load conditions. Therefore, the first thing we do is to put in a feedback loop that modifies the actuating signal whenever the load, and hence the speed of the motor, changes. We then choose the gain values of the power amplifier and the feedback tachometer in order to meet the required accuracy of this system.

285

This approach, then, represents the preliminary design phase for the control system. The next step, as implied at the end of the example, is to examine how the system responds to changes in the motor load. Is the response sluggish, or does it overshoot and oscillate? What is the duration of the settling time and, most important, is the system stable? It is not unusual at this stage to find that the stability or the response of the control system is not what we desire. Thus, although we fulfilled the primary specification of speed control and steady-state accuracy, it may still be necessary to correct the transient behavior or even the stability of the system before it can be of any use to us. It is at this stage that we must examine the various methods for improving the response of a control system.

If the control system is otherwise stable and needs improving only in terms of the system response (e.g., reducing or removing the steady-state error, or increasing the speed of response), then we may consider using a two- or three-term controller.

If, however, the preliminary design for the control system proves to be unstable or close to instability, or if the system tends to instability when one is trying to improve the system response, then a compensating device must be added to the control system. Such a device modifies the gain and phase of the system in order to improve the gain and phase margin.

In this chapter we will examine the use of term controllers and compensation devices, using practical examples to illustrate the merits of each type of device.

10-2 TWO- AND THREE-TERM CONTROLLERS

10-2.1 general

These types of controllers are cascade control devices, which means that they are inserted into an existing control loop to form part of the forward gain of the control system. As its name implies, the controller is comprised of two or three of the following control terms:

1. Proportional control
2. Integral control
3. Differential control

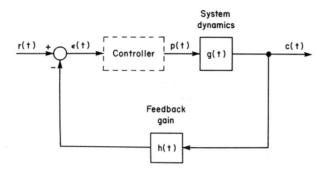

figure 10-1 the location of a two- or three-term controller in a control loop

If the controller contains all three terms, it is more often referred to as a P.I.D. (proportional, integral, differential) controller.

These devices are widely used in industry wherever process control is involved and are usually employed as part of the control circuitry on process control valves and hydraulic actuators.

The input to such a controller is usually the error signal ϵ, as shown in Figure 10-1. By adjusting the magnitude of the control terms the actuating signal, p (i.e., controller output), can be modified so as to produce the required system response. Before demonstrating how such controllers are used in practical situations, we will take a closer look at the individual terms of the controller in order to explain why each term is used and how it affects the response of the control system.

10-2.2 proportional control

This type of control is the most straightforward of the three terms, where the output of the controller varies directly as the input (or system error); i.e.,

$$p = K\epsilon$$

where K is the gain of the proportional term. Increasing K will increase the loop gain of the system and can therefore be used to increase the speed of the system response and to reduce the magnitude of any steady-state errors. Proportional control alone, however, is often not good enough, because increasing K not only makes the system more sensitive but also tends to destabilize the system. Consequently, the amount by which K can be increased is limited, and this limit may not be high enough to achieve the desired response.

In fact, when trying to adjust the gain K, we have conflicting requirements. On the one hand, we want to reduce any steady-state errors as much as possible, but to attempt this by increasing the gain of the system is likely to cause the response to oscillate, thereby prolonging the settling time. On the other hand, the response to any change of the input command should be as fast as possible but with little overshoot or oscillation. The fast response can also be achieved by increasing K, but once again the increase is likely to destabilize the system.

To resolve the conflicting requirements made on the system gain, we need a controller that has:

1. A gain term that is high at very low frequencies (i.e., the steady-state condition) in order to reduce system error.

2. A gain term that is high at high frequencies (i.e., immediately following an input change, when the rate of change of the transient is fastest) in order to ensure a rapid response.

3. At mid-frequencies (i.e., during the latter part of the transient response and before the onset of steady-state conditions), the gain should be low enough to ensure that the response does not overshoot excessively and that any tendency to oscillate is quickly damped.

A Bode plot of the composite gain of such a controller is shown in Figure 10-2, but how do we construct a controller having this characteristic? Examination of Figures 8-11, 8-12, and 8-13 shows that it is

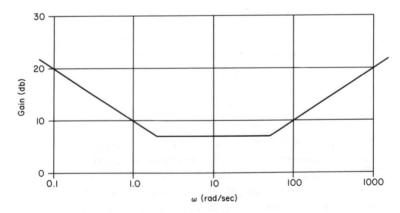

figure 10-2   idealized composite gain of the controller

possible to construct such a controller by adding to the proportional controller the two remaining control terms mentioned in Section 10-2.1, i.e., integral control and differential control.

10-2.3 integral control

The prime purpose for introducing an integral term into a controller is to remove any steady position errors. Integral action achieves this aim by introducing a gain term which effectively has infinite value at zero frequency, i.e., the steady-state condition (see Figure 8-13).

This control function is usually used together with derivative and proportional control, although in cases where speed of response and instability are not a problem, integral and proportional control alone may suffice. This type of two-term controller has to be used with caution, as the addition of an integral term to the control loop has a destabilizing effect on the system (i.e., by introducing an extra phase lag).

This type of controller can be pneumatic, hydraulic, electromechanical, or electronic. Figure 10-3 shows a typical proportional-plus-integral electronic controller in the form of an operational amplifier and its associated networks.

Circuit design and component selection of this and other circuits will be covered in Chapter 11 together with their feedback analysis. For the present we will content ourselves with merely stating the relationship between the input and output voltages of the circuit; i.e.,

$$v_0 = -\left[\frac{R_f}{R_1} \cdot v_i + \frac{1}{R_1 C} \int v_i \, dt\right] \qquad (10.1)$$

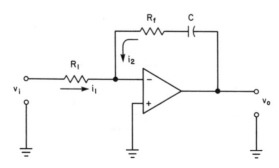

figure 10-3 an op-amp integrator performing proportional-plus-integral action

It will be seen from equation (10.1) that this circuit contains a proportional gain term, K_1 say, that depends upon the ratio R_f/R_1 and an integral control term with a time constant, τ_I, that varies as $R_1 C$; thus let

$$K_1 = \frac{R_f}{R_1} \quad \text{(proportional gain)} \tag{10.2}$$

$$K_2 = 1/\tau_I = 1/R_1 C \quad \text{(integral gain)} \tag{10.3}$$

If we ignore the negative sign in equation (10.1), the block diagram for this circuit is shown in Figure 10-4.

From Figure 10-4(b) the transfer function for this type of controller is

$$\frac{P(s)}{\mathcal{E}(s)} = K_1 + \frac{K_2}{s} \tag{10.4}$$

or

$$\frac{P(s)}{\mathcal{E}(s)} = \frac{K_1}{s} \left(s + \frac{K_2}{K_1} \right) \tag{10.5}$$

The transfer function can be plotted in root locus form, as in Figure 10-5(a), or in Bode plot form, as in Figure 10-5(b).

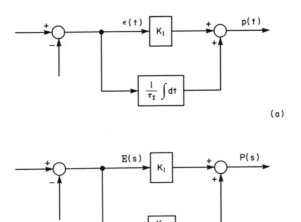

(a)

(b)

figure 10-4   (a) proportional-plus-integral action in the time domain; (b) the same action in the s domain

By choosing the circuit parameters carefully, the zero or corner frequency ω_c can be located so that integral action is only effective for low frequencies with the constant gain of the proportional controller, providing stability to the system at intermediate and higher frequencies. From Figure 10-5(b) we can see that by increasing K_2 (i.e., by reducing τ_I) we can increase the effect of the integral action which causes ω_c to increase, thereby raising the effective value of the loop gain, at higher frequencies, and also increasing the phase lag of the control system with a consequent destabilizing effect. In terms of the root locus diagram, increasing K_2 causes the zero to be driven further along the real axis, away from the pole at the origin, thereby reducing the restraining, or cancelling, action of the zero on the pole.

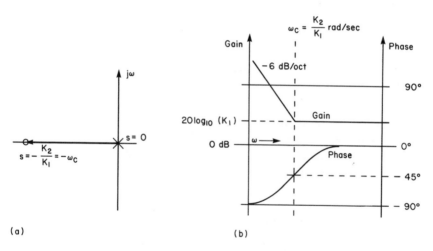

(a) (b)

figure 10-5 (a) root locus diagram; (b) bode diagram of a proportional-plus-integral controller

10-2.4 derivative control

Derivative action or rate control is used in the controller to speed up the transient response of the control system. Derivative action is always accompanied by proportional control; integral is used only when necessary.

The inclusion of derivative action in the controller has a stabilizing effect on the control system by virtue of the addition of phase lead to the control loop gain (i.e., reducing the phase lag of the gain).

Once again, proportional-plus-derivative controllers may be manufactured as pneumatic, hydraulic, electromechanical, or electronic de-

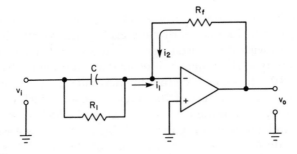

figure 10-6   an op-amp differentiator performing proportional-plus-derivative action

vices, and Figure 10-6 illustrates a typical electronic controller in the form of an operational amplifier and its associated networks.

The output voltage of this circuit depends upon the input voltage in the following manner.

$$v_0 = -\left[\frac{R_f}{R_1}v_i + R_f C\frac{dv_i}{dt}\right]$$ (10.6)

The proportion gain of this circuit depends upon the ratio R_f/R_1 (in the same way as the proportion-plus-integral controller), and the derivative term causes the output voltage to vary directly as the rate of change of the input signal. Once again, let

$$K_1 = \frac{R_f}{R_1} \text{ (proportional gain)}$$ (10.7)

and $$K_3 = \tau_D = R_f C \text{ (derivative gain)}$$ (10.8)

The circuit is presented in block diagram form in Figure 10-7 [where the negative sign of equation (10.6) has been dropped for convenience].

The transfer function for a proportional-plus-derivative controller, as can be seen from Figure 10-7(b), is

$$\frac{P(s)}{\mathcal{E}(s)} = K_1 + K_3 s$$ (10.9)

or $$\frac{P(s)}{\mathcal{E}(s)} = K_3\left(s + \frac{K_1}{K_3}\right)$$ (10.10)

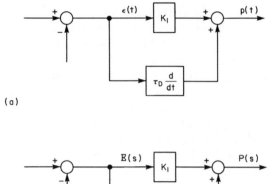

(a)

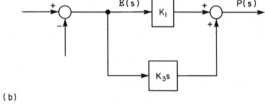

(b)

figure 10-7 (a) proportional-plus-derivative action in the time domain, (b) the same action in the *s* domain

The root locus and Bode diagrams for this controller are drawn in Figure 10-8.

The root locus in Figure 10-8(a) shows only a single zero as the contribution of the controller to the overall system root locus.

The action of this single zero will be to draw any of the existing root loci, located between the zero and the origin, further to the left,

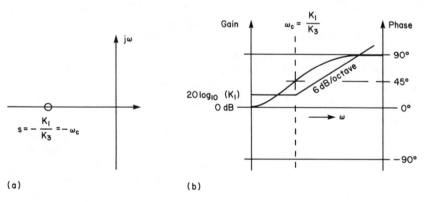

(a)   (b)

figure 10-8 (a) root locus diagram, (b) bode diagram of a proportional-plus-derivative controller

thereby increasing the system damping. Indeed, by locating the zero on an existing locus joining two poles it is possible to prevent the formation of closed-loop complex poles in certain control systems.

As K_3 is increased, the derivative action increases and ω_c gets smaller, moving the zero closer to the origin. As the zero approaches the origin, the distance that existing root loci can be drawn to the left is reduced; hence the stabilizing effect of the zero is also reduced. Indeed, as the zero gets close to the origin, it is likely to pass to the right-hand side of an existing locus, and its effect then will be to draw any nearby loci on its left-hand side towards the right, thereby reducing the system damping.

The Bode diagram of Figure 10.8(b) shows that for the correct choice of ω_c the controller introduces high gain at high frequencies with the added advantage of a phase lead of $+90$ deg at frequencies greater than $10\omega_c$. Thus derivative action will produce increased sensitivity at higher frequencies, producing a fast response to large rates of change of the input. Unfortunately, derivative control is susceptible to system noise, which is characterized by small but extremely rapid perturbations superimposed on the input signal. At frequencies below the corner frequency, ω_c, the controller tends to become insensitive to changes in the input, and the controller gain is entirely dependent upon the value of the proportional action (K_1).

10-2.5 proportional, derivative, integral controllers

As mentioned earlier, these controllers are widely used for controlling the response of pneumatic control valves and hydraulic actuators. The derivative action is used to increase the speed of response, while the integral action prevents steady-state errors from occurring in the process flow rate or actuator position.

The integral action of the PID controller is usually used when the controller is trying to maintain the process variable at its nominal working value and where changes in the process variable only occur as a result of changes in the load.

If the input command to a PID controller is changed significantly (e.g., on start-up, when the command may be to open a valve to its nominal flow position), the integral action of the controller is usually turned off or suppressed until the process variable is close to its nominal value. If the integral action is not suppressed, then the large change of input into the integral controller causes large oscillations to be superimposed onto the response of the control system. The oscillating response interacts with the other two control elements, and the result is a very cyclic system response with a very long settling time. A general rule governing the use of integral action in a cascade controller is that it should only be used when steady-state errors exist that cannot be tolerated, and even then the amount of integral action used should be just enough to remove steady-state errors without causing the steady response to oscillate. Where steady-state errors either do not exist or can be tolerated, then a two-term proportional-plus-derivative controller will prove to be sufficient.

An electronic circuit that performs PID control is shown in Figure 10-9, where the output voltage v_0 is given by

$$v_0 = -\left[\frac{R_1C_1 + R_fC_f}{R_1C_f} \cdot v_1 + \frac{1}{R_1C_f} \int v_1 \, dt + R_fC_1 \frac{dv_1}{dt} \right] \quad (10.11)$$

The block diagram of this circuit is shown in Figure 10-10.

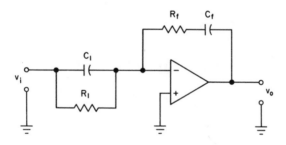

figure 10-9 an op-amp band stop filter performing proportional-plus-integral-plus-derivative action

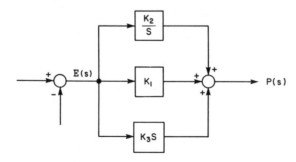

figure 10-10 proportional-plus-integral-plus-derivative action in block diagram form

where
$$K_1 = \frac{R_1C_1 + R_fC_f}{R_1C_f} \text{ (proportional control)} \qquad (10.12)$$

$$K_2 = 1/\tau_I = 1/R_1C_f \text{ (integral control)} \qquad (10.13)$$

$$K_3 = \tau_D = R_fC_1 \text{ (derivative control)} \qquad (10.14)$$

The controller transfer function is

$$\frac{P(s)}{\mathcal{E}(s)} = \frac{K_3s^2 + K_1s + K_2}{s} \qquad (10.15)$$

or
$$\frac{P(s)}{\mathcal{E}(s)} = \frac{K_3}{s} \cdot (s + a)(s + b) \qquad (10.16)$$

where a and b are zeros with values

$$a = \frac{1}{2K_3}[-K_1 + \sqrt{K_1^2 - 4K_2K_3}] \qquad (10.17)$$

$$b = \frac{1}{2K_3}[-K_1 - \sqrt{K_1^2 - 4K_2K_3}] \qquad (10.18)$$

from which it can be seen that if

$$4K_2K_3 > K_1^2$$

the zeros become a complex conjugate pair which will tend to reduce

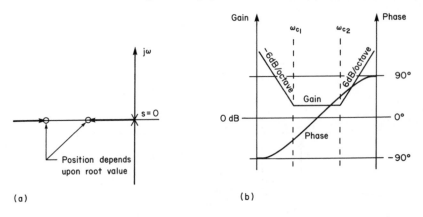

(a)   (b)

figure 10-11 (a) the root locus diagram, (b) the Bode diagram for the band stop filter performing proportional-plus-integral-plus-derivative action

the damping of the control system. Figure 10-11(a) is the root locus of the PID controller for the case when

$$4K_2K_3 < K_1^2$$

The Bode diagram for this controller is drawn in Figure 10-11(b).

The choice of ω_{c1} and ω_{c2} for the PID controller is governed by the same criteria described in Sections 10-2.3 and 10-2.4. At this stage it would be profitable for us to examine a couple of examples and see how we go about choosing the controller gains and time constants in a practical problem. Before doing this, however, we should make some mention of the various terminology associated with PID controllers when used by the process control industry.

10-2.6 process control terminology

PID controllers are available from a number of instrument manufacturers for use in the process control industry. Whether the controller is electronic, pneumatic, hydraulic, or electromechanical, the specifications for all these controllers use a common process control terminology. The relevant terminology relating to PID controllers is listed and described below.

Proportional Band (or Throttling Range): The proportional band is another way of measuring the proportional gain of the controller.

Remembering that the input to the controller is the error signal and that this signal is obtained by subtracting the value of the output, or controlled variable, from the demanded value, we define the proportional band of a controller as the change in the value of the controlled variable (measured as a percentage of its full-scale value) that is necessary to cause full travel of the final control element.

This definition is best understood by means of an example.

EXAMPLE 10.1

Figure 10-12 shows a hot water boiler heated via a steam heat exchanger. The temperature of the hot water depends upon the amount of steam allowed through the steam valve, and the stem movement of the steam valve is in turn controlled by an electropneumatic controller. The demanded temperature for hot water is T_{wD} and the actual temperature, T_w, is measured and fed into the con-

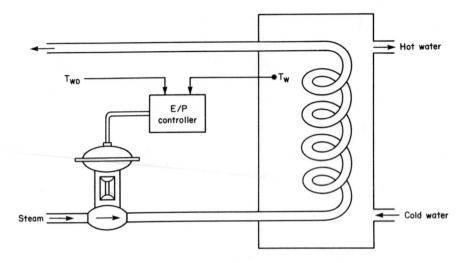

figure 10-12   steam-heated water boiler

troller via a thermocouple. The full stem travel of the valve is 1 in. from fully closed to fully open.

Suppose that when the valve (the final control element) is fully open the temperature of the hot water is 80°C (full-scale value) and that T_w is required to be 50°C. If we now adjust the proportional gain of the controller so that a drop in T_w of 8°C causes the valve to open 0.1 in., then we can say that a 10 percent (8°C) change of the full-scale value of the controlled variable (T_w) causes a 10 percent (0.1 in.) movement of the valve. In other words, a 100% change of the output variable causes full travel of the valve. Hence from the previous definition the proportional band for this case is 100 percent. Of course, the above is only another way of saying that the proportional gain between controller input and valve stem movement is unity.

Suppose now we adjust the proportional gain so that 0.1 in. stem movement is caused when T_w drops by only 4°C. In this case, full stem movement would be caused by a change in the controlled variable of only 40°C, i.e., 50 percent of the full-scale value. Thus the proportional band is said to be 50 percent and the gain between controller input and the valve stem is 2.

At the other end of the scale, if T_w changes by its full-scale value of 80°C and the valve opens only ½ in. (50 percent of its full travel) then a 200 percent change in the controlled variable would be required to cause full stem travel, and the proportional band is said to be 200 percent.

In fact, if this gain is defined as K_c, then

$$\text{Proportional band} = \frac{1}{K_c} \times 100 \text{ percent} \qquad (10.19)$$

and the narrower the proportional band, the higher the loop gain.

Rate Control: Rate control or rate action is just another name for derivative control.

Automatic Reset: Automatic reset or just simply reset is another name for integral control; it gets the name from the ability of the integral controller to reset the controlled variable to the demanded value following a process disturbance.

Reset Rate or Repeats Per Minute: The reset rate or repeats per minute is another way of specifying τ_I or K_2 [from equation (10.13)] with respect to K_1.

If the value of K_1 is chosen so that a change in T_w of, say, $\Delta T°C$ causes the stem to move 0.2 in. and the same change in T_w causes the integral action to move the stem at 0.4 in. per minute, then the integral action will, in one minute, cause the stem to move twice as far as the total movement caused by the proportional action. In this case the reset rate is said to be two per minute. If r is the reset rate, and τ_I is expressed in minutes, then

$$K_1 \Delta T = 0.2 \text{ in.}$$

and from equation (10.13)

$$K_2 \Delta T = \frac{\Delta T}{\tau_I} = 0.4 \text{ in. per minute}$$

Therefore,

$$r = K_2 \Delta T \cdot \frac{1}{K_1 \Delta T} \text{ per minute}$$

or

$$r = \frac{K_2}{K_1} \text{ per minute} \qquad (10.20)$$

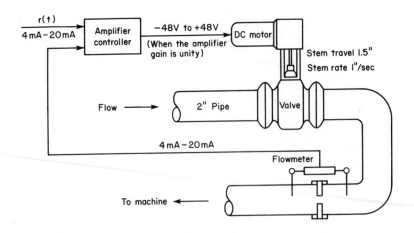

figure 10.13 small batch process control system. (non-si units have been specified, because such units are current in the process control industry. If the reader wishes to make the proper conversions, 1 gallon = 3.7854 litres, a 2-inch nominal pipe is 50 mm (hard conversion), and 1 inch = 0.0254 m).

We now turn our attention to examining how we would design two- and three-term controllers for practical problems.

10-2.7 a worked example—proportional-plus-differential controller

Consider Figure 10-13, in which a 2-inch motor-operated control valve is used to provide flow control to a hydraulic machine downstream of the valve.

In "Standby" mode the machine requires 20 gallons per minute (gpm) ± 5 percent, which is set by a 6 mA signal into the input of the controller. On "Load" the machine requires 100 gpm ± 2 percent, which is set by a 14 mA signal into the controller. The change from 6 mA to 14 mA is extremely fast (step function), and the feedwater flow is required to respond quickly without undue overshoot.

The motor on the control valve is driven by a 250 VDC motor, and the valve stem has a 1½-inch stroke. The characteristics of the valve can be assumed to be linear in that the flow rate of water through the valve varies directly with the valve stem movement. At the fully open position the flow through the valve is 150 gpm.

The amplifier/controller shown in Figure 10-13 comprises standard units which allow a number of 4 mA to 20 mA input signals but

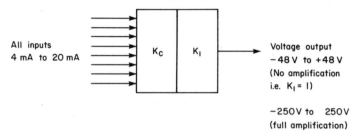

figure 10-14 multi-current input amplifier controller

develops a single voltage output, as shown in Figure 10-14. The amplifier gain, K_1, for this case, is initially set at unity.

The only inputs to the controller are $r(t)$ and the flowmeter signal. If both signals are 4 mA, the output developed by the controller is 0 V; if $r(t)$ is increased to 20 mA the output becomes +48 V. The flow meter (differential pressure transducer and square root extractor) produces a current signal of 4 mA when the flow is 0 gpm and a signal of 20 mA when the flow is 160 gpm; (however, this latter flow rate can never be achieved, since the maximum flow with the valve fully open is only 150 gpm.) If, now, $r(t)$ is 4 mA while the flowmeter signal is 20 mA the controller output goes to −48 V and whenever the two signals have the same value the controller output is 0 V. The flowmeter transducer is located in close proximity to the valve; hence transport lag can be neglected. (Transport lag is the time between the valve's operating and the transducer's sensing the change.)

With +48 volts applied to the motor the valve stroke is in the "Up" direction to open the valve; with −48 volts applied, the valve stroke is in the "Down" direction to close the valve. Either of these two voltages will operate the valve stem at a rate of 1 inch per second.

The response time of the motor-valve, obtained from the manufacturer's specifications, is 550 ms. Response time is the time taken for the stem speed to rise from 10 percent to 90 percent of its top speed.

The control system specifications are

1. Following a step change from "Standby" mode to "Load" mode (i.e., 20 gpm to 100 gpm or an input change of 8 mA), the valve must deliver 100 gpm ± 2 percent within one second of the step being applied. In other words, t_s, the Settling Time (or the time taken for overshoot or oscillations to reduce to less than ±2 percent of the steady output), must be less than one second.

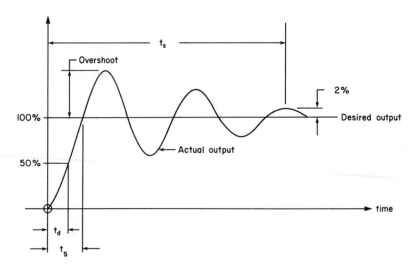

figure 10-15 typical response of a system to a step input

2. The System Rise Time (t_{rs}) must be less than 700 ms. (Rise time is the time taken for the flow rate to rise to the steady-state value.) Alternatively, the response time (t_r) must be less than 400 ms.

3. The percentage overshoot of the flow rate shall not exceed 10 percent.

4. The System Delay Time, t_d, must be less than 200 ms. (Delay time is the time taken for the flow rate to rise to 50 percent of the required change.)

Settling time, rise time, and delay time are shown graphically in Figure 10-15.

One further requirement should be noted: The open-loop phase margin must be at least 60 deg.

solution:

Step 1. The system schematic diagram must be converted to a block diagram so that the appropriate analysis can be made. Figure 10-16 shows the block diagrams for this control system, where the transfer function of each dynamic component is shown incorporated in each block.

Step 2. Determine the numerical values of all the block diagram parameters. The input and feedback signals shown in Figure 10-16 are specified as having a range from 0 mA to 16 mA,

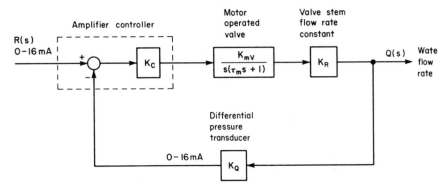

figure 10-16 block diagram of the process control system of figure 10-13

whereas in reality the range would span 4 mA to 20 mA. (That is, the 4 mA bias has been subtracted from the 4 mA to 20 mA range. The span in both cases is precisely the same, namely 16 mA.)

The proportional constant, K_c, is

$$K_c = \frac{48 \text{ V}}{16 \text{ mA}} = 3 \text{ V/mA} \qquad (10.21)$$

The motor-valve constant, K_{mv}, must relate the input voltage to the output stem rate; hence

$$K_{mv} = \frac{1 \text{ in./s}}{48 \text{ V}} = 0.020833 \text{ in./volt-s} \qquad (10.22)$$

The motor-valve time constant, τ_{mv}, is determined from a knowledge of the motor-valve's specified response time, $t_r = 550$ ms. The rise time is shown graphically in Figure 10-17.

Because the speed change is essentially exponential, the time constant, τ_{mv}, may be calculated in the following manner by using the equations:

$$90 \text{ percent full speed} = \text{full speed}(1 - e^{-t_2/\tau_{mv}})$$

$$10 \text{ percent full speed} = \text{full speed } (1 - e^{-t_1/\tau_{mv}})$$

Hence $\qquad \dfrac{1}{9} = \dfrac{e^{-t_2/\tau_{mv}}}{e^{-t_1/\tau_{mv}}} = e^{-(t_2-t_1)/\tau_{mv}}$

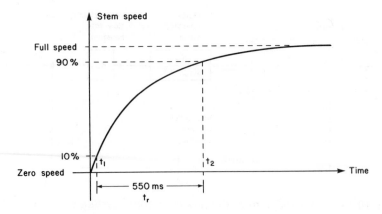

figure 10-17 the speed response of the motor-valve unit showing the response time $t_r = 550$ ms.

or
$$0.1111 = e^{-0.55/\tau_{mv}}$$

which yields
$$\tau_{mv} = 0.25 \text{ s}$$

The valve stem/flow rate constant, K_R, relates the valve stem displacement to the actual flow rate of the water. The value of this constant is

$$K_R = \frac{150 \text{ gpm}}{1\frac{1}{2} \text{ in.}} = 100 \text{ gpm/in.} \qquad (10.23)$$

Finally, the gain of the flow meter is determined by relating the flow rate to the current output in the following manner.

$$K_Q = \frac{16 \text{ mA}}{160 \text{ gpm}} = 0.1 \text{ mA/gpm} \qquad (10.24)$$

Step 3. Redraw the system block diagram, including all numerical values, as shown in Figure 10-18.

Step 4. (a) Determine the time response of the original, uncompensated system to a step input change of 8 mA (20 gpm to 100 gpm).

(b) Determine the stability of the system with a root locus diagram.

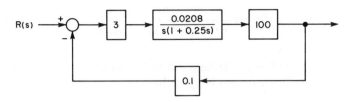

figure 10-18 the process control system block diagram, including all numerical values

    (c) Establish the gain/phase relationship by means of an open-loop Bode diagram.

    Note that it is not necessary to perform all three of these analyses when actually deciding on the gain of a controller. We use the three techniques here in order to show how each method may be used and leave it to the individual to decide which method he prefers.

(a) The Time Domain Response
    The control ratio of the system illustrated in Figure 10-18 is determined by the equation

$$\frac{Q(s)}{R(s)} = \frac{G(s)}{1 + G(s)H(s)} \qquad (10.25)$$

where
$$G(s) = \frac{25}{s(s + 4)} \qquad (10.26)$$

Hence
$$G(s)H(s) = \frac{2.5}{s(s + 4)} \qquad (10.27)$$

and
$$\frac{Q(s)}{R(s)} = \frac{25}{s^2 + 4s + 2.5} \qquad (10.28)$$

    Now the input to the control system is a step demand to increase the flow rate by 80 gpm (20 gpm to 100 gpm) and is represented at the input by a step change in the current signal of 8 mA (6 mA to 14 mA in the actual control system).

Therefore,
$$R(s) = \frac{8}{s} \qquad (10.29)$$

and
$$Q(s) = \frac{200}{s(s^2 + 4s + 2.5)} \qquad (10.30)$$

or
$$Q(s) = \frac{200}{s(s + 0.775)(s + 3.225)} \qquad (10.31)$$

Equation (10.31) has the same form as $F(s)$ in 15(b) of Table 5-1, and therefore the flow rate $q(t)$ is given by

$$q(t) = \frac{200}{0.775 \times 3.225}\left[1 + \frac{0.775\,e^{-(3.225t)} - 3.225\,e^{-(0.775t)}}{3.225 - 0.775}\right] \qquad (10.32)$$

i.e., $q(t) = 80 + 25.31\,e^{-3.225t} - 105.31\,e^{-0.775t}$ $\qquad (10.33)$

Equation (10.33) is plotted for various values of t in Figure 10-19. Two points are very obvious from examination of the curve in Figure 10-19.

First, because we have considered the input demand signal to change from 0 mA to 8 mA (rather than the correct change from 6

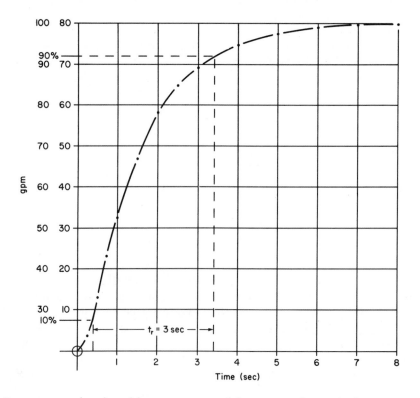

figure 10-19   the closed-loop response of the system shown in figure 10-13 to a step function change of 20 gpm to 100 gpm

mA to 14 mA), the output response calculated in equation (10.32) will start at zero and rise to 80 gpm.

To obtain the true flow rate, 20 gpm should be added to $q(t)$ in equation (10.33), as shown on the far left-hand side of Figure 10-19.

Second, the system is very sluggish and, as shown in Table 10-1, it does not meet the specifications laid down earlier.

TABLE 10-1—PARAMETER SPECIFICATIONS

PARAMETER	REQUIRED VALUE	ACTUAL VALUE
t_r	< 400 ms	3000 ms
t_d	< 200 ms	1200 ms
t_s	< 1000ms	7000 ms

(b) The root locus diagram is developed from the open-loop transfer function in equation (10.27); i.e.,

$$G(s)H(s) = \frac{2.5}{s(s + 4)} \qquad (10.27)$$

The root locus diagram, plotted in Figure 10-20, shows the system to be rigidly stable regardless of the system gain.

(c) The Bode diagram, plotted in Figure 10-21, shows that the un-compensated system adequately meets the phase margin specification. It is important that, regardless of changes made to the system, this condition should not be affected.

It is obvious from Table 10-1 that some type of improvement is needed in the system response. Because the system is initially quite stable, we may consider using a two-term controller. It is important to note that a three-term controller is not necessary here, because integral control would be redundant. That is, we have a type 1 control system with a step input and hence no steady-state errors. Let us look, therefore, at what can be achieved with a proportional-plus-derivative cascade controller, and let us start by using only proportional control, adding derivative control later on, to see what effect adjusting the controller gains has on the system response. In this way we will get a better understanding of the action of each of the control terms and how we arrive at a final choice for the control gains

Step 5. To remove the sluggishness, the system requires more sensitivity, which can easily be accomplished by increasing the

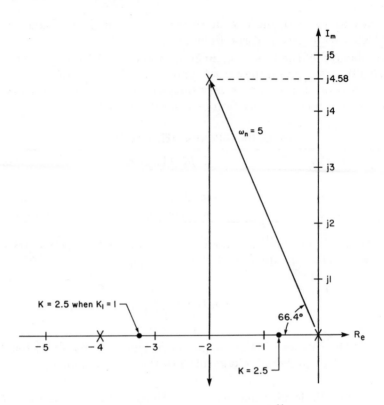

figure 10-20 the root locus diagram for $G(s)H(s) = \dfrac{K}{s(s + 4)}$

amplifier control gain K_1 and hence the loop gain of the system. The control system with proportional gain controller is shown in Figure 10-22. Because sluggishness indicates an overdamped condition, we should aim for a system damping coefficient, ζ, which is less than unity.

From equations (10.25) through (10.28) the control system transfer function now becomes

$$\frac{Q(s)}{R(s)} = \frac{25K_1}{s^2 + 4s + 2.5K_1} \qquad (10.34)$$

Also, from equation (10.29),

$$R(s) = \frac{8}{s}$$

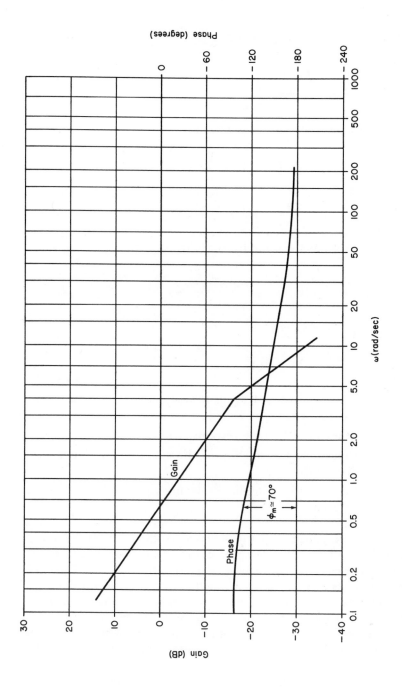

figure 10-21 bode diagram of $G(\omega)H(\omega) = \dfrac{0.625}{j\omega(0.25j\omega + 1)}$

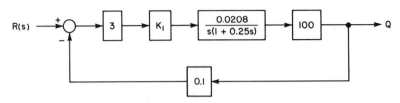

figure 10-22 a simple method used to change the system sensitivity

Therefore, $$Q(s) = \frac{200K_1}{s(s^2 + 4s + 2.5K_1)} \qquad (10.35)$$

Equation (10.35) has the same form as $F(s)$ in 17 of Table 5-1; i.e.,

$$F(s) = \frac{A\omega_n^2}{s(s^2 + 2\zeta\omega_n s + \omega_n^2)} \qquad (10.36)$$

where $\omega_n = \sqrt{2.5K_1}$ rad/s

and $\zeta = \dfrac{4}{\sqrt{10K_1}}$

Step 6. Let us repeat Step 4 again to assess the effect, on the response, of a change in K_1.

(a) Determine the time response.

(b) Determine the stability of the system with a root locus diagram.

(c) Establish the gain/phase relationship by means of a Bode diagram.

(a) The time response of $q(t)$ is obtained from the inverse Laplace transform of equation (10.36). Before evaluating $q(t)$, however, we should establish a practical range of values for K_1. Table 10-2 lists values of ζ and ω_n as a function of K_1; a useful value for ζ is usually somewhere between 0.4 and 0.8. It would seem, therefore, that if we selected four values of K_1, from $K_1 = 2$ to $K_1 = 10$, we could plot the most appropriate step responses of the system. The four step response curves are shown in Figure 10-23.

On examination of Figure 10-23, the system response time (t_r) shows a remarkable improvement with successive increases of

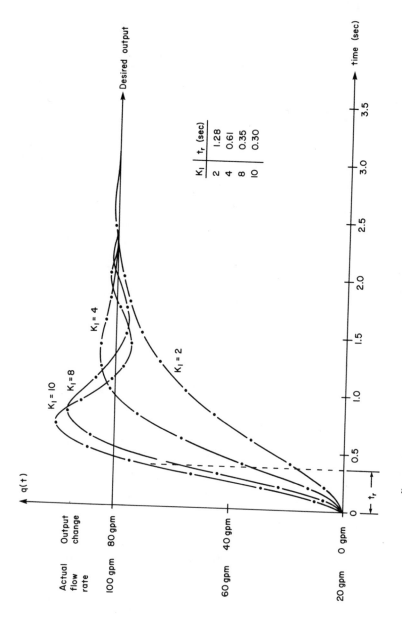

figure 10-23   step response for various values of K_1

gain K_1. However, with K_1 set at a value that enables the system to meet the specification $t_r \leqslant 400$ ms, the overshoot becomes intolerable and the settling time far exceeds the specification of 1 second.

(b) The root locus diagram for the system is precisely the same as that shown in Figure 10-20 in that the system is still rigidly stable, although oscillations may be present, depending upon the value of K_1. The system is oscillatory for any value of K_1 exceeding 1.6 (see Table 10-2).

(c) Figure 10-24 is the system Bode diagram when $K_1 = 10$. As K_1 increases, the gain versus frequency line moves up the graph, and the *gain crossover frequency* ω_{gc} (the point at which the gain curve crosses the 0 db line) moves to the right. When $K_1 = 10$, ω_{gc} is 5

TABLE 10-2—DAMPING COEFFICIENT AND RESONANT FREQUENCY AS A FUNCTION OF GAIN

K_1	$\zeta = \dfrac{4}{\sqrt{10K_1}}$	$\omega_n = \sqrt{2.5K_1}$ (rad/s)
1.5	1.03	1.94
1.6	1	2
1.7	0.97	2.06
2	0.89	2.24*
3	0.73	2.74
4	0.63	3.16*
5	0.57	3.54
6	0.52	3.87
7	0.48	4.18
8	0.45	4.47*
9	0.42	4.74
10	0.4	5 *
12	0.37	5.48
14	0.34	5.92
16	0.32	6.32
18	0.3	6.71
20	0.28	7.07
25	0.25	7.91
30	0.23	8.66
35	0.21	9.35
40	0.2	10

* Values of K_1 used in plotting the step response.

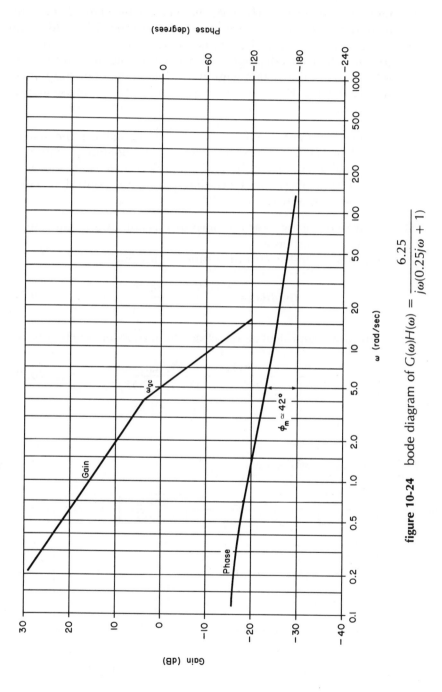

figure 10-24 bode diagram of $G(\omega)H(\omega) = \dfrac{6.25}{j\omega(0.25j\omega + 1)}$

rad/s and the phase margin, ϕ_m, has been reduced to 42 deg. This degradation of phase margin must be improved by adding an extra stabilizing control term, i.e., derivative control. Derivative control will give us the high gain needed during the transient but will not affect the system gain unduly when the open loop phase shift is 180 deg. Hence we can expect to return to a phase margin similar to that of the original control system, i.e., 70 deg.

Step 7. Because derivative control will tend to stabilize the system, let us retain the present sensitivity by holding the proportional gain at $K_1 = 10$, and add a derivative control term to the controller. The block diagram for the control system is now shown in Figure 10-25.

Step 8. We now assess how the derivative action affects the control system's response.

The new open-loop transfer function becomes

$$G(s)H(s) = \frac{0.832K_3(s + 3K_1/K_3)}{s(s + 4)} \qquad (10.37)$$

from which we can calculate the closed-loop transfer function to be

$$\frac{Q(s)}{R(s)} = \frac{8.32K_3(s + 3K_1/K_3)}{s^2 + (4 + 0.833K_3)s + 2.5K_1} \qquad (10.38)$$

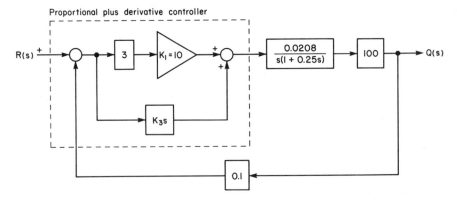

figure 10-25 the original system with increased gain and derivative compensation

Hence $\qquad\qquad \omega_n = \sqrt{2.5K_1} = 5 \text{ rad/s}$ $\qquad\qquad$ (10.39)

and $\qquad\qquad \zeta = \dfrac{4 + 0.833K_3}{2\sqrt{2.5K_1}}$ $\qquad\qquad$ (10.40)

If we use a to denote the zero in equation (10.37),

then $\qquad\qquad a = -3K_1/K_3 = -30/K_3$ $\qquad\qquad$ (10.41)

and $\qquad\qquad \zeta = \dfrac{4 - 2.5K_1/a}{2\sqrt{2.5K_1}} = 0.4 - 2.5/a$ $\qquad\qquad$ (10.42)

As rules of thumb we would like ζ to be less than unity, so as not to be too sluggish and to have phase margin greater than 60 deg, so that oscillations and overshoot are not too great. From equations (10.41) and (10.42),

if $\qquad\qquad \zeta \leqslant 1$

then $\qquad\qquad K_3 \leqslant 7.2$ $\qquad\qquad$ (10.43)

and $\qquad\qquad a \leqslant -4.1667$ $\qquad\qquad$ (10.44)

Because the specifications for the performance of the control system are written in terms of time response parameters, we find that the Bode diagram and root locus methods do not offer the most efficient techniques for designing the control system. We will, however, demonstrate how both these methods can be used to get close approximation to the final solution.

Using the Bode Diagram: If we also require of the phase margin, ϕ_m, in Figure 10-24 that

$$\phi_m \geqslant 60°$$ $\qquad\qquad$ (10.45)

then we can see from that figure that the break point in the phase curve must begin at

$$\omega_b = 1.5 \text{ rad/s}$$ $\qquad\qquad$ (10.46)

i.e., where the phase curve intersects -120 deg (or $\phi_m = 60$ deg). Thus the zero in equation (10.37) must be chosen so that

$$a = -10\omega_b$$

i.e., $\qquad\qquad a = -15 \text{ rad/s} \qquad\qquad\qquad (10.47)$

Therefore, from equations (10.41) through (10.44) the useful ranges for K_3, a, and ζ are

$$2 \leqslant K_3 \leqslant 7.2 \qquad\qquad\qquad (10.48)$$

$$-15 \leqslant a \leqslant -4.1667 \qquad\qquad\qquad (10.49)$$

$$0.57 \leqslant \zeta \leqslant 1 \qquad\qquad\qquad (10.50)$$

Probably the first value we would try for ζ is 0.707, which is the so-called optimum damping factor for a second-order control system with no zeros present in the open-loop transfer function. Note that in our control system, $G(s)H(s)$ does contain a zero; as we stated earlier, the optimum performance of our control system can only be judged from how well the final time response fits the specifications stated at the beginning of this section. However, before solving for the time response, let us look at how we would decide upon a value of K_3 (or a) using only the root locus method.

Root Locus Approach: We wish to locate the zero a such that

$$\zeta = 0.7071$$

Referring to Figure 10-26 and the rules of Chapter 7, because we have an open-loop pole at $s = -4$ and at $s = 0$, we must locate the open-loop zero, a, on the negative real axis of the diagram such that

$$\theta_1 + \theta_2 - \theta_3 = 180° \qquad\qquad\qquad (10.51)$$

We know θ_1 to be 135 deg because we have chosen

$$\zeta = 0.7071 = \cos 45° \qquad\qquad\qquad (10.52)$$

We also know that

$$\omega_n = \sqrt{2.5K_1} = 5 \text{ rad/s}$$

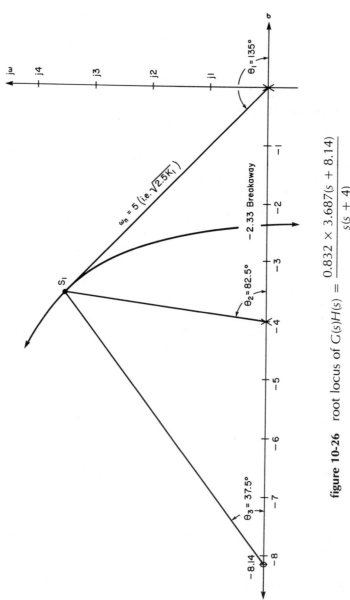

$\theta_1 = 135°$

$\omega_n = 5\ \left(\text{i.e. } \sqrt{2.5K_1}\ \right)$

-2.33 Breakaway

$\theta_2 = 82.5°$

S_1

$\theta_3 = 37.5°$

-8.14

figure 10-26 root locus of $G(s)H(s) = \dfrac{0.832 \times 3.687(s + 8.14)}{s(s + 4)}$

Hence the closed-loop poles are

$$s_1 = 5(\cos \theta_1 + j \sin \theta_1)$$

or
$$s_1 = 3.536 + j\,3.536 \qquad\qquad (10.53)$$

and
$$s_2 = 5(\cos \theta_1 - j \sin \theta_1)$$

or
$$s_2 = 3.536 - j3.536 \qquad\qquad (10.54)$$

By drawing a line between s_1 and the open-loop pole at $s = -4$ we can measure θ_2 to be

$$\theta_2 = 82.5° \qquad\qquad (10.55)$$

Knowing θ_1 and θ_2, we can calculate θ_3 from equation (10.51); i.e.,

$$\theta_3 = -(180° - 13.5° - 82.5°)$$

i.e.,
$$\theta_3 = 37.5° \qquad\qquad (10.56)$$

Consequently, we must locate the open-loop zero on the negative real axis of Figure 10-26 so that the line drawn between the zero and s_1 intersects the real axis at an angle of 37.5 deg.

Drawing this line on the diagram, we find that the line intersects the real axis at

$$a = -8.1 \text{ rad/s}$$

Therefore from equation (10.41), when

$$\zeta = 0.7071 \qquad\qquad (10.57)$$

we find
$$a = -8.1 \text{ rad/s} \qquad\qquad (10.58)$$

and
$$K_3 = 3.687 \qquad\qquad (10.59)$$

To check if these values fulfill the specifications, we must examine the time response for the control system.

Time Response Solution: From equations (10.29) and (10.38) the closed-loop transfer function is

$$\frac{Q(s)}{R(s)} = \frac{8.333K_3(s + 3K_1/K_3)}{s^2 + (4 + 0.833K_3)s + 2.5K_1} \qquad\qquad (10.38)$$

and
$$R(s) = \frac{8}{s} \qquad (10.29)$$

Therefore,

$$Q(s) = \frac{8 \times 8.333 K_3 (s + 3K_1/K_3)}{s(s^2 + (4 + 0.833 K_3)s + 2.5 K_1)} \qquad (10.60)$$

The inverse Laplace transform for this equation can be derived by partial fractions using the Laplace transforms in Table 5-1. From the inverse Laplace transform of equation (10.60), the flow rate following an input step change of 8 mA is

$$q(t) = 80 \left[1 + \frac{(\omega_n^2/a^2 - 2\zeta\omega_n/a + 1)^{1/2} e^{-\zeta\omega_n t}}{\sqrt{1 - \zeta^2}} \sin(\omega_n \sqrt{1 - \zeta^2} \cdot t - \phi) \right] \text{gpm}$$
$$(10.61)$$

where ζ, a, and ω_n are defined in equations (10.39) through (10.42).

$$\phi = \tan^{-1} \left[\frac{\sqrt{1 - \zeta^2}}{\omega_n/a - \zeta} \right] \qquad (10.61a)$$

and 80 gpm is the resulting steady-state change in flow rate.

The function in equation (10.61) has been plotted in Figure 10-27 for the cases considered when one is using the Bode diagram and root locus techniques, i.e.,

$$K_1 = 10; \qquad K_3 = 1.95 \qquad (\phi m = 60°)$$

and
$$K_1 = 10; \qquad K_3 = 3.687 \qquad (\zeta = 0.7071)$$

Let us see if either of these two curves fulfills the specifications.
To restate the specifications, we require:

1. That t_D, the time to reach 50 percent of the steady flow rate, be less than 0.2 s.
   — Both curves fulfill this requirement.
2. That t_r, the rise time, is less than 0.7 s.
   — Both curves fulfill this requirement.
3. That the percentage overshoot shall be less than 10 percent.
   — When $K_3 = 1.95$ (phase margin of 60 deg) the overshoot is approximately 13 percent, so this value of K_3 is insufficient. Increasing K_3 reduces the overshoot, and when K_3 is 3.687, overshoot is only 7 percent.

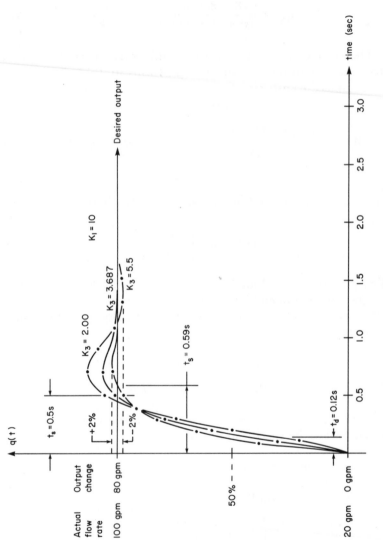

figure 10-27 step response-flow control system for various values of K_3

4. That t_s, the settling time (or the time for overshoot or oscillations to decay to less than 2 percent of the steady value), shall be less than one second.

— With both values of K_3, the settling time exceeds one second, although in the case of $K_3 = 3.687$ ($\zeta = 0.7071$) the excess is marginal.

Thus neither of these two cases exactly fulfill our specification, and they indicate that the amount of derivative action should be increased. By plotting the flow rate for one or two values of K_3 in the region of $t = 0.75$ s (i.e., around the peaks of the other two curves), we come up with values of K_3 in the area of 5.5 which yield small overshoot (less than 2 percent) with settling times less than one second and delay times less than 0.2 second. In fact, when

$$K_3 = 5.5 \qquad (10.62)$$

we find
$$\text{Delay time, } t_d = 0.12 \text{ s}$$
$$\text{Rise time, } t_{rs} = 0.59 \text{ s}$$
$$\text{Settling time, } t_s = 0.5 \text{ s}$$
$$\text{Percentage overshoot} = 1.84 \text{ percent}$$

which fulfills all the specifications. Also for this value of K_3 the open-loop zero is located at

$$a = -5.45 \text{ rad/s} \qquad (10.63)$$
and that
$$\zeta = 0.86 \qquad (10.64)$$

The flow rate response for this value of K_3 is also plotted in Figure 10-27 for comparison with the other two response curves.

Finally, Figure 10-28 illustrates the effect of each successive stage of controller action.

10-2.8 a worked example—proportional-plus-integral control

If the motorized control valve of the previous example were replaced by a pneumatic control valve, we would have a very different problem to deal with. The problem would change because the transfer function for the pneumatic control valve is of type 0 (approximated by a first-order lag in the next example); hence the open-loop transfer function of the flow control system will also be type 0 and steady-state errors will accompany any step change in flow rate demand.

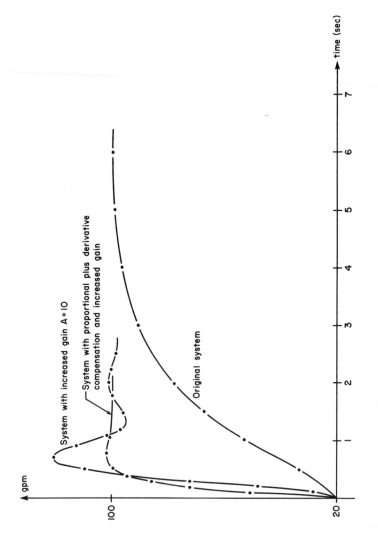

figure 10-28 the step response of the system at various stages of compensation

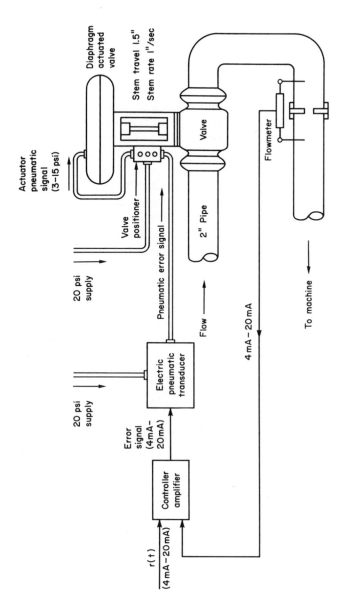

figure 10-29 small batch process control system

However, results similar to those obtained in Example 10-2.7 can be achieved with a pneumatic control valve, provided that an integral control term is used in the controller. Consider the control system as modified in Figure 10-29.

In this system we have an amplifier controller at the input, as in the last system. It sums both the demand signal and the feedback signal as before. However, this time we are using a current output into an electropneumatic transducer (EPT). The EPT operates with an input range of 4 mA to 20 mA and an output range of 3 psi to 15 psi. The gain is naturally 0.75 psi/mA. The valve positioner receives the pneumatic error signal which modifies the 20 psi supply to between 3 psi and 15 psi, which is then delivered to the diaphragm on top of the valve. Because the pressure range is 12 psi (from 3 psi to 15 psi) and the stem movement is 1½ in., the gain of the diaphragm-actuated valve is 0.125 in/psi. The pneumatic valve of this example is considered to have the same time constant as the previous motorized valve, namely, 0.25 s, and the flow meter is also unchanged from the last example.

The operational specifications are

1. Zero steady-state error.
2. Percentage overshoot to be less than 2.5 percent.
3. Settling time, t_s, to be less than 1 s.
4. Rise time, t_r, to be less than 0.7 s.

solution

Step 1. The system schematic diagram must be converted to a block diagram with appropriate symbols for each transfer function, as shown in Figure 10-30.

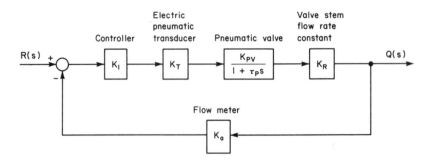

figure 10-30 block diagram of the process control system of figure 10-29

Step 2. Redraw the block diagram, including all the numerical values of the transfer functions; this is done in Figure 10-31. To begin with, let $K_1 = 1$, and $K_2 = K_3 = 0$, as shown in Figure 10-31.

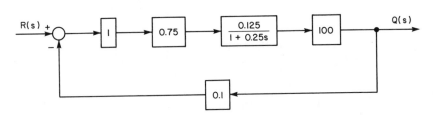

figure 10-31 the process control system block diagram, including all numerical values

Step 3. As before, let us investigate the operation of the unimproved system, and then we will know how to change it to meet the operational requirements. Because this system is a type 0 control system, there will be a steady-state error in response to a step input, so as a first step we ask, what is the steady-state error?

Step 4. (a) Steady-state error
From Chapter 9, the steady-state error, $\epsilon_{ss}(t)$ is

$$\epsilon_{ss}(t) = \lim_{s \to 0} s \cdot \mathcal{E}(s) = \lim_{s \to 0} s \cdot \frac{R(s)}{1 + G(s)H(s)} \qquad (10.65)$$

where $$R(s) = \frac{8}{s}$$

From the block diagram of Figure 10-30 the open-loop transfer function is easily obtained, namely,

$$G(s)H(s) = \frac{K_1 \, K_T \, K_{pv} \, K_R \, K_Q}{\tau_p s + 1} \qquad (10.66)$$

hence $$G(s)H(s) = \frac{0.9375}{0.25s + 1} \qquad (10.67a)$$

Substituting $R(s)$ and $G(s)H(s)$ into equation (10.65) yields

$$\epsilon_{ss}(t) = \lim_{s \to 0} \left[s \cdot \frac{\dfrac{8}{s}}{1 + \dfrac{0.9375}{0.25s + 1}} \right]$$

or

$$\epsilon_{ss}(t) = \lim_{s \to 0} \left[\frac{8}{1 + \dfrac{0.9375}{0.25s + 1}} \right]$$

If we let $s = 0$, the steady-state error is

$$\epsilon_{ss}(t) = \frac{8}{1.9375} = 4.129 \text{ mA}$$

Because 16 mA (4 mA and 20 mA) at the input represents a flow rate of 160 gpm (0 to 160 gpm), then a 4.129 mA error represents a 41.29 gpm error at the output. This is shown graphically in Figure 10.32.

The error at the output is ridiculously large and is intolerable under any circumstances. Before we proceed to improve the system, let us examine the phase and gain relationship. As shown in the Bode diagram of Figure 10.33, the phase margin is large, $\phi_m = 155$ deg; in fact, the phase margin, regardless of the open-loop gain, can

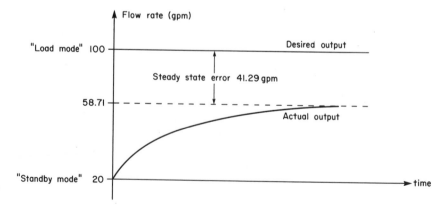

figure 10-32 the output flowrate response to a step input demand showing the steady-state error at 41.29 gpm.

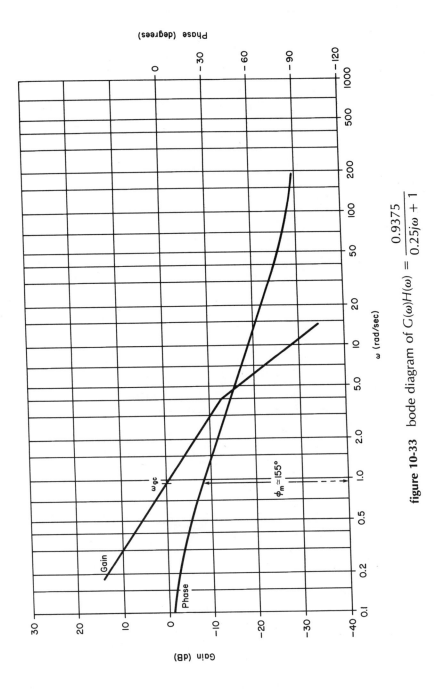

figure 10-33 bode diagram of $G(\omega)H(\omega) = \dfrac{0.9375}{0.25j\omega + 1}$

never be less than 90 deg, because $G(s)H(s)$ contains only a single pole.

Step 5. Again, as a first step in improving this system, let us increase K_1 and examine the output response. At each successive increase we have recalculated the steady-state error, as shown in Table 10-3. (N.B. To prevent saturation K_1 must be <1.7).

TABLE 10-3—STEADY-STATE OUTPUT ERROR FOR VARIOUS VALUES OF K_1

K_1	ERROR IN GPM
1	41.29
3	20.98
10	7.71
30	2.75

The response of the system to the input demand becomes more rapid as K_1 increases, which can be seen in Figure 10-34.

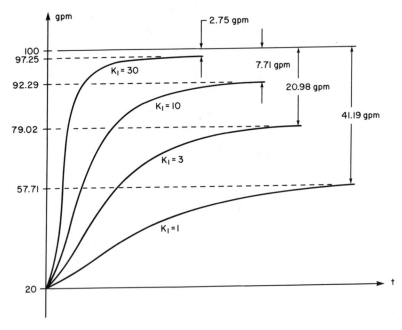

figure 10-34 the response of the system with increasing values of gain K_1 showing the output error

It is obvious that if we could increase the gain we can meet three of the four specifications laid down for operation of the system. However, although $\epsilon_{ss}(t)$ decreases as K_1 increases, the steady-state error will never be zero.

Step 6.   By adding integral action to the controller we can remove the steady-state error completely. Unfortunately, integral action will tend to make the system more sluggish; however, by optimal selection of the integral constant K_2 we can still meet the operating requirements.

When integral action is added to the controller, the resulting block diagram is as depicted in Figure 10.35.

Step 7.   Let us verify that this system has zero steady-state error. The input is still

$$R(s) = \frac{8}{s}$$

So the open-loop transfer function now becomes

$$G(s)H(s) = \frac{0.9375(K_1 s + K_2)}{s(0.25s + 1)} \qquad (10.68)$$

Thus,

$$\epsilon_{ss}(t) = \lim s \cdot \frac{\dfrac{8}{s}}{1 + \dfrac{0.9375\,(K_1 s + K_2)}{s(0.25s + 1)}}$$

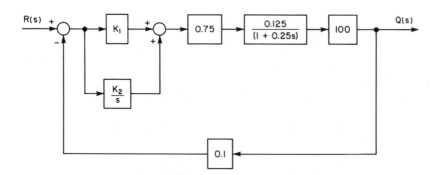

figure 10-35 flowrate control system with proportional-plus-integral action

Hence $\quad \epsilon_{ss}(t) = \dfrac{8}{1 + \dfrac{0.9375K_2}{0}} = 0$

Step 8. Let us now try to find values for K_1 and K_2 that will achieve the specifications for the time response. This is now quite simple, because we did most of the work for this example while solving the previous example. Let us explain, if $G(s)H(s)$ in equation (10.68) is rewritten as

$$G(s)H(s) = \frac{3.75K_1(s + K_2/K_1)}{s(s + 4)} \qquad (10.69)$$

we see, in comparing equations (10.69) and (10.37), that the form of the open-loop transfer functions in both examples is identical. Therefore, by suitable choice of K_1 and K_2 we can obtain a time response that is also identical to the previous example and, more importantly, one that we know will fulfill the specifications. Thus from equations (10.37), (10.41), and (10.63), if we choose K_1 and K_2 so that

$$K_1 = 1.222 \qquad (10.70)$$

$$K_2 = 6.667 \qquad (10.71)$$

we will arrive at a time response solution for the flow rate identical to that given by equations (10.61), (10.39), (10.63), and (10.64). This response was illustrated in Figure (10-27), for the case $K_3 = 5.5$ and is repeated in Figure 10-36 to show the effect of adding the proportional-plus-integral controller.

10-3 COMPENSATION TECHNIQUES

After a control system is designed, if it is found to be marginally or inherently unstable, a compensating device can often be added to the control to improve the gain and phase margin of the frequency response.

Compensation techniques can be applied in two ways:

1. By a cascade device in the forward gain of the control system; these devices may be phase-lead, phase lag, or phase-lag-lead networks.

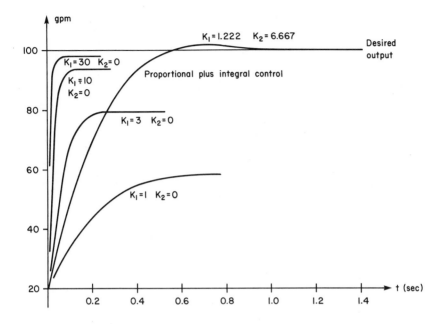

figure 10-36 time response of pneumatic flow control valve with proportional and proportional-plus-integral control

2. Minor loop or rate feedback, where the addition of an inner feedback loop provides stability to the control system.

10-3.1 cascade compensation devices

Phase-Lead Network: Phase-lead networks are used to reduce the phase lag and improve the phase margin of a control system. As its name implies, a cascade phase-lead network will shift the phase of the actuating signal (network output) so that it leads the error signal (network input).

This network is often formed by using passive components, as shown in Figure 10-37. Similar networks can be formed by using amplifiers; they behave in the same way as our passive network.

The transfer function can easily be developed from the block diagram; i.e., a block diagram reduction leads to

$$\frac{E_0(s)}{E_i(s)} = \frac{R_2(R_1Cs + 1)}{R_1 + R_2(R_1Cs + 1)} \qquad (10.72)$$

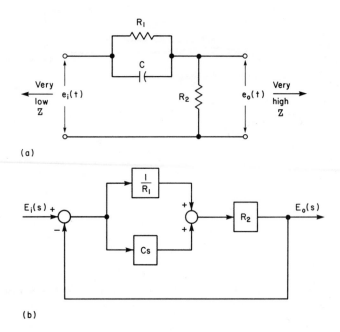

figure 10-37 (a) phase-lead circuit configuration; (b) phase-lead network block diagram.

or

$$\frac{E_0(s)}{E_i(s)} = \frac{R_2}{R_1 + R_2} \cdot \frac{(R_1Cs + 1)}{\dfrac{R_2}{R_1 + R_2} \cdot R_1Cs + 1} \qquad (10.73)$$

which simplifies to

$$\frac{E_0(s)}{E_i(s)} = \alpha \cdot \frac{\tau_{LD}s + 1}{\alpha\tau_{LD}s + 1} \qquad (10.74)$$

where the circuit gain

$$\alpha = \frac{R_2}{R_1 + R_2} \qquad (10.75)$$

and the lead time constant

$$\tau_{LD} = R_1Cs \qquad (10.76)$$

The frequency response and the locations of the pole and zero of equation (10.74) are shown in Figure 10-38

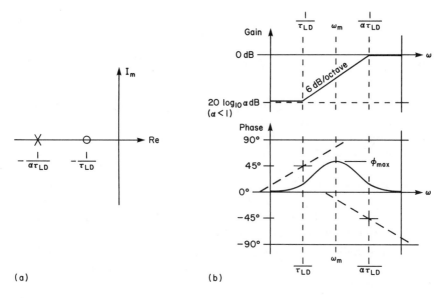

figure 10-38 (a) the pole-zero configuration of the phase-lead network, (b) the bode diagram of the same network

Figure 10.38(a) shows one pole and one zero located at $-\dfrac{1}{\alpha\tau_{LD}}$ and $-\dfrac{1}{\tau_{LD}}$, respectively. If R_2 is reduced, the pole will move to the left, and if R_2 is increased, the pole will move to the right. If R_1 is adjusted, both the pole and zero location will change. If the circuit is introduced into a control system that requires some stabilization, a reasonable result can be achieved by proper selection of R_1, R_2, and C.

The Bode diagram shows that for frequencies less than $\dfrac{1}{\tau_{LD}}$ the circuit introduces attenuation of $20 \log \alpha$ db and does not affect the system phase shift. At frequencies above $\dfrac{1}{\alpha\tau_{LD}}$ there is little attenuation or phase shift. However, at the *geometric mean frequency* ω_m (located on the log of frequency scale, halfway between the lower and upper corner frequencies) there is one-half the maximum attenuation developed and a maximum phase lead of ϕ_{max}.

It is useful to be able to calculate both the mean frequency ω_m and the maximum phase lead introduced into a system. These two parameters are determined by:

$$\omega_m = \frac{1}{\tau_{LD}\sqrt{\alpha}} \tag{10.77}$$

and
$$\sin \phi_{\max} = \frac{1 - \alpha}{1 + \alpha} \tag{10.78}$$

Phase-lead compensation will improve the rise time response to a step input and will usually reduce the overshoot, but it increases the closed-loop bandwidth, which is a disadvantage, since it makes the system more susceptible to noise.

Phase-Lag Network: Phase-lag networks are usually used to attenuate the open-loop gain prior to the crossover frequency and thereby to improve the gain margin. This network shifts the phase of the actuating signal so that it lags the error signal.

Just as before, the transfer function can easily be developed from the block diagram of Figure 10-39(b). The reduction of the block diagram yields

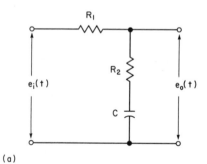

(a)

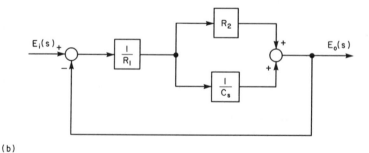

(b)

figure 10-39 (a) phase-lag circuit configuration, (b) phase-lag network block diagram

$$\frac{E_0(s)}{E_i(s)} = \frac{R_2Cs + 1}{R_1Cs + R_2Cs + 1} \qquad (10.79)$$

or

$$\frac{E_0(s)}{E_i(s)} = \frac{\dfrac{R_2}{R_1 + R_2}(R_1 + R_2)Cs + 1}{(R_1 + R_2)Cs + 1} \qquad (10.80)$$

which simplifies to

$$\frac{E_0(s)}{E_i(s)} = \frac{\alpha\tau_{LG}s + 1}{\tau_{LG}s + 1} \qquad (10.81)$$

where the high frequency gain α is given by

$$\alpha = \frac{R_2}{R_1 + R_2}$$

and the lag time constant is

$$\tau_{LG} = (R_1 + R_2)C$$

The frequency response and the location of the pole and zero of equation (10.81) are shown in Figure 10-40.

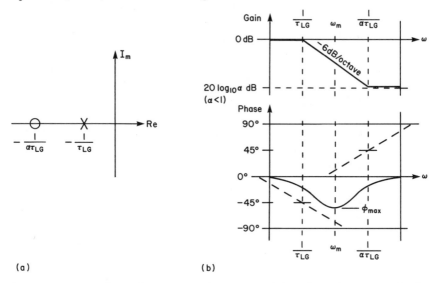

(a) (b)

figure 10-40 (a) the pole-zero configuration of the phase-lag network, (b) the bode diagram of the same network

Figure 10-40(a) shows a single pole and zero located at $-\dfrac{1}{\tau_{LG}}$ and $-\dfrac{1}{\alpha\tau_{LG}}$ respectively. Changing the values of either R_1 or R_2 will cause the pole and the zero to move their positions, permitting adjustment of the control system root locus to meet a particular stability requirement.

The Bode diagram shows two corner frequencies, as before, occurring at $\dfrac{1}{\tau_{LG}}$ and $\dfrac{1}{\alpha\tau_{LG}}$. Below the lower corner frequency the network has essentially unity gain and 0 deg phase shift. However, above the upper corner frequency the network attenuates to a maximum of 20 log α db, and again the phase shift is approximately 0 deg. At the mean frequency ω_m the circuit develops one-half its maximum attenuation and a maximum phase lag of ϕ_{max}.

The mean frequency is calculated as before.

$$\omega_m = \frac{1}{\tau_{LG}\sqrt{\alpha}} \qquad (10.82)$$

and the maximum phase lag angle is determined by the relationship

$$\sin \phi_{max} = \frac{\alpha - 1}{\alpha + 1} \qquad (10.83)$$

Phase-lag compensation reduces the overshoot and makes the system more stable. The bandwidth of the closed-loop system is reduced, which tends to increase the rise time.

Phase Lag-Lead Networks: A phase lag-lead network will cause the phase of the actuating signal to lag the phase of the error signal at low frequencies and to lead it at high frequencies.

A circuit and block diagram for such a network is shown in Figure 10-41.

From the block diagram the transfer function for this circuit is

$$\frac{E_0(s)}{E_i(s)} = \frac{\tau_1\tau_2 s^2 + (\tau_1 + \tau_2)s + 1}{\tau_1\tau_2 s^2 + (\tau_1 + \tau_2 + \tau)s + 1} \qquad (10.84)$$

where
$$\tau_1 = R_1 C_1 \qquad (10.85)$$

$$\tau_2 = R_2 C_2 \qquad (10.86)$$

$$\tau = R_1 C_2 \qquad (10.87)$$

Figure 10-42 shows the Bode diagram for this network.

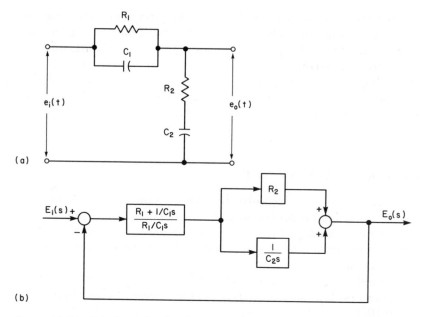

figure 10-41   (a) phase lag-lead circuit, (b) equivalent block diagram

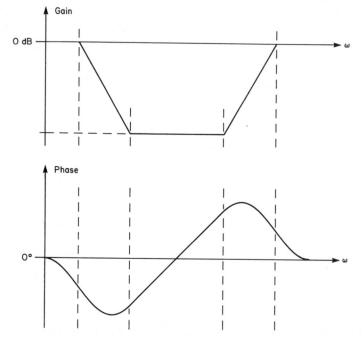

figure 10-42   bode diagram of phase lag-lead network

Notice from the Bode diagram that at low frequencies the gain of the network is approximately unity and that the phase of the output lags the input. As the frequency increases, the gain falls off, until at some intermediate frequency the gain is a minimum but the phase shift is now 0 deg. As we approach higher frequencies, the phase of the output leads the input and then reduces again to 0 deg, while the gain increases, returning to its original unity value.

10-3.2 minor loop or rate feedback

Rate feedback is a pure derivative feedback action in that the output from such an element is proportional to the rate of change (time derivative) of its input signal. When used as a feedback device, this allows the derivative of the output to be compared at the summing junction, with the reference input to the system.

The most common feedback device of this type is a tachogenerator. Simple schematic diagrams of two types of tachogenerator are illustrated in Figure 10-43.

The tachogenerators are often connected directly to a motor shaft and simply translate this output speed into a directly proportional voltage. A block diagram of this action is shown in Figure 10-44, where K_T is the tachogenerator gain constant measured in volt-seconds per radian, although manufacturers' data sheets usually specify it in volts per 1000 rpm.

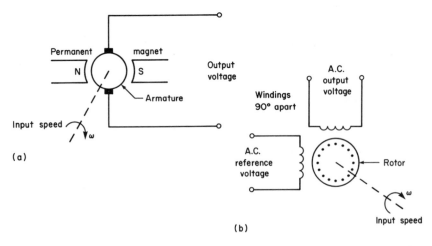

figure 10-43 schematic diagram of, (a) permanent magnet dc tachogenerator, (b) two-phase ac induction tachogenerator

figure 10-44 a block diagram representing the speed/voltage translation of a tachogenerator.

From Figure 10-44

$$\frac{d\theta}{dt} \cdot K_T = e_0 \qquad (10.88)$$

The Laplace transform of this equation is

$$K_T s\theta(s) = E_0(s) \qquad (10.89)$$

which yields the transfer function,

$$\frac{E_0(s)}{\theta(s)} = K_T s \qquad (10.90)$$

10-3.3 a worked example—phase lead compensation

As a final example of system compensation, consider the tracking antenna of Figure 10-45. When the signal echo is received, part of the signal processing causes a voltage to be developed and fed through an amplifier to a control motor. The control motor rotates the antenna dish through gearing in the direction of the target motion. The motion of the antenna dish is sensed by a precision potentiometer, the output of which is fed back to the input of the amplifier. Although we shall consider only one system, two are really necessary: one to rotate the antenna in elevation (up and overhead) and the other to rotate the antenna in azimuth (from side to side).

A system schematic diagram is shown in Figure 10-46(a) and its equivalent block diagram in Figure 10-46(b).

The component parameters are to be determined from a knowledge of the system operation.

Antenna: The antenna is initially required to scan 120 deg back and forth, each 120 deg taking 1 second. When the target is acquired, the antenna must track it at a maximum rotation of about 11½ deg per second to an accuracy of at least 20 minutes of arc $\left(\frac{1°}{3}\right)$. The antenna

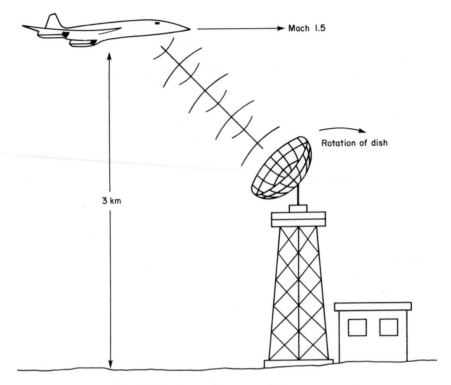

Mach 1.5

Rotation of dish

3 km

figure 10-45   a radar tracking antenna

will, therefore, present a load inertia (J_L) and a maximum load torque (T_L) to the motor through the gearing.

> *Antenna Data*
> | | |
> |---|---|
> | Mass of dish, counterweight, and gearing | 180 kg |
> | Effective distance of mass from pivot point | 0.4 m |
> | Load inertia, J_L | 28.8 Kg.m² |
> | Maximum load torque, T_L | 612 Nm |

Control Motor: The motor characteristics are as follows:

> Size, 1.75 HP, 150 volt dc
> | | |
> |---|---|
> | Maximum speed (2000 rpm) | 209.44 rad/sec |
> | Continuous starting torque (T_{ST}) | 13.6 Nm |
> | Motor inertia (J_m) | 0.014 Kg.m² |

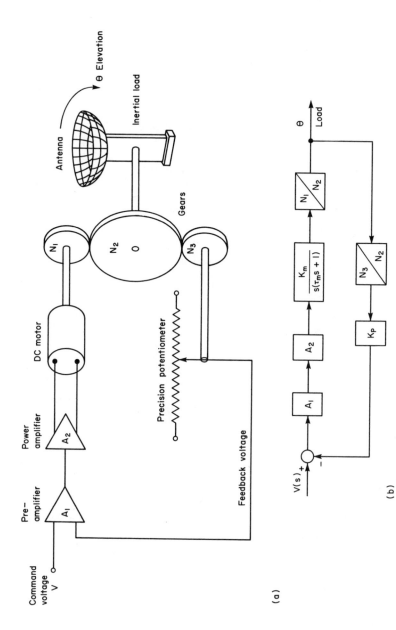

figure 10-46 (a) target acquisition system schematic diagram, (b) simplified block diagram

Motor Transfer Function: The motor gain constant, K_m, and the motor time constant, τ_m, are calculated in the following way:

Motor constant $\qquad K_m = \dfrac{\text{no load speed}}{\text{motor voltage}}$

hence $\qquad K_m = \dfrac{209.44}{150 \text{ V}}$ rad/sec

or $\qquad K_m = 1.4$ rad/volt-second

Time constant $\qquad \tau_m = \dfrac{\text{(total inertia) (no load speed)}}{\text{(starting torque)}}$

or $\qquad \tau_m = \dfrac{J_T \omega_0}{T_{ST}}$

where $\qquad \omega_0 = 209.44$ rad/sec

$\qquad T_{ST} = 13.6$ Nm

and $\qquad J_T = J_m + J_L'$

$$J_L' = \text{reflected load inertia} = J_L \left(\frac{N_1}{N_2}\right)^2$$

or $\qquad J_L' = \dfrac{28.8}{45^2} = 0.014$ Kg.m^2

Hence $\qquad \tau_m = \dfrac{(0.028)\,(209.44)}{13.6}$

or $\qquad \tau_m = 0.43$ s

Gear Ratios: The motor/load gear ratio is chosen from the ratio of motor starting torque to load torque:

$$\frac{N_1}{N_2} = \frac{13.6 \text{ Nm}}{612 \text{ Nm}} = \frac{1}{45}$$

The feedback pot. to load gear ratio is selected as a 1:1 ratio.

$$\frac{N_2}{N_3} = \frac{1}{1}$$

Feedback Potentiometer: This is a precision potentiometer with a maximum rotation of 356 deg. The transfer function is determined from a
knowledge that 12 volts of input drive signal will rotate the antenna, in
its reciprocating mode, to an angle of 60 deg. The feedback potentiometer must, therefore, produce 12 volts for 60 deg rotation. Hence,

$$K_p = 11.46 \text{ volts/radian}$$

Amplifiers: The gain of the pre-amp is adjusted initially to develop the
motor voltage from the 12 volt input; hence,

$$A_1 = 12.5$$

The power amplifier has a gain of unity and is used to provide current
demand at the motor voltage, in this case at least 20 A at start.

$$A_2 = 1$$

The numerical transfer functions can now be introduced into the system
block diagram, as illustrated in Figure 10-47.

The only operational specifications that we must accommodate are

1. The steady-state error of the system must not be greater than $\frac{1°}{3}$
   (0.00582 rad) when following a target at a maximum constant rate
   of 11½°/second (0.2 rad/s). This is equivalent to an input signal
   change of 2.292 V/s.
2. The system must be stable with an open-loop phase margin not
   less than 40 deg.

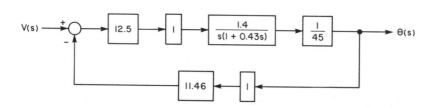

figure 10-47 the system block diagram

Step 1. Using the specified rate input 2.292 V/s, determine the steady-state error of the uncompensated system.

The steady-state error, $\epsilon_{ss}(t)$, is

$$\epsilon_{ss}(t) = \lim_{s \to 0} [s\mathcal{E}(s)] = \lim_{s \to 0} \left[s \cdot \frac{R(s)}{1 + GH(s)} \right]$$

where

$$R(s) = \frac{2.292}{s^2}$$

and

$$G(s)H(s) = \frac{4.46}{s(0.43s + 1)}$$

Hence,

$$\epsilon_{ss}(t) = \lim_{s \to 0} \left[s \cdot \frac{\dfrac{2.292}{s^2}}{1 + \dfrac{4.46}{s(0.43s + 1)}} \right]$$

or

$$\epsilon_{ss}(t) = \lim_{s \to 0} \left[\frac{\dfrac{2.292}{s}}{s + \dfrac{4.46}{(0.43s + 1)}} \right]$$

The steady-state error is, therefore,

$$\epsilon_{ss}(t) = \frac{2.292}{4.46} = 0.514 \text{ V}$$

Now, because 12 volts represents 60 deg, then 0.514 volts represents 2.57 deg (0.0449 rad). It is obvious that the system has far too large a tracking error and something must be done to reduce it from a little over 2½ deg to ⅓ deg or less. Since ⅓ deg (0.00582 rad) represents a steady-state error voltage of 0.0667 volt, the simplest solution would seem to be to increase the open-loop gain $G(s)H(s)$ from 4.46 to 34.38; then the steady-state error would be

$$\epsilon_{ss}(t) = \frac{2.292}{34.38} = 0.0667 \text{ V} = \frac{1°}{3}$$

This is easily done in practice by increasing the gain of the preamp from 12.5 to 96.4, which yields the block diagram in Figure 10-48.

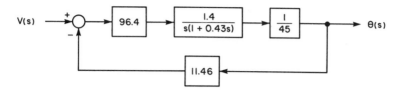

figure 10-48 the system with increased gain to reduce the steady-state error

Step 2. By means of an open-loop Bode diagram, determine the phase
margin of
(a) The uncompensated system.
(b) The system with increased gain.

Figure 10-49 shows that the original system has a phase margin of
41 deg (ϕ_{m_A}), which meets the original specification. However, when the
gain is increased to 34.38, the gain plot moves up by about 17.7 db and
the phase margin is reduced to 25 deg (ϕ_{m_B}).

Once again, we can see that by a simple adjustment of gain we
cannot meet all the system requirements, and we are constrained to
introduce a little more sophistication in an attempt to compensate the
system.

Step 3. At this point we have two options: We could introduce an integral controller action, which would eliminate the error completely but because of the marginal stability of this system would
probably destabilize the system. Alternatively, we could introduce a phase-lead network to increase the phase margin at the
gain crossover point (8.4 rad/s) in Figure 10-49. The second
option is the more logical choice and offers more chance of
success.

The phase-lead network is inserted into the system between the
pre-amp and the power amplifier, but for simplicity it is shown to be
located between the amplifying section and the motor in Figure 10-50.

The compensating network has a transfer function of

$$G_N = \alpha \cdot \frac{\tau_{LD}s + 1}{\alpha\tau_{LD}s + 1} \qquad (10.91)$$

Unfortunately, the dc gain of the network, α, will upset the overall
gain of the system unless we offset its effect by adjusting the amplifier
gain by $\dfrac{1}{\alpha}$, as shown in Figure 10-51.

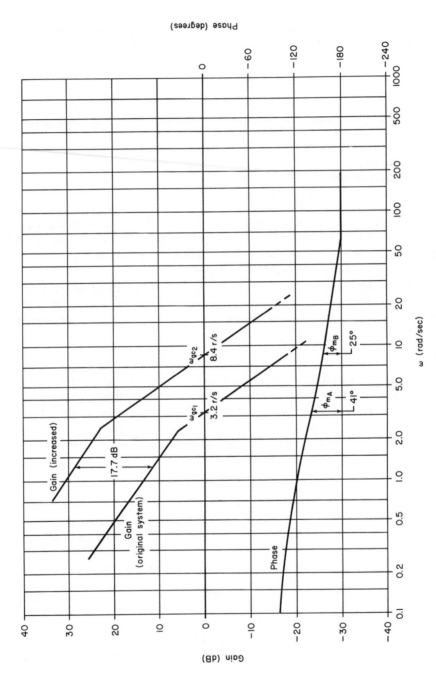

figure 10-49 bode diagram of the original system and the system when the gain is increased

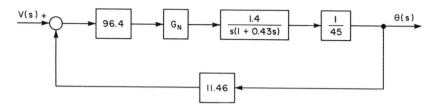

figure 10-50   the original system with increased gain of 96.4 and a phase-lead network, G_N, inserted for compensation

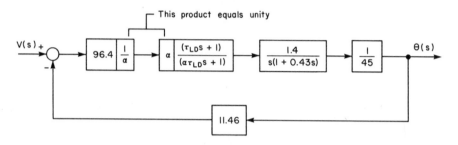

figure 10-51   adjustment of the amplifier gain to compensate for α

All we need to do now is to calculate α by using equation (10.75) and τ_{LD} by using equation (10.76).

In order not to operate the system at the limit of the specification, let us consider that the required phase margin is 45 deg and 40 deg is the absolute minimum. At the gain crossover frequency (from the increased gain curve in Figure 10-49), $\omega_{gc} = 8.4$ rad/s, and at this point we want to increase the phase by $\phi_{\max}$, where

$$\phi_{\max} = \phi_m - \phi_{m_B} \qquad (10.92)$$

where
$$\phi_{\max} = \text{maximum phase shift (leading)}$$
$$\phi_m = \text{desired phase margin (45 deg)}$$
and
$$\phi_{m_B} = \text{actual phase margin (25 deg)}$$

Hence we want

$$\phi_{\max} = 45° - 25° = 20° \qquad (10.93)$$

Since
$$\sin(\phi_{max}) = \frac{1-\alpha}{1+\alpha} \qquad (10.77)$$

then
$$\sin 20° = \frac{1-\alpha}{1+\alpha}$$

Therefore, $\qquad\qquad \alpha = 0.49$

Since
$$\omega_m = \frac{1}{\tau_{LD}\sqrt{\alpha}} = \omega_{gc} \qquad (10.78)$$

and $\qquad\qquad \omega_m = 8.4 \text{ rad/s}$

hence $\qquad\qquad \tau_{LD} = 0.17 \text{ s}$

We have now determined the phase-lead network transfer function

$$G_N = 0.49 \frac{(0.17s + 1)}{(0.0833s + 1)} \qquad (10.94)$$

The block diagram of the newly compensated control system is shown in Figure 10-52, and the Bode diagram is plotted in Figure 10-53 for

$$G(\omega)H(\omega) = \frac{34.38(0.17j\omega + 1)}{j\omega(0.0833j\omega + 1)(0.43j\omega + 1)} \qquad (10.47)$$

or
$$G(\omega)H(\omega) = \frac{7.25(j\omega + 5.88)}{j\omega(j\omega + 12)(j\omega + 2.32)} \qquad (10.47a)$$

At the gain crossover frequency $\omega_{gc} = 12.4$ rad/s, shown in Figure 10-53, the phase margin of the compensated system is measured as $\phi_{mc} = 40$ deg, indicating that the system meets the operating specifications.

Step 4. The root locus diagram is shown in Figure 10-54. At the open-loop gain of 34.38 the damping ratio $\zeta = 0.32$ and the system resonant frequency $\omega_n = 11.86$ rad/s.

The process of compensation for this antenna system was reasonably simple. All that was needed was that the forward loop gain be increased by a factor of 7.7 and that a phase-lead network be introduced into the forward path.

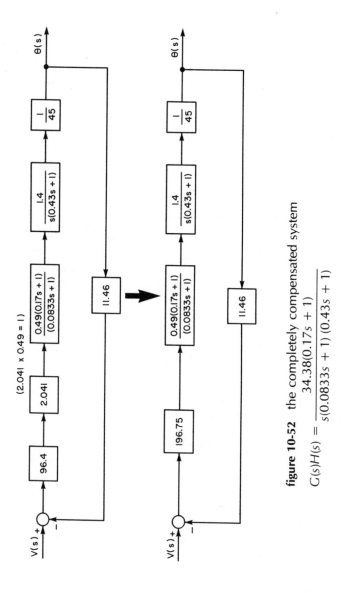

figure 10-52 the completely compensated system

$$G(s)H(s) = \frac{34.38(0.17s + 1)}{s(0.0833s + 1)(0.43s + 1)}$$

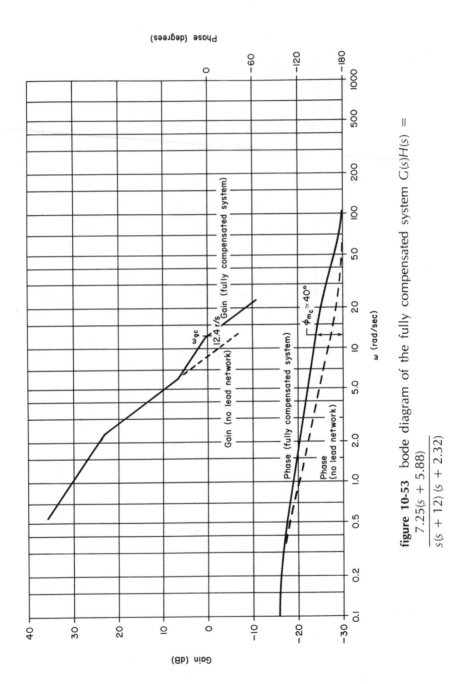

figure 10-53 bode diagram of the fully compensated system $G(s)H(s) =$

$$\frac{7.25(s + 5.88)}{s(s + 12)(s + 2.32)}$$

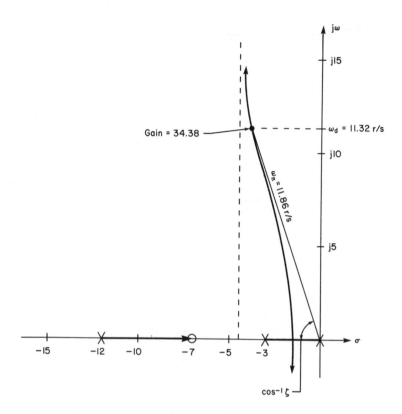

figure 10-54 the root locus diagram of the compensated system

10-4 SUMMARY

It should be emphasized that design of real systems is not quite as simple as suggested in this chapter. For clarity it was necessary to omit some very tangible real-life problems. However, the object was to introduce the logical step-by-step procedure of system compensation. When an engineering system malfunctions, servicing is made much easier if it is understood why various parts of the system have been incorporated and how they are meant to interact with the rest of the system.

PROBLEMS

Section 10-2.3

1. For a proportional-plus-integral controller, where $K_1 = 4$ and $K_2 = 5$ s^{-1},

   (a) Determine the magnitude of the poles and zeros.
   (b) Sketch the Bode diagram.
   (c) Using the block diagram of any type 0 system, show that this controller action eliminates the steady-state position error.

Section 10-2.4

2. For a proportional-plus-derivative controller, where $K_1 = 4$ and $K_3 = 0.8$ sec,

   (a) Determine the magnitude of the poles and zeros.
   (b) Sketch a Bode diagram.
   (c) Using the block diagram of any second-order system with $\zeta < 1$, show that this controller action increases the system damping.

Section 10-3.1

3. (a) For the phase-lead circuit of Figure 10-37(a), where $R_1 = 1$ K, $R_2 = 10$ K, and $C = 0.1$ μF, determine
      (i)   The dc gain (α).
      (ii)   The transfer function.
      (iii)   The maximum phase lead (ϕ_{max}).
      (iv)   The geometric mean frequency (ω_m) at which ϕ_{max} is introduced.

   (b) For the phase-lag circuit of Figure 10-39(a) when $R = 1$ K, $R_2 = 10$ K, and $C = 0.1$ μF, determine
      (i)   The high frequency gain (α).
      (ii)   The transfer function.
      (iii)   The maximum phase lag (ϕ_{max}).
      (iv)   The mean frequency (ω_m) at which ϕ_{max} is introduced.

4. A process control system has a forward loop transfer function of

$$G(s) = \frac{0.3125}{(0.25s + 1)\,(0.0625s + 1)}$$

and unity feedback. Design a cascade device to be introduced into the forward loop to achieve

(a) $\zeta = 0.8$.
(b) $\omega_d = 6$ rad/s.
(c) Minimum steady-state position error.

5. The schematic diagram below shows an arsenic trisulfide lens directing the sun's rays onto a gimballed mirror. The mirror reflects the focused beam onto and through a quadrature array detector mounted on a solar spectrometer. Infrared in the beam passes through the hole, but the red light excites the quadrature array detector and produces 0.1 V/deg of mirror rotation from the array. The difference amplifier has a gain of unity, and the power amplifier has a gain of A_2. The dc motor drives the mirror through a 50:1 gear reduction.

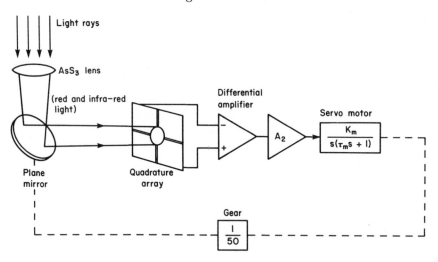

If the motor transfer function parameters are K_m = 2.4 rad/volt-s and τ_m = 0.1 s, determine the optimum value for A_2 and design a compensator which will allow the system when it is airborne to track input sunlight when it is rotating at 18 deg/ s with an error no greater than 10 minutes of arc. The system phase margin should be at least 60 deg. and the damping ratio limited such that $0.4 < \zeta < 0.7$.

6. Suppose that the output flow rate for the system shown in Figure 10-13 is required to change in linear fashion from 20 gpm to 100 gpm in 5 seconds with an error no greater than 1 gpm. Redesign the compensator to meet the original specifications and to accommodate this new one.

7. Repeat Problem 6 above for the system shown in Figure 10-29.

8. A unity feedback control system is shown below.

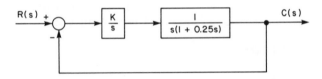

Design a cascade controller so that the system is able to follow an acceleration input of 0.1 unit/ s^2 with a maximum steady-state error of 1 percent and the phase margin of the system is no less than 60 deg.

part

practical design
of OP-AMP
circuity
and modelling of
control systems

4

system modelling: part i—the operational amplifier

I took a kettle large and new
fit for the deed I had to do...

Lewis Carroll, Through the Looking Glass

11-1 INTRODUCTION

In the engineering industry, the assembly and testing of prototype equipment is a necessary and very important part of the production process. However, with the advent of computers, much of the information that would previously be gathered from prototype testing can now be obtained via computer simulations, using a mathematical model of the equipment. Such system modelling can be performed by using a digital or analog or hybrid computer.

Digital computer simulations will not be considered here, and in this chapter we will confine ourselves to learning how to construct and breadboard the major circuits used in an analog computer. In Chapter 12 we will go on to describe how we use these circuits (in an analog computer) to "model" a control system. First, however, we must learn how to build electronic circuits that will perform the dynamic operations typically encountered in a control system block diagram.

If we build circuits that simulate each block and then connect the circuits in the same way as in the block diagram, we will produce an analog model of the control system. A more detailed description of the procedure for constructing analog models is given in Chapter 12.

The circuits, which we will use to simulate various dynamic operations, each contain an operational amplifier, an input, an output, and feedback resistors and capacitors; in addition, each circuit requires a power supply.

Before the design techniques are demonstrated, it would be wise to familiarize ourselves with operational amplifier (op-amp) circuits at the block diagram level.

11-2 THE DIFFERENTIAL AND OPERATIONAL AMPLIFIERS

The heart of the op-amp is a high gain differential amplifier. If the op-amp is in IC (integrated circuit) form there will be other circuits, operating in concert with the differential amplifier, which are used for signal conditioning and biasing. The basic differential amplifier is shown in Figure 11-1(a) and (b). The circuit diagram shows six essential terminals: two for input (1 and 2), two for output (3 and 4), and two for power supply voltages (V_{CC} and $-V_{EE}$). The block diagram shows only the input and output terminals.

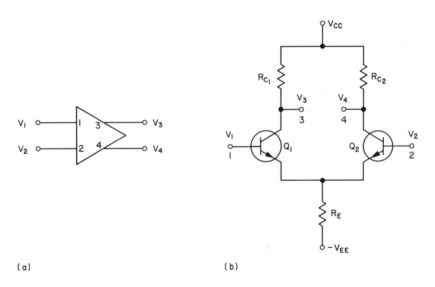

(a)   (b)

figure 11-1 the differential amplifier: (a) block diagram, (b) simple circuit diagram

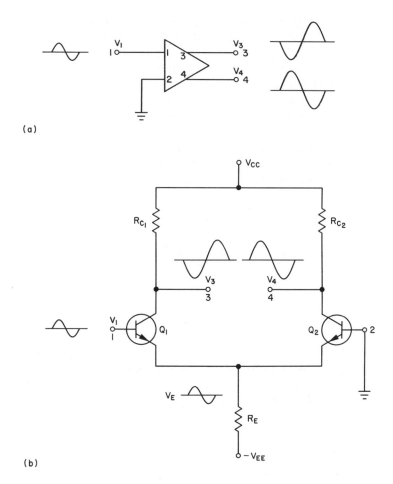

figure 11-2 (a) the block diagram with signal applied at input 1 with input 2 grounded, (b) the circuit diagram showing the same condition

The operation is quite simple; consider Figure 11-2. In the diagram input 2 has been grounded and a signal is applied at input 1. When the voltage rises on the base of Q_1, the current through Q_1 increases, causing the collector of Q_1 to fall and the voltage across the common emitter resistor, R_E, to rise. V_3 is, therefore, an amplified and inverted (opposite in phase) form of the input to Q_1. The base of Q_2 is fixed, but the voltage at the emitter of Q_2 rises; therefore, the base to emitter voltage on Q_2 drops; hence the collector current decreases. As the collector current through Q_2 decreases, the collecter voltage rises and an amplified, noninverted form of the input to Q_1 appears at V_4.

If the input ground and signal were interchanged, the output voltages would be interchanged, but otherwise identical.

If terminal V_4 were removed and the amplifier made to have very large gain, in the region of 200,000 or more, we would have a basic op-amp. Figures 11-3(a) and (b) show the inverting and noninverting effects of the two inputs to the differential amplifier in relation to one output. Figure 11-3(c) shows the block diagram of the basic op-amp with its single output and two inputs. One input is always marked minus (−) indicating that any signal connected to it will be inverted at the output; the other input is always marked plus (+), indicating that any signal connected to it will not be inverted at the output.

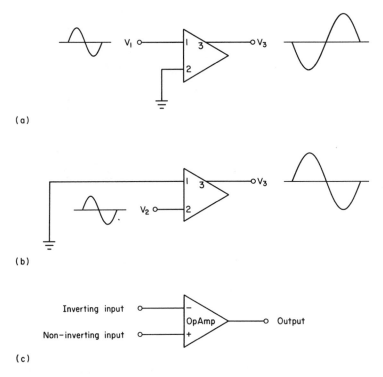

(a)

(b)

(c)

figure 11-3 (a) the differential amp with inverted output, (b) the differential amp with noninverted output, (c) the basic op-amp showing the inverting and noninverting inputs

The op-amp has some very important characteristics that are useful to remember. In fact, a number of them are used to greatly simplify circuit analysis when one is deriving the operational circuit relationships. A few of the more important characteristics and typical values are as follows:

1. Very high differential open loop gain (> 100 db).
2. High differential input resistance (2 MΩ).
3. Low output resistance (75 Ω).
4. Wide power bandwidth (50 KHz).
5. Fast slew rate (0.5 V/μs).

11-3 SIMPLE MATHEMATICAL OPERATIONS

Because op-amp circuits can, among other things, add, subtract, multiply, divide, integrate, or differentiate electrical input signals, they can be used to perform almost any mathematical operation. They are made to perform these operations by the appropriate connection of resistors and capacitors to the external pins of the IC package. Examples of typical IC op-amp packages are shown in Figures 11-4(a) and (b). Let us now look at some of the mathematical operations that can be performed by using op-amps.

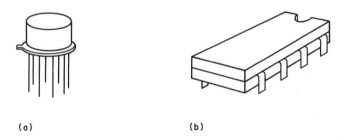

(a)   (b)

figure 11-4 typical IC op-amp packages: (a) TO-5 style, (b) dual-in-line style

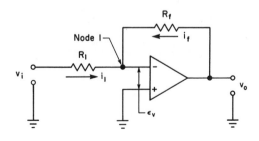

figure 11-5 the inverting amplifier

11-3.1 scale changing or constant gain multiplier

Sometimes it is necessary to multiply a parameter by a constant value. For example, $y = 5x$ indicates that the independent variable x is always multiplied by the constant 5 to yield the dependent variable y. The constant is positive in this case, but it can also be negative. Multiplication in this way is accomplished by the simplified circuit shown in Figure 11.5.

As we said earlier, the op-amp has very high input resistance, and for analysis purposes it is considered that negligible current flows into the inverting input; hence the steady-state error voltage ϵ_v is approximately equal to zero. This being the case, there is a virtual ground at node 1. Applying Kirchhoff's current law to this circuit, we can say

$$i_1 + i_f = 0 \qquad (11.1)$$

which yields

$$\frac{v_i - \epsilon_v}{R_1} + \frac{v_0 - \epsilon_v}{R_f} = 0 \qquad (11.2)$$

Since
$$\epsilon_v = 0,$$

then
$$\frac{v_0}{v_i} = -\frac{R_f}{R_1} \qquad (11.3)$$

If, for example, $R_f = 5R_1$, then

$$v_0 = -5v_i \qquad (11.4)$$

Hence the output voltage is five times larger than the input voltage but negative (or opposite in phase); i.e., it is an inverting amplifier. The block diagram for this feedback circuit is illustrated in Figure 11-6.

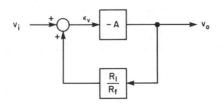

figure 11-6 the block diagram for the inverting amplifier circuit

Block diagram reduction of Figure 11-6 yields

$$\frac{v_0}{v_i} = \frac{-A}{1 - (-A\frac{R_1}{R_f})} \qquad (11.5)$$

but since the op-amp has a very large forward loop gain $-A$ (the negative sign means inverting) then equation (11.5) reduces to

$$\frac{v_0}{v_i} = \frac{-A}{A\frac{R_1}{R_f}} \qquad (11.5a)$$

which further reduces to equation (11.3), namely,

$$\frac{v_0}{v_i} = -\frac{R_f}{R_1} \qquad (11.3)$$

The noninverting amplifier is shown in simplified circuit form, together with its block diagram in Figures 11-7(a) and (b).

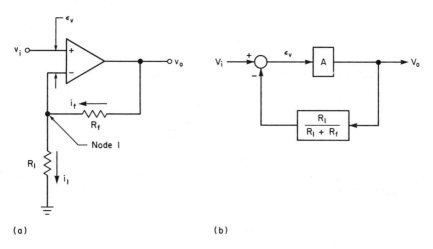

(a)   (b)

figure 11-7 the noninverting amplifier: (a) simplified circuit, (b) block diagram

Due to the very high input resistance, the error voltage is once again approximately equal to zero, which allows us to write the equation

$$v_i - \epsilon_v = v_{R_1} \qquad (11.6)$$

Hence
$$v_i = v_{R_1} \qquad\qquad (11.6a)$$

Applying Kirchoff's current law at node 1 yields

$$i_f - i_1 = 0 \qquad\qquad (11.1)$$

Hence

$$\frac{v_0}{R_1 + R_f} - \frac{v_{R_1}}{R_1} = 0 \qquad\qquad (11.7)$$

Therefore,

$$\frac{v_0}{v_{R_1}} = \frac{R_1 + R_f}{R_1} \qquad\qquad (11.7a)$$

and since

$$v_i = v_{R_1}$$

then

$$\frac{v_0}{v_i} = 1 + \frac{R_f}{R_1} \qquad\qquad (11.8)$$

Block diagram reduction of Figure 11-7(b) yields

$$\frac{v_0}{v_i} = \frac{A}{1 + \left(A \dfrac{R_1}{R_1 + R_f}\right)} \qquad\qquad (11.9)$$

but since the op-amp has very large forward loop gain A (positive value indicates noninverting), equation (11.9) reduces to

$$\frac{v_0}{v_i} = \frac{A}{A \dfrac{R_1}{R_1 + R_f}} \qquad\qquad (11.9a)$$

which further reduces to equation (11.8), namely,

$$\frac{v_0}{v_i} = 1 + \frac{R_f}{R_1} \qquad\qquad (11.8)$$

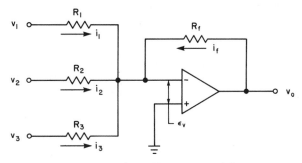

figure 11-8 the summing amplifier

11-3.2 addition

Addition of signals is accomplished by a circuit similar to the one illustrated in Figure 11-8. By using the previously stated assumptions, we have

$$i_1 + i_2 + i_3 + i_f = 0 \qquad (11.10)$$

Hence

$$\frac{v_1}{R_1} + \frac{v_2}{R_2} + \frac{v_3}{R_3} + \frac{v_0}{R_f} = 0 \qquad (11.11)$$

Therefore,

$$v_0 = -\left[\frac{R_f}{R_1}v_1 + \frac{R_f}{R_2}v_2 + \frac{R_f}{R_3}v_3\right] \qquad (11.12)$$

and if we let $R_1 = R_2 = R_3 = R_f$, the output is simply the inverted sum of all the input signals, namely,

$$v_0 = -[v_1 + v_2 + v_3] \qquad (11.12a)$$

11-3.3 subtraction

The difference of two variables represented by the signals, v_1 and v_2 can be accomplished by a circuit very similar to the one shown in Figure 11-9. Because v_1 does not equal v_2, then v_A will not equal v_B, and hence an extremely small voltage, v_{in}, will be sensed between the two inputs of the amplifier. However, as the input resistance of the op-amp is very

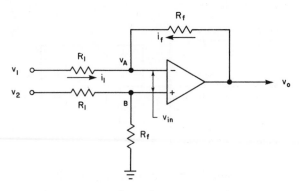

figure 11-9 the subtractor circuit

high, we can still assume that no current flows into either of the op-amp inputs; hence

$$i_1 + i_f = 0 \qquad (11.1)$$

Therefore,

$$\frac{v_1 - v_A}{R_1} + \frac{v_0 - v_A}{R_f} = 0 \qquad (11.13)$$

Hence,

$$v_A = \frac{R_f v_1 + R_1 v_0}{R_f + R_1} \qquad (11.14)$$

Now if we consider the effect of v_2 at the input of the amplifier, we can state that

$$v_B = v_2 \frac{R_f}{R_1 + R_f} \qquad (11.15)$$

In addition, we can also state that

$$v_{in} = v_B - v_A \qquad (11.16)$$

and since $$v_0 = A v_{in} \qquad (11.17)$$

then $$v_0 = A(v_B - v_A) \qquad (11.18)$$

If we combine equations (11.14) and (11.15) and substitute appropriately into equation (11.18), we obtain

$$v_0 = A \left[\left(\frac{v_2 R_f}{R_1 + R_f} \right) - \left(\frac{R_f v_1 + R_1 v_0}{R_f + R_1} \right) \right] \qquad (11.19)$$

Solving for v_0 in equation (11.19) and knowing that $AR_1 \gg (R_1 + R_f)$ yields

$$v_0 = \frac{R_f}{R_1} (v_2 - v_1) \qquad (11.20)$$

If $R_1 = R_f$, then the output is simply the difference of the two input voltages.

11-3.4 integration

A circuit that will integrate an input voltage signal is shown in its simplest form in Figure 11-10(a); the equivalent block diagram is given in Figure 11-10(b). The feedback resistor of the inverting amplifier has been replaced by a capacitor, C_f. The input resistor, R, remains unchanged. Using this circuit, we find that the output voltage, v_0, is the time integral of the input voltage, v_i.

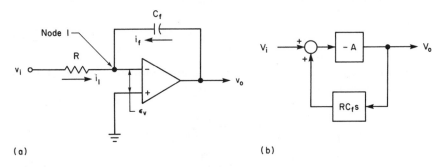

(a)   (b)

figure 11-10 the integrator: (a) simple schematic diagram, (b) block diagram

For simple analysis we can once again assume that

$$i_1(t) + i_f(t) = 0 \qquad (11.1)$$

because of the virtual ground at node 1, also

$$i_1(t) = \frac{v_i(t)}{R}$$

and
$$i_f(t) = C_f \cdot \frac{dv_0(t)}{dt}$$

Substituting these values into equation (11.1), we get

$$\frac{v_i(t)}{R} + C_f \cdot \frac{dv_0(t)}{dt} = 0 \qquad (11.21)$$

Therefore,

$$-\frac{v_i(t)}{RC_f} = \frac{dv_0(t)}{dt} \qquad (11.21a)$$

If we integrate both sides of equation (11.21a) with respect to time, we obtain

$$v_0(t) = -\frac{1}{RC_f} \int^t v_i(t)\, dt + V_c \qquad (11.22)$$

where V_c is the initial charge on the capacitor C and the product RC_f is the time constant of the circuit (usually symbolized by τ). Equation (11.22) is perhaps more properly stated as

$$v_0(t) = -\frac{1}{\tau} \int_{t_0}^t v_i(t)\, dt + V_c(t_0) \qquad (11.22a)$$

If we approach the analysis by block diagram reduction, Figure 11.10(b) will reduce to one block in the following way.

$$\frac{V_0(s)}{V_i(s)} = \frac{-A}{1 - (-ARC_f s)} \qquad (11.23)$$

Hence
$$\frac{V_0(s)}{V_i(s)} = \frac{-A}{1 + ARC_f s} \qquad (11.23a)$$

Since A is extremely large, then $ARC_f \gg 1$ and equation (11.23a) becomes

$$\frac{V_0(s)}{V_i(s)} = -\frac{A}{ARC_f s} \qquad (11.24)$$

which in turn reduces to

$$\frac{V_0(s)}{V_i(s)} = -\frac{1}{RC_f s} \qquad (11.24a)$$

Hence $$V_0(s) = -\frac{V_i(s)}{RC_f s} \qquad (11.24b)$$

or $$v_0(t) = -\frac{1}{RC_f} \int^t v_i(t)\, dt \qquad (11.22)$$

EXAMPLE 11.1

Determine the output voltage of the integrator shown in the accompanying diagram when the input is a square wave $\pm$ 2 volts in magnitude and 5 KHz frequency.

1. $R = 1$ MΩ, $C = 1$ μF
2. $R = 10$ kΩ, $C = 0.022$ μF

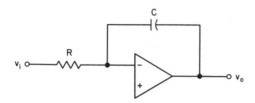

solution:

The input wave form is

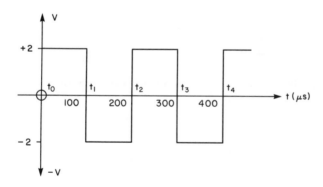

Provided that $v_i(t)$ is continuous between t_i and t, then

$$v_0(t) = -\frac{1}{T}\int_{t_i}^t v_i(t)\, dt + V_c(t_i)$$

Case (i)

$$T = (1\ M\Omega)(1\ \mu F) = 1\ s$$

then
$$v_0(t) = -1 \int_{t_i}^{t} v_i(t)\, dt + V_c(t_i)$$

and when
$$0 \leqslant t < 100 \ \mu s$$
$$v_0(t) = - \int_0^{t} 2\, dt + V_c(t_0)$$

Hence
$$v_0(t) = -2t + V_c(t_0)$$

where $V_c(t_0)$ is the charge on the capacitor C at time t_0. The factor -2 indicates that the input voltage is changing at -2 volts per second. If the initial charge on the capacitor is zero, then

$$v_0(t_0) = -2 \cdot 0 + 0 = 0 \ V$$

and
$$V_c(t_1) = v_0(t_1) = -2 \cdot 100 \times 10^{-6} + 0 = -200 \ \mu V$$

Now when
$$100 \ \mu s < t < 200 \ \mu s$$

the input changes to -2 volts; hence

$$v_0(t) = - \int_{t_1}^{t} - 2\, dt + V_c(t_1)$$

Therefore,
$$v_0(t) = 2(t - t_1) + v_0(t_1)$$

Therefore,
$$v_0(t_2) = 2(10^{-4}) + (-200 \ \mu V) = 0$$

when $t = 300 \ \mu s$, the integrator output will have once again ramped downward to a value of

$$v_0(t_3) = -200 \ \mu V$$

Thus the output of the integrator will be

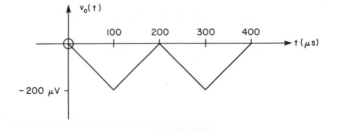

Case (ii)

For $R = 10$ KΩ and $C = 0.022$ μF, the output voltage is

$$v_0(t) = -4545.45 \int^t v_i \, dt$$

Using the same reasoning as in Case 1 we can see that

when $\qquad$ $t = 0$, $\qquad$ $v_0 = 0$ V

when $\qquad$ $t = 100\mu$s $\qquad$ $v_0 = -0.91$V

when $\qquad$ $t = 200\mu$s $\qquad$ $v_0 = 0$ V

when $\qquad$ $t = 300\mu$s $\qquad$ $v_0 = -0.91$ V

Thus the new output waveform is

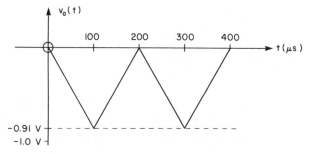

Notice that both outputs are inverted triangular waves and that with a large value of RC (time constant) the rate of change of the output is small, 2 volts per second in Case 1. When the RC is reduced, the rate of change of the output increases [> 4500 volts per second in Case 2].

EXAMPLE 11.2

For the circuit shown below, if three dc voltages are applied simultaneously, determine the output voltage rate of change and the output voltage after 100 μs.

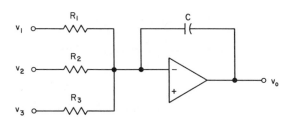

$$v_1 = 2 \text{ volts}, \qquad v_2 = 7 \text{ volts}, \qquad v_3 = 12 \text{ volts}$$

$$R_1 = 12 \text{ K}\Omega, \qquad R_2 = 10 \text{ K}\Omega, \qquad R_3 = 5.6 \text{ K}\Omega$$

$$C = 0.22 \ \mu F$$

Consider that the initial charge on C is zero and let $t_1 = 10^{-4}$ s.

solution:

$$v_0(t_1) = - \left[\frac{1}{R_1 C} \int_0^{t_1} v_1 \, dt + \frac{1}{R_2 C} \int_0^{t_1} v_2 \, dt + \frac{1}{R_3 C} \int_0^{t_1} v_3 \, dt \right]$$

Hence $\quad v_0(t_1) = - \left[378.97 \int_0^{t_1} 2 \, dt + 454.55 \int_0^{t_1} 7 \, dt + 811.69 \int_0^{t_1} 12 \, dt \right]$

Therefore, $\qquad v_0(t_1) = - \left[757.94 \, t_1 + 3181.85 t_1 + 9740.28 t_1 \right]$

or $\qquad\qquad v_0(t_1) = -13680 t_1$

Therefore, Output rate of change $= -13.68$ KV/second and $v_0(t_1)$, the output magnitude at $t = 100 \ \mu s$, is -1.37 V.

EXAMPLE 11.3
Consider that a simple integrator with input resistor R and feedback capacitor C has an input voltage of $E \sin (\omega t)$ Determine the output voltage and comment upon some of its properties.

solution:
The circuit diagram is as follows:

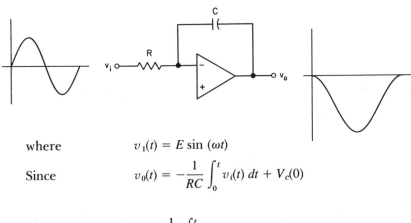

where $\qquad\qquad v_1(t) = E \sin (\omega t)$

Since $\qquad\qquad v_0(t) = -\frac{1}{RC} \int_0^t v_i(t) \, dt + V_c(0)$

then $\qquad\qquad v_0(t) = -\frac{1}{RC} \int_0^t E \sin (\omega t) \, dt + V_c(0)$

which yields $v_0(t) = -\dfrac{1}{RC} \cdot E \cdot \dfrac{1}{\omega} \cdot \left[-\cos \omega t \right]_0^t + V_c(0)$

or $v_0(t) = \dfrac{E}{\omega RC} \left[\cos \omega t - 1 \right] + V_c(0)$

Comments

1. If the capacitor is initially discharged, the output is always below the zero voltage line. (The output can be centered about the zero voltage line either by adding a bias voltage at the summing junction of the next stage or by starting with an initial charge on the capacitor of $E/(\omega RC)$ volts.)

2. The output is at the same frequency (ω) as the input.

3. The output is phase-shifted by 90 deg leading.

4. The maximum amplitude of the output is determined by the unitless product of frequency and time constant, ωRC.

5. The time constant is fixed when R and C are chosen; the output maximum amplitude is then dependent upon and inversely proportional to frequency.

6. This last characteristic makes the integrator a good noise suppressor.

11-3.5 differentiation

Although this circuit is not used in analog computers because of its inherent instability and susceptibility to noise, it is often used, in a more sophisticated version than Figure 11-11(a), as a high pass filter. The block diagram equivalent of the simplified circuit is illustrated in Figure 11-11(b).

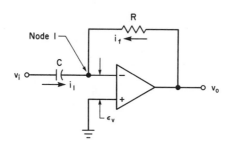

(a)

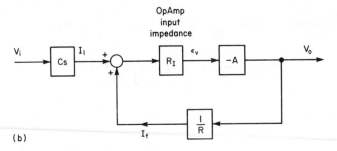

figure 11-11   the differentiator: (a) simple schematic diagram, (b) block diagram

Summing the currents at node 1 yields

$$i_1(t) + i_f(t) = 0 \qquad (11.1)$$

Because $\epsilon_v = 0$, we can therefore state that

$$C\frac{dv_i(t)}{dt} + \frac{v_0(t)}{R} = 0 \qquad (11.25)$$

Hence
$$v_0(t) = -RC\frac{dv_i(t)}{dt} \qquad (11.25a)$$

or
$$v_0(t) = -\tau\frac{dv_i(t)}{dt} \qquad (11.26)$$

where $\tau = RC$ and is the time constant of the circuit.

The block diagram of Figure 11-11(b) reduces in the usual way, so that

$$\frac{V_0(s)}{V_i(s)} = \frac{-AR_I}{1 - \left(-A \cdot \dfrac{R_I}{R}\right)} \cdot sC \qquad (11.27)$$

$$\frac{V_0(s)}{V_i(s)} = \frac{-AR_I}{1 + AR_I/R} \cdot sC \qquad (11.27a)$$

A and R_I are both very large; therefore, $\dfrac{AR_I}{R} \gg 1$ and equation (11.27a)

becomes
$$\frac{V_0(s)}{V_i(s)} = \frac{-AR_I(s)}{\dfrac{AR_I}{R}} \qquad (11.28)$$

Therefore,
$$\frac{V_0(s)}{V_i(s)} = -RCs \qquad (11.28a)$$

or
$$V_0(s) = -RCsV_i(s) \qquad (11.28b)$$

Hence
$$v_0(t) = -RC\frac{dv_i(t)}{dt} \qquad (11.25a)$$

EXAMPLE 11.4

Repeat Example 11.3 for a differentiator circuit.

solution:

A simple differentiator has the following configuration:

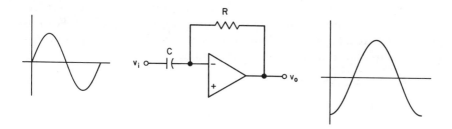

where
$$v_i(t) = E\sin(\omega t)$$

Since
$$v_0(t) = -RC\frac{dv_i(t)}{dt}$$

then
$$v_0(t) = -RC\frac{d(E\sin \omega t)}{dt}$$

which yields
$$v_0(t) = -E\omega RC\cos(\omega t)$$

Comments:

1. The output is centered about the zero voltage line.
2. The output is at the same frequency (ω) as the input.
3. The output is phase-shifted by 90 deg lagging.
4. The maximum amplitude of the output is dependent upon the unitless product of the frequency and time constant (ωRC).
5. The time constant is fixed when R and C are chosen; therefore, the maximum amplitude of the output is directly proportional to the input frequency.
6. This last characteristic makes the differentiator very susceptible to noise, but it is a good high pass filter.

11-4 OP-AMP CIRCUIT DESIGN

Designing and "breadboarding" op-amp circuits is both fascinating and instructive. In the rest of this chapter we will look at how to breadboard the major op-amp circuits used in an analog computer when designing and modelling control systems.

The two main op-amp circuits found in any analog computer are

1. A variable gain inverting amplifier. (A special case of this type of circuit is when the gain of the amplifier is −1 and the circuit is then referred to simply as an inverter).
2. A variable gain inverting integrator.

Both of these circuits may be combined if desired; a more detailed description of combination circuits will be given in Chapter 12. For the present, let us take a closer look at how we would breadboard an inverting amplifier and an inverting integrator.

11-4.1 the inverting amplifier

Consider an amplifier circuit that has the following requirements:

Voltage gain	−50
Maximum input signal	20 mV(rms)
Input resistance of circuit	> 1 kΩ
Minimum load resistance	1 kΩ
Op-amp used Fairchild	μA741C (see Appendix V)

The circuit diagram is shown in Figure 11-12.

We will follow a step-by-step approach in determining the values of all the resistors in this circuit.

Step 1. The circuit requires a power supply whose voltage must be less than the absolute maximum value specified in the data sheet; e.g., from the μA741 data sheet,

Maximum supply voltage:	±18 Vdc
Recommended safe voltage:	±15 Vdc

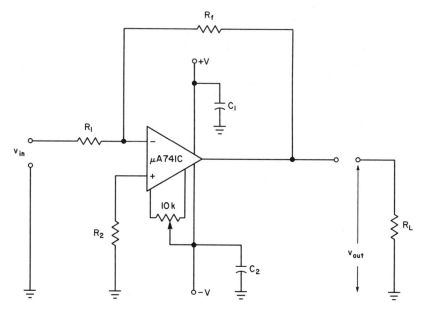

figure 11-12 a practical op-amp amplifier circuit

A well-regulated power supply can easily provide a voltage to within ± 1 percent of the full voltage range. However, spurious high frequency voltages are often impressed on the power supply voltage, causing unwanted interference; it is, therefore, necessary to provide a pathway around the op-amp circuit by using bypass capacitors. Typical values for such bypass capacitors are 0.01 μF to 0.1 μF. The power supply circuit connections will be as shown schematically in Figure 11-13.

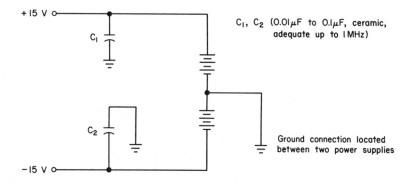

figure 11-13 construction of an op-amp power supply from two dc sources

Step 2. The value of the input resistor, R_1, depends upon the input voltage signal, V_i, and the input bias current, I_b. Because the input transistors of the first differential stage of the op-amp need to be forward-biased they draw a small current from the source. This input bias current causes a voltage drop across the input resistors and must be allowed for in the design. In practice, the voltage drop should not be more than 5 percent of the input signal; i.e., for

$$\text{Input signal} = 20 \text{ mV}$$

then 5 percent of input, $\qquad V_{R_1} = 1 \text{ mV}$

Also from the data sheet the maximum input bias current is 500 nA, and since

$$I_b R_1 = V_{R_1}$$

where V_{R_1} is the voltage drop across R_1,

then $\qquad\qquad R_1 = \dfrac{1 \times 10^{-3}}{500 \times 10^{-9}}$

i.e., $\qquad\qquad R_1 = 2000 \ \Omega$

Step 3. The feedback resistor, R_f, can be calculated from a knowledge of the gain required. The required closed-loop voltage gain A_{VCL} was given as -50.

Since $\qquad\qquad A_{VCL} = -\dfrac{R_f}{R_1}$ $\qquad\qquad\qquad$ (11.3)

then $\qquad\qquad -50 = -\dfrac{R_f}{R_1}$

Hence $\qquad\qquad R_f = 100 \text{ k}\Omega$

Step 4. The two input transistors of any op-amp can never be made identical, so they both draw different amounts of input bias current, I_b. The difference between these two values is called *input offset current*. The smaller the value, the better is the matching in the first differential stage. The offset current produces an offset voltage across the input resistors which can, however, be effectively reduced by connecting resistor R_2 to ground, as illustrated in Figure 11-12. The value of R_2 is given by

$$R_2 = \frac{R_1 R_f}{R_1 + R_f}$$

Therefore
$$R_2 = \frac{(2k)(100k)}{2k + 100k}$$

hence
$$R_2 = 1.96k\Omega$$

Since a 2 kΩ resistor is a standard value, it should be used rather than using something extravagant like a trim potentiometer to achieve 1.96 kΩ.

Step 5. It is essential that there be no output voltage when the input signal is zero. To achieve this and also to reduce offest further, the manufacturer specifies the use of a 10 kΩ potentiometer connected across the two offset null terminals of the op-amp. The slider of the potentiometer should be connected to the negative power supply terminal, V^-, as shown in Figure 11-12.

Step 6. The bandwidth of the circuit can now be calculated. The bandwidth is the range of frequency from dc (or close to it) to some upper limit, in which the gain does not drop off by more than 3 db. The upper frequency limit (obtained from the data sheet graph on open-loop voltage gain versus frequency) is given by

$$f_u = \frac{f_{op} A_{VOL} R_1}{R_f} \qquad (11.30)$$

where A_{VOL} is the open-loop voltage gain of the op-amp
and f_{op} is the first-pole frequency of the op-amp.

Thus f_{op} is the frequency at which the gain curve begins to fall away at 6 db/octave from the maximum open-loop gain level A_{VOL}.

From the data sheet: $f_{op} = 8$ Hz (approximately)

$$A_{VOL} = 3 \times 10^5 \text{ (approximately)}$$

Hence
$$f_u = \frac{8 \times 3 \times 10^5 \times 2}{100} \text{ Hz}$$

Therefore,
$$f_u = 48 \text{ kHz}$$

The closed-loop bandwidth of the amplifier is, therefore, approximately 48 kHz.

Step 7. Because the steady-state error voltage between the inverting and noninverting input terminals is practically zero, a virtual ground is said to exist at the inverting input. Hence the circuit input resistance is effectively equal to R_1, as shown in Figure 11-14.

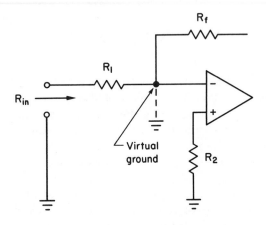

figure 11-14   the effective circuit input resistance, R_1

A more accurate value for R_{in} is given by

$$R_{in} = R_1\left(1 + \frac{R_f}{A_{VOL}R_1}\right) \qquad (11.31)$$

which in our case yields a difference of only 0.33 Ω and can, therefore, be neglected. Thus R_{in} is 2 kΩ and well within the circuit specification of 1 kΩ minimum.

Step 8. The closed-loop output resistance, R_{out}, is determined from the following relationship:

$$R_{out} = \frac{R_0}{1 + A_{VOL}\left(\dfrac{R_1}{R_1 + R_f}\right)} \qquad (11.32)$$

where R_0 is the op-amp's open-loop output resistance and is given as 75 Ω in the data sheet.

Hence
$$R_{out} = \frac{75}{1 + 3 \times 10^5 \left(\dfrac{2k}{2k + 100k} \right)}$$

Therefore, $\qquad R_{out} = 0.01 \ \Omega$

Usually R_{out} is considered zero for all practical purposes.

Step 9. When using a power supply voltage of ± 15 Vdc and a load $\geqslant 2$ kΩ, the manufacturer's data sheet recommends that the output voltage be confined to the range ± 13 volts. Because our circuit has a maximum input of 20 mVrms and the closed-loop gain is -50, the output voltage range will be

$$\text{Output voltage range} = \pm \frac{\sqrt{2} \times 20 \times 50}{1000}$$

or
$$= \pm 1.414 \ \text{V}$$

which is well within the recommended range.

Step 10. Naturally, the output power delivered to the load depends upon the effective value of load resistance (or input resistance of the next stage in the circuit).

There is, however, a maximum value of power which can be drawn from this circuit. The method of calculating the output power, P_{out}, is very simple; P_{out} is determined by subtracting the op-amp power consumption from the op-amp internal power dissipation. Both values are listed in the data sheet. From the data sheet:

Maximum internal power dissipation	500 mW
Maximum op-amp power consumption	85 mW
Maximum power out (P_{out})	415 mW

Since the minimum value of load resistance is given as 1 kΩ, then the maximum power demand made upon this circuit will be

$$P_0 = \frac{V_0^2(\text{rms})}{1 \ \text{k}\Omega} = \frac{1}{1000} = 1 \ \text{mW}$$

and is well under the maximum available power. Thus the completely designed inverting amplifier, having a gain of -50 and a bandwidth of 48 kHz (-3 db), will be as shown in Figure 11-15.

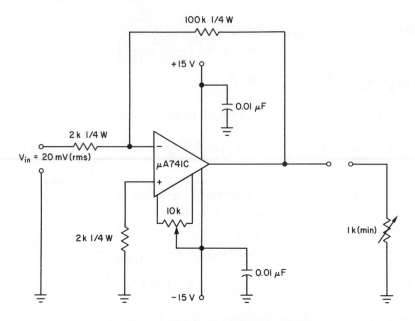

figure 11-15   the completely designed inverting amplifier

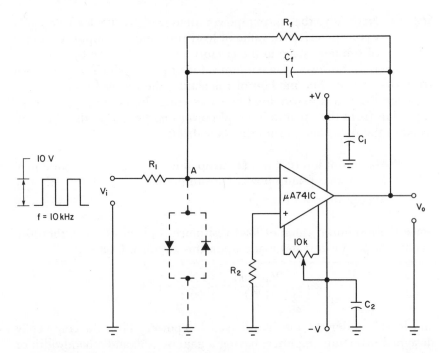

figure 11-16   a practical op-amp integrator circuit

11-4.2 inverting integrator

Figure 11-16 shows a breadboard design for an inverting integrator. Some designers suggest using diodes connected between points A and ground to protect the op-amp input against power-up transients and other fast-rising input signals that may be connected to the circuit. When such signals are applied, they rise to maximum voltage long before the output has time to change, and the inbalance at the inverting input may, on occasion, damage the op-amp. The large differential input voltage characteristic, $\pm$ 30 Vdc of the μA741C will often preclude the necessity of using such diodes. Under normal circumstances the diodes will not conduct, because the virtual ground at point A is not high enough to cause conduction. However, unusually high signals at A will be shunted around the op-amp via the diode. The integrator is required to perform to the following specifications:

1. Input signal, 10 KHz. Alternating square waveform, $\pm$5V peak to peak.
2. Output signal to be centered about zero volts.
3. Integrator time constant to be not greater than 0.1 ms $\pm$ 2.5 percent.
4. Slew rate of output signal to be less than 0.5 V/μs.

As before, we will determine the values of all the resistors and capacitors on a step-by-step basis.

Step 1. The power supply voltage and bypass capacitors are determined in the same manner as Step 1 in section 11.4.1. Remember that the ground connection located between the two halves of the power supply must also be connected to the circuit ground; otherwise the circuit will be inoperative.

Step 2. Input offset voltage, V_{io}, and input offset current, I_{io}, cause an error voltage at the output. For this effect to be minimized the value of R_1 is calculated from these two parameters, both of which are specified in the manufacturer's data sheet.

$$R_1 = \frac{V_{io}}{I_{io}} \qquad (11.33)$$

where V_{io} is the maximum input offset voltage (6 mV from the data sheet) and I_{io} is the maximum input offset current (200 nA from the data sheet).

Hence
$$R_1 = \frac{6 \text{ mV}}{200 \text{ nA}} = 30 \text{ k}\Omega$$

Although 30 kΩ is an available value of resistance, the more common, standard value resistor is 27 kΩ.

Step 3. For the fixed frequency integrator, the time constant R_1C_f is chosen to be equal to the period of the input frequency, T (see Figure 11-17). Also, in order to center the output waveform about the zero volt line, the initial charge on the capacitor is set at $\frac{V_{O_{p/p}}}{2}$, i.e., 1.285 V.

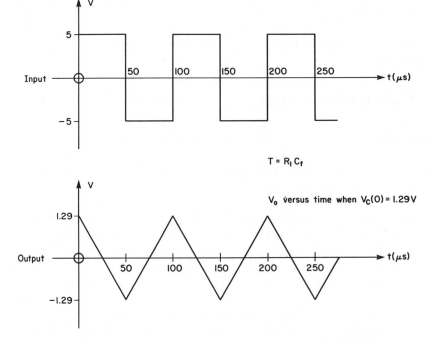

figure 11-17 input/output waveforms of the integrator circuit

Since
$$T = \frac{1}{f} \tag{11.34}$$

where
$$f = 10 \text{ KHz}$$

then
$$C_f = \frac{1}{R_1 \cdot f} \tag{11.35}$$

or
$$C_f = \frac{1}{27K \cdot 10K} = 3704 \text{ pF}$$

3600 pF is a standard value, and we would use it rather than the calculated value.

Step 4. R_f provides direct current feedback from the output, which stabilizes the integrator and reduces dc drift. R_f also sets the low frequency limit of the integrator (see frequency/gain curve,) below which, the output, instead of being a linear ramp, will tend to droop. Thus R_f is usually chosen such that

$$R_f = 10R_1 \qquad (11.36)$$

i.e., $\qquad\qquad\qquad R_f C_f \simeq 10T$

Hence in this case, $\qquad\qquad R_f = 270 \text{ k}\Omega$

and the low frequency limit is

$$f_L = \frac{1}{2\pi \, R_f C_1} \text{ Hz} \qquad (11.37)$$

hence $\qquad\qquad\qquad f_L = 163.7 \text{ Hz}$

Step 5. R_2 is again calculated by use of equation (11.29), which yields:

$$R_2 = 24.55 \text{ k}\Omega$$

The nearest standard value would be 22 kΩ.

Step 6. The peak-to-peak output of the integrator may be calculated by integrating the peak value input voltage over one-half the time period, T, and multiplying by the gain. Hence,

$$V_{O\,p/p} = \frac{1}{R_1 C_f} \int_0^{T/2} V_{\text{IN}} \text{ (peak) } dt \qquad (11.38)$$

Therefore, $\qquad\qquad V_{O\,p/p} = \frac{V_{\text{IN}} \text{ (peak)}}{R_1 C_f} \left(\frac{T}{2}\right)$

Substituting values for the various parameters yields

$$V_{O\,p/p} = \frac{(5)(50 \times 10^{-6})}{(27 \times 10^3)(3600 \times 10^{-12})}$$

or $$V_{0\ p/p} = 2.57 \text{ V}$$

The output is, therefore, a center-zero, inverted triangular waveform which is 2.57 volts peak to peak with an initial peak value of $V_C(0) = 1.285$ V.

Step 7. To determine the maximum slew rate at the output, we note that maximum slew rate occurs when the input changes sign, i.e., when the rate of change of the output goes from:

$$\frac{2.57}{50\ \mu s} \quad \text{to} \quad \frac{-2.57}{50\ \mu s}$$

The maximum slew rate is thus

$$\frac{2.57 \times 10^6}{50} - \left(\frac{2.57 \times 10^6}{50} \right) \qquad (11.39)$$

i.e., 0.103 V/μs which is well within the specification of 0.5 V/μs.

The final version of the integrator is shown in Figure 11-18, together with all the values of the resistors and capacitors.

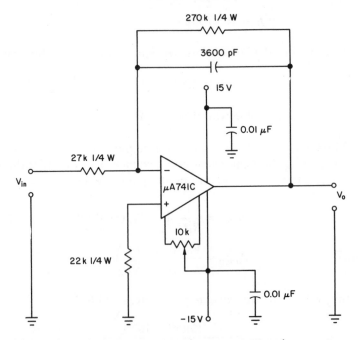

figure 11-18 the integrator circuit with component values

The values of R_1 and C_f can be varied, within reason, while the $R_1 C_f$ product is maintained reasonably constant. With the circuit in its present configuration, the lowest value of R_1 for which offset errors would be minimum is established if the typical instead of the maximum value of V_{io} is used. The data sheet specifies $V_{io} = 2$ mV typically, which would result in a value of $R_1 = 10$ kΩ. However, if R_2 were not present, the bias current, I_b, would be more of a problem and R_1 would be calculated by using the input offset voltage, V_{io}, and the bias current, I_b so that

$$R_1 = \frac{V_{io}}{I_b} = \frac{6 \text{ mV}}{500 \text{ nA}} = 4 \text{ k}\Omega$$

The nearest standard value resistor of 3.9 kΩ would be sufficient for R_1 in this case.

11-5 SUMMARY

There are a large number of operational amplifier IC packages on the market today. Some are single op-amps, while some are dual. They cover a wide range of frequency operation and, like any other electronic device, have different operating characteristics. Most have short circuit protection and are very forgiving if not handled too carelessly.

When the circuits discussed here are combined with more complex systems, care has to be taken to keep stage loading to a minimum and to prevent unwanted ground noise from causing distortion and loss of signal.

System modelling is an excellent way for one to begin developing confidence and experience at using op-amp circuits while, at the same time, cultivating an understanding of feedback control systems.

PROBLEMS

Section 11-4   Design the following circuits, using a μA741C op-amp.

1. A summing amplifier with three sinusoidal input signals,

where $V_1 = 10$ mVrms; $G_1 = 10$

$$V_2 = 20 \text{ mVrms}; \qquad G_2 = 40$$

$$V_3 = 100 \text{ mVrms}; \qquad G_3 = 40$$

$$R_{in} > 1 \text{ k}\Omega; \qquad R_L > 2 \text{ k}\Omega$$

Calculate the amplifier output impedance, R_o, the bandwidth (BW) and the peak-to-peak value of the output voltage, $V_{o_{p/p}}$;

2. A subtractor for two sinusoidal input signals,

where $V_1 = 50 \text{ mVrms}$

$$V_2 = 140 \text{ mVrms}$$

$$G = 100$$

$$R_{in} > 1 \text{ k}\Omega; \qquad R_L > 2\text{k}\Omega$$

Calculate R_o, BW, and $V_{o_{p/p}}$;

3. An integrator where

$V_{in} = 5 \sin (\omega t) \text{ V}$
and $300 \text{ rad/s} < \omega < 750 \text{ rad/s}$

4. A differentiator where V_{in} is a triangular waveform with a slope of $\pm 0.1 \text{ V}/\mu s$ maximum. V_o is to be a square wave $\pm 10 \text{ V}$.

system modelling: Part II —the analog computer

"Contrariwise," continued Tweedledee, "if it was so, it might be; and if it were so, it would be; but as it isn't, it ain't. That's Logic."

Lewis Carroll Through the Looking Glass

12-1 INTRODUCTION

As an alternative to building all the integrators and amplifiers, when modelling a control system, we could model the control system directly, using an analog computer. An analog computer comprises a patchboard that provides connections to large numbers of integrators, amplifiers, and potentiometers. Adjusting the values of the potentiometers enables us to vary the gains of the amplifiers and the time constants of the integrators. Each amplifier and integrator is connected to a summing junction, allowing several input signals to be connected to these devices. The variables of a control system are represented on an analog computer by the output voltages of the amplifiers and integrators. Thus by use of patchcords and with suitable scaling of the input and output voltages, the integrators and amplifiers can be connected together to synthesize any linear dynamic control system.

12-2 ANALOG COMPUTER SYMBOLS AND COMPONENTS

The commonly used mathematical operations available on an analog computer are

1. Addition
2. Inversion
3. Multiplying by a constant
4. Integration

12-2.1 addition

Summing amplifiers and the addition of several input signals were described in Section 11-3.2; Figure 12-1 shows the symbol used to represent a summing amplifier (of unspecified gain) with three input signals. This symbol is used to denote

$$y = -(G_1 x_1 + G_2 x_2 + G_3 x_3) \qquad (12.1)$$

figure 12-1 summing signals on an analog computer

where y is the inverted sum of the three input signals x_1, x_2, and x_3 when multiplied by their respective gains G_1, G_2, and G_3. These gains can be varied by use of input potentiometers, as we will describe in Section 12-2.3.

In the circuit x_1, x_2, and x_3 are all voltages measured with respect to ground and would be the electrical analogs of some physical variables.

12-2.2 inversion

A special case of equation (12.1) is when

$$G_1 = G_2 = G_3 = 1$$

so that

$$y = -(x_1 + x_2 + x_3) \qquad (12.2)$$

In this case the amplifier is termed an *inverter* and can be used as a summing junction or, as is more often the case, to change the polarity of a signal for feedback purposes.

The computer symbol for an inverter is shown in Figure 12-2.

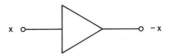

figure 12-2   the analog computer symbol for an inverter

12-2.3 multiplying by a constant

At various points within the computer circuit it is necessary to develop the electrical equivalent of the coefficients of the independent variables found in the system equations. For example, in the equation

$$y = Kx$$

K is the coefficient of the independent variable x, and in the equation

$$y = C\frac{dx}{dt}$$

C is the coefficient of the derivative of the independent variable.

For the constant-coefficient, linear differential equations dealt with in this text, such coefficients as K and C would be fixed throughout a computer simulation (although their values may be changed from simulation to simulation if, for example, we were trying to determine an optimal value for a particular coefficient).

The coefficients K and C are imposed on the voltage signals for x and dx/dt by use of potentiometers, as is shown symbolically in Figure 12-3.

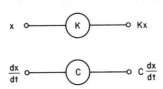

figure 12-3   multiplying a variable by a constant coefficient

Potentiometers used for this purpose usually range from 1 KΩ to 50 KΩ, depending on various factors, which will become apparent as we develop the computer circuits.

Potentiometers (pots), in an analog computer, may be ungrounded or have one side of the pot grounded. (They are more often used in the grounded mode.) Figure 12-4 shows both the grounded and ungrounded connections, together with their computer equivalent symbols.

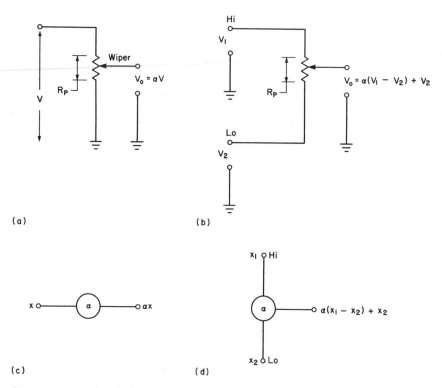

(a)   (b)

(c)   (d)

figure 12-4 unloaded potentiometers: (a) grounded, (b) ungrounded, (c) symbol for grounded pot, (d) symbol for ungrounded pot

It is obvious from Figure 12-4 that a pot can only provide a multiplication factor which lies between zero and unity; i.e.,

$$0 < \alpha < 1$$

But often we require coefficients such as K and C to be greater than unity and it becomes necessary to factorize the coefficient into two parts; one part, which is less than unity, is used as a pot setting, and the second part, which is greater than unity, is incorporated into the gain of the following amplification stage. For example, if the coefficient happened to be 25, then $K = 25$ is too large for a pot setting, so 25 is factorized thus:

$$25 = 0.5 \times 50$$

i.e.,
$$\frac{\text{coefficient}}{\text{value}} = \frac{\text{pot}}{\text{setting}} \times \frac{\text{gain of}}{\text{next stage}}$$

When one is setting up the analog circuits, it is a good idea to give the initial pot settings a value of about 0.5, because if the coefficient needs adjustment at a later time, there will be 50 percent wiper arm movement on either side of the original setting. In the case just considered, it would be possible to adjust the value of K from zero to 50, which provides plenty of working room.

12-2.4 integration

The theory of the integrating amplifier has been covered in Chapter 11. The computer symbol for an integrator is shown in Figure 12-5; note the difference between the integrator and the inverter symbol.

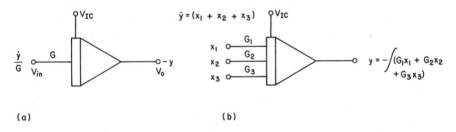

(a)   **(b)**

figure 12-5   the integrator symbol: (a) single input, (b) multiple inputs

In analog simulations the input to an integrator is always arranged to be a derivative of the dependent variable, y, but never the variable itself. In Figure 12-5(a), for instance, the input is $\frac{dy}{dt}$ (the first derivative of y with respect to time). In a computer diagram the derivatives are often simplified thus:

$$\frac{dy(t)}{dt} = \dot{y}$$

or
$$\frac{d^2x(t)}{dt^2} = \ddot{x}$$

The gain G represents the coefficient of integration [i.e., $\dfrac{1}{RC}$ in equation (11.22)]. Similarly, in Figure 12-5(b)

$$G_1 = \frac{1}{R_1 C_f}, \qquad G_2 = \frac{1}{R_2 C_f}, \qquad \text{and} \qquad G_3 = \frac{1}{R_3 C_f}$$

and V_{IC} represents the initial condition voltage that is sometimes "dumped" across the capacitor, by a relay, just prior to integration commencing. The integrator symbol in Figure 12-5(a) represents the following operation:

$$v_0(t) = -\frac{1}{R_1 C_f} \int_{t_0}^{t} v_{in} \cdot dt + V_{IC} \qquad (12.3)$$

or symbolically

$$y(t) = -G \int_{t_0}^{t} \dot{x}\, dt + y_0 \qquad (12.4)$$

where y_0 is the initial condition value of y.

When using an integrator we can, therefore, integrate, invert, and multiply by a constant and, if necessary, add initial conditions.

12-3 PROBLEM SOLVING ON AN ANALOG COMPUTER

An analog computer can be used for solving a wide variety of problems, including problems involving nonlinear functions and time-dependent coefficients. However, in this text we are concerned only with how to use an analog computer to solve linear, constant-coefficient, differential equations, i.e., those equations describing the behavior of linear, time-invariant, control systems.

Usually the problem to be solved is to measure, or record, the response of a control system to the application of a particular input signal. We can solve this problem by modelling the control system on the analog computer, applying the relevant input signals and recording the particular output voltage representing the control system response. Amplifiers, integrators, and potentiometers are used to construct the "model," and the system differential equation dictates the way in which these components are connected together to simulate the control system.

EXAMPLE 12.1

To demonstrate this process, let us begin with a simple system and translate it into an analog computer diagram. Consider the following first-order differential equation which describes the natural phenomenon of decay, where the rate of decay of the quantity x is directly proportional to the magnitude of x, i.e.,

$$\frac{dx(t)}{dt} = -Kx(t) \qquad (12.5)$$

Equation (12.5) may, for example, represent a capacitor discharging, or radioactive decay of an isotope, or even a cup of coffee cooling to room temperature.

The solution to equation (12.5), if x is unity when t is zero, is easily obtained as follows:

$$\frac{dx(t')}{dt'} = -Kx(t')$$

or $$-\frac{1}{K} \int_{1}^{x(t)} \frac{dx(t')}{x(t')} = \int_{0}^{t} dt'$$

Hence $$-\frac{1}{K} \operatorname{Ln}[x(t)] = t$$

i.e., $$\operatorname{Ln}[x(t)] = -Kt$$

Therefore, $$x(t) = e^{-Kt} \qquad (12.6)$$

The solution to equation (12.5) is shown graphically in Figure 12-6.

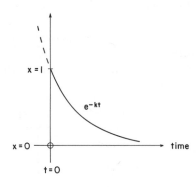

figure 12-6   simple exponential decay $x(t) = e^{-Kt}$

How do we model equation (12.5) on an analog computer?

1. Note all the variables in the equation, i.e., x and dx/dt.

2. Write the equation so that the term having the highest derivative of x is the only term on the left-hand side of the equation (in this case, because there are only two terms, one the derivative, the equation is unaltered).

$$\dot{x} = -Kx \qquad (12.7)$$

3. Draw an integrator with the input given by equation (12.7) and an output equal to the integral of the left-hand side of equation (12.7), i.e.,

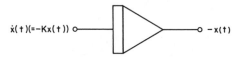

figure 12-7 integration of the derivative $\dot{x}(t)$ to produce the variable $x(t)$

4. Introduce successive integrators until $x(t)$ is obtained (no further integrators needed in this case).

5. Combine the outputs of all the integrators, with suitable potentiometer settings, to provide an input to the first integrator in accordance with equation (12.7).

6. Repeat step 5 for all successive integrators (not necessary in this case).

7. Set the magnitude of the initial output voltage of each integrator to take account of the initial values of the system variables (set to unity in this case).

Consequently, we must take the output of our only integrator [i.e., $-x(t)$] and multiply it by a gain K to satisfy equation (12.7). This product must then be fed into the input to the first integrator, as shown in Figure 12-8.

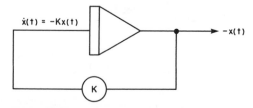

figure 12-8 a model of $\dot{x}(t) = -Kx(t)$

If we now want a signal $x(t)$ rather than $-x(t)$, we must use an inverter to follow the $-x(t)$ signal. However, in this case, because there is no input reference signal, the negative sign is relative and Figure 12-8 could have also been drawn as shown in Figure 12-9.

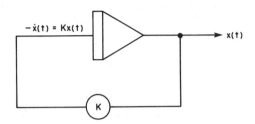

figure 12-9 a model of $\dot{x}(t) = -Kx(t)$

If we now add to Figure 12-9 the symbol for providing initial conditions for $x(t)$, and we make the initial value unity, then we have, as in Figure 12-10, the final version of the model for equation (12.5), exhibiting a response given by equation (12.6).

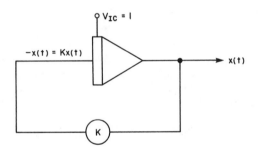

figure 12-10 a model of $\dot{x}(t) = -Kx(t)$; $x(0) = V_{IC} = 1$

When the computer is turned on, $x(t)$ will begin with a value of $V_{IC} = 1$, and as $x(t)$ is fed back to the input of the integrator via K, the magnitude of $x(t)$ begins to reduce exponentially.

EXAMPLE 12.2

As an example of a system described by a second-order differential equation, consider the simplified automobile suspension system discussed in Section 3-6, Chapter 3. The particular system is illustrated in Figure 3-16, and the system equation is defined as

$$M\frac{d^2x(t)}{dt^2} + B\frac{dx(t)}{dt} + Kx(t) = F(t) \qquad (3.42)$$

or, more simply,

Step 1.

$$M\ddot{x} + B\dot{x} + Kx = F \qquad (12.8)$$

Step 2. To determine the basic computer diagram, we solve equation (12.8) for $\ddot{x}$, which yields

$$\ddot{x} = -\frac{B}{M}\dot{x} - \frac{K}{M}x + \frac{F}{M} \qquad (12.9)$$

Step 3 and 4. Figure 12-11 shows the simplest connection of integrators and inverters that could generate the operations described in equation (12.9).

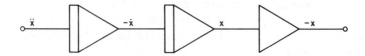

figure 12-11   the basic solution of equation (12.9)

Step 5 and 6. It is now only necessary to introduce the coefficients of the variables into feedback loops and apply the input F through a pot representing its coefficient, $\frac{1}{M}$. Figure 12-12 shows the completed analog computer diagram.

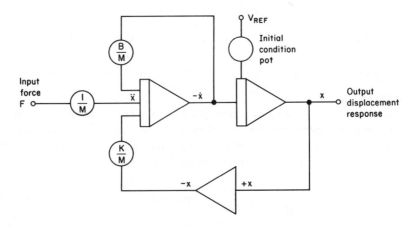

figure 12-12   the complete basic computer program diagram

Regardless of the complexity of the problem, the steps to programming a solution are quite straightforward and can be summarized thus:

1. Describe the system by its linear differential equations.
2. Use the simplified coding for the derivatives and solve for the highest derivative.
3. Assemble in-line computer symbol diagrams representing integration and inversion from the highest derivative down to its variable.
4. Generate the coefficients by means of potentiometers, and close each loop back to the input.
5. Provide initial conditions by means of potentiometers and voltages if necessary.

12-4 AMPLITUDE AND TIME SCALING

12-4.1 introduction

The output from an analog computer model may be a voltage representing the response of the modelled control system. However, the amplitude of the output voltage, regardless of the magnitude of the physical quantity that it represents, has to be within the voltage range of the final op-amp. Needless to say, the output voltages of all the other op-amps in the model must also be constrained to operate within the op-amp limits. It is necessary, therefore, to introduce amplitude scaling factors, so that the voltage range of an amplifier can be made to represent the working range of the variable being synthesized without causing amplifier saturation.

Output signals from an analog computer are usually displayed graphically on a chart recorder, an X-Y recorder, or an oscilloscope. Depending on the control system being modelled, the response time may be very long or it may be extremely short; in either case, observing the response can be difficult or inconvenient. It would be more convenient if the response time of the model were adjusted so that the response was more easily observed and recorded. Adjusting the response time of the model in this way is termed *time scaling*.

Two display instruments used effectively by one of the authors and many of his students are

1. Mark 220 Recorder, Gould Inc., Cleveland, Ohio.
2. 5111 Storage Oscilloscope, Tektronix Inc., Beaverton, Oregon.

The Mk220 is a chart recorder which, at its fastest speed of 125 mm/s, can easily record rise times of 20 ms duration and sine waves of 25 hz (≈ 160 rad/s) frequency. The speed of response of this instrument is limited only by the mechanical inertia of the pen and its electromechanical drive system. On the other hand, the great advantage of a chart recorder is that it produces a permanent record.

Compared to the chart recorder, the oscilloscope is capable of displaying much faster response times, i.e., rise times down to 100 μs and sine waves with frequencies up to 10 kHz (63 kr/s). However, in order to obtain a permanent recording from an oscilloscope, we must photograph the trace (either as it happens or from the semi-permanent recording obtained when the oscilloscope is in the "storage mode"). The disadvantage of obtaining permanent records in this way is that resolution is lost on the photograph; hence the accuracy of the recording suffers.

12-4.2 a worked example—modelling and scaling of spring restrained mass, with damping

The System Equation Amplitude and time scaling will be demonstrated by means of the automobile suspension system described in Chapter 3 and shown for convenience in Figure 12-13. Figure 12-13 is a much simplified model but is very suitable as an introductory example. In this example, instead of considering the application of a force from above, as we did in Chapter 3, we will consider that the road applies a step displacement, x_1, in the upward direction, as it might do if the vehicle were to run over a curb. When this happens, the spring will be compressed, pushing the mass of the vehicle up into the air. The vehicle will then bounce around a little until it resumes running at a steady height above the new ground level.

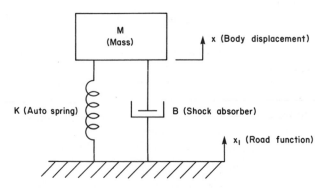

figure 12-13   the simplified model of an auto suspension system

The vehicle parameters are

$$\text{Mass, } M = 400 \text{ Kg}$$

$$\text{Damping coefficient, } B = 3000 \frac{\text{N·s}}{\text{m}}$$

$$\text{Spring coefficient, } K = 15000 \text{ N/m}$$

$$\text{Maximum spring compression, } x_{max} = 0.45 \text{ m}$$

$$\text{Maximum body velocity, } \dot{x}_{max} = 2.3 \text{ m/s}$$

$$\text{Applied step function, } x_1 = 0.13 \text{ m}$$

The sum of all the forces acting on the system equals the product of the mass and the acceleration. Identifying all the forces following the displacement, x_1, we find that:

$$\text{The restoring force due to the spring, } F_K = -K(x - x_1)$$

$$\text{The restraining force due to the damper, } F_B = -B\frac{d(x - x_1)}{dt}$$

Since x_1 is a fixed distance and hence constant, for $t > 0$, then

$$\frac{dx_1}{dt} = 0$$

and

$$F_B = -B\frac{dx}{dt}$$

We can now write the system equation as follows:

$$F_K + F_B = M\frac{d^2(x - x_1)}{dt^2} = M\frac{d^2x}{dt^2} \qquad (12.10)$$

Hence

$$Kx_1 - Kx - B\frac{dx}{dt} = M\frac{d^2x}{dt^2} \qquad (12.11)$$

Using derivative coding yields

$$Kx_1 - Kx - B\dot{x} = M\ddot{x} \qquad (12.12)$$

Solving for the highest derivative yields

$$\ddot{x} = -\frac{B}{M}\dot{x} - \frac{K}{M}(x - x_1) \qquad (12.13)$$

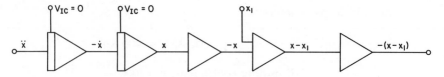

figure 12-14 basic solution of equation (12.13)

Figure 12-14 shows the simplest connection of op-amps that could generate the operations described in equation (12.13). The initial value of x (displacement of the mass before the application of the step function) has been arbitrarily taken to be zero. Similarly, $\dot{x}_0$, the initial vertical velocity of the mass, is also assumed to be zero.

From Figure 12-14 we can easily derive the complete but unscaled analog computer diagram, as shown in Figure 12-15.

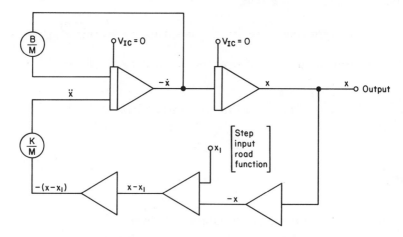

figure 12-15 the complete but unscaled analog computer diagram of equation (12.13)

Amplitude Scaling We now apply amplitude scaling factors into equation (12.13) so that at no time will the op-amps be driven into saturation. If the maximum voltage swing of the op-amps is ± 10 volts, then for our example the maximum spring compression of 0.45 m can be represented by the 10-volt saturation limit of the amplifier. With this scaling, any change in the value of x will produce a voltage swing of less than ± 10 volts. Hence in terms of the displacement, x,

$$\left| \frac{x}{x_{\max}} \right| < 1$$

and in terms of the scaled voltage signal [x],

$$-10 \text{ volts} \leq [x] = \left(\frac{x}{x_{\max}}\right) \times 10 \leq 10 \text{ volts}$$

Similarly, the maximum displacement rate, $\dot{x}_{\max}$, will be represented by 10 volts so that any change in the value of $\dot{x}$ will produce a voltage swing less than ± 10 volts. Thus, once again, in terms of $\dot{x}$,

$$\left|\frac{\dot{x}}{\dot{x}_{\max}}\right| < 1$$

and in terms of the scaled voltage signal, $[\dot{x}]$,

$$-10 \text{ volts} \leq [\dot{x}] = \left(\frac{\dot{x}}{\dot{x}_{\max}}\right) \times 10 \leq 10 \text{ volts}$$

We can now proceed with amplitude scaling of equation (12.13) as follows: First we rewrite the equation, thus:

$$\frac{d}{dt}\dot{x} = -\frac{B}{M}\dot{x} - \frac{K}{M}(x - x_1) \qquad (12.13a)$$

We now multiply and divide each expression of equation (12.13a) by the maximum value of the appropriate variable, thus leaving the equation unchanged, i.e.,

$$\frac{d}{dt}(\dot{x})\frac{\dot{x}_{\max}}{\dot{x}_{\max}} = -\frac{B}{M} \cdot (\dot{x}) \cdot \frac{\dot{x}_{\max}}{\dot{x}_{\max}} - \frac{K}{M} \cdot (x - x_1) \cdot \frac{x_{\max}}{x_{\max}} \qquad (12.14)$$

which yields

$$\frac{d}{dt}\left(\frac{\dot{x}}{\dot{x}_{\max}}\right) \cdot \dot{x}_{\max} = -\frac{B}{M}\left(\frac{\dot{x}}{\dot{x}_{\max}}\right) \cdot \dot{x}_{\max} - \frac{K}{M}\left(\frac{x - x_1}{x_{\max}}\right) \cdot x_{\max} \qquad (12.14a)$$

Dividing both sides of equation (12.14a) by $\dot{x}_{\max}$ yields

$$\frac{d}{dt}\left(\frac{\dot{x}}{\dot{x}_{\max}}\right) = -\frac{B}{M}\left(\frac{\dot{x}}{\dot{x}_{\max}}\right) - \frac{K}{M}\left(\frac{x - x_1}{x_{\max}}\right) \cdot \frac{x_{\max}}{\dot{x}_{\max}} \qquad (12.15)$$

If we now substitute the parameter values, we obtain

$$\frac{d}{dt}[\dot{x}] = -\frac{3000}{400}[\dot{x}] - \frac{15,000}{400} \cdot \frac{0.45}{2.3}([x] - [x_1]) \qquad (12.16)$$

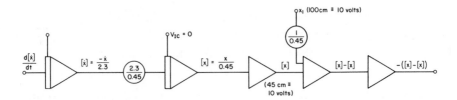

figure 12-16 the basic computer solution of equation (12.16)

In the same way that we used equation (12.13) to give us Figire 12-14, we have shown in Figure 12-16 the simplest connection of op-amps that will generate the operations described in equation (12.16).

We now feed the relevant loops back to the input to achieve the complete, amplitude-scaled computer solution, as shown in Figure 12-17.

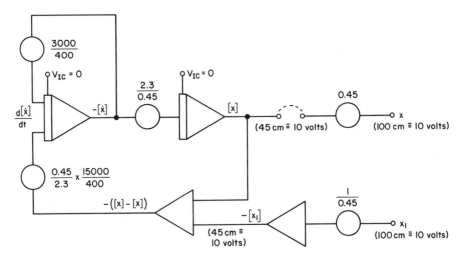

figure 12-17 the complete, amplitude-scaled computer solution

In Figure 12-17 there are four potentiometers with factors greater than unity; if they are all set in the neighborhood of 0.5, the following stage gains can be adjusted to produce the appropriate factor. For example,

$$\frac{B}{M} = \frac{3000}{400} = 7.5 = (0.5) \cdot (15)$$

$$= \text{(pot setting)} \cdot \text{(gain)}$$

$$\frac{K}{M} \times \frac{x_{\max}}{\dot{x}_{\max}} = \frac{15,000}{400} \times \frac{0.45}{2.3} = 7.337 = (0.489) \cdot (15)$$

$$= (\text{pot setting}) \cdot (\text{gain})$$

$$\frac{\dot{x}_{max}}{x_{max}} = \frac{2.3}{0.45} = 5.11 = (0.511) \cdot (10)$$

$$= (\text{pot setting}) \cdot (\text{gain})$$

$$\frac{1}{x_{max}} = \frac{1}{0.45} = 2.22 = (0.444)\,(5)$$

$$= (\text{pot setting}) \cdot (\text{gain})$$

Figure 12-18 shows the computer diagram with the potentiometer settings and gains that were calculated above.

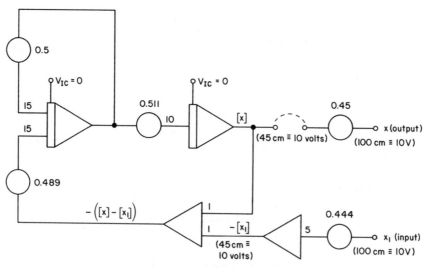

figure 12-18   the amplitude-scaled computer diagram showing all pot settings and stage gains

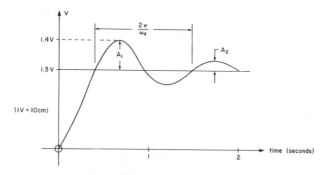

figure 12-19   the response of the analog computer to a 1.3-volt step input at x_1

If a step input measuring 1.3 volts (equivalent to 13 cm) is applied at x_1 (INPUT) in Figure 12-18, the response at x (OUTPUT) is a voltage, illustrated in Figure 12-19, which rises to a maximum of 1.4 volts. The output of the second integrator is obviously 3.14 volts maximum or

$$\frac{1.4 \text{ V}}{0.45}$$

It is worth noting that if we obtain the Laplace transform of equation (12.13), we obtain

$$s^2 \cdot X(s) = -\frac{B}{M} \cdot s \cdot X(s) - \frac{K}{M} \cdot (X(s) - X_1(s)) \tag{12.17}$$

which yields

$$\frac{X(s)}{X_1(s)} = \frac{K/M}{s^2 + B/M \cdot s + K/M} \tag{12.18}$$

and since $X_1(s) = \dfrac{0.13}{s}$, then

$$X(s) = \frac{0.13 \times 37.5}{s(s^2 + 7.5s + 37.5)} \tag{12.19}$$

from which we obtain

$$\omega_n = 6.12 \text{ rad/s } (0.97 \text{ Hz})$$

$$\zeta = 0.612$$

$$\omega_d = 4.84 \text{ rad/s } (0.77 \text{ Hz})$$

Of course, we could quite easily measure and calculate all of these terms from a chart recording of the response; e.g., if Figure 12-19 represents the response, then ω_d can be directly measured as shown, and ω_n is calculated from

$$\omega_n = \frac{\omega_d}{\sqrt{1 - \zeta^2}} \tag{12.20}$$

where

$$\zeta = \frac{\dfrac{\pi}{2} \text{Ln} \left(\dfrac{A_1}{A_2} \right)}{\sqrt{1 + \dfrac{\pi}{2} \text{Ln} \left(\dfrac{A_1}{A_2} \right)^2}} \tag{12.21}$$

A_1 = amplitude of the first overshoot and A_2 = amplitude of the second overshoot.

Time Scaling It is obvious, from a brief examination of Figure 12-19, that the settling time of the step response is of the order of five seconds or less. Thus, for this example, the response can be easily observed and recorded, without having to apply time scaling to equations (12.13) and (12.16). However, in order to demonstrate how to use time scaling on an analog computer, we will suppose that we wish to slow down the response of the analog model (which we would do if we were particularly interested in getting a better look at a certain portion of the response curve, e.g., in order to measure the rate of rise of the response).

The time frame of the analog model can be slowed down or speeded up as wished, in accordance with the actual system being modelled. For example, if we were simulating a geological event where ω_n or ω_d represented thousands of years per cycle, then the computer time must be shortened or speeded up.

Real time t is never changed; obviously, it is only computer time T that is changed by an appropriate scaling factor γ; i.e.,

Time scale factor,

$$\gamma = \frac{\text{time taken by the computer model}}{\text{time taken by the actual process being modelled}}$$

or
$$\gamma = \frac{T}{t} \qquad (12.22)$$

If real events are slowed down, then

$$\gamma_1 = \frac{T_1}{t}$$

where
$$T_1 > t$$

Hence
$$\gamma_1 > 1 \qquad (12.22a)$$

If real events are speeded up, then

$$\gamma_2 = \frac{T_2}{t}$$

where
$$T_2 < t$$

Hence
$$\gamma_2 < 1 \qquad (12.22b)$$

Real time t can, therefore, be expressed in terms of γ, the time scale factor, and T, computer time; i.e.,

$$t = \frac{T}{\gamma} \qquad (12.22c)$$

For our purposes we want to slow down the response of the analog model, so we are interested in the case where $\gamma > 1$. Suppose we wish to improve our measurement of "the rate of rise of the response" by a factor of 10; in this case, we want to slow down the analog model by a factor 10, which means the time scaling factor is

$$\gamma = 10 \qquad (12.23)$$

Because γ affects only the time-dependent terms of the system differential equation, we now apply the time scaling factor to each of the derivative terms of equation (12.13) [and, consequently, equation (12.16)]. Thus the first derivative of the unscaled variable $x(t)$ becomes, after application of equations (12.22c) and (12.23):

$$\frac{dx(t)}{dt} \rightarrow \frac{dx(T)}{d\left(\dfrac{T}{\gamma}\right)} = \gamma \cdot \frac{dx(T)}{dT}$$

Hence
$$\frac{dx(T)}{dT} = \frac{1}{\gamma} \frac{dx(t)}{dt} = 0.1 \frac{dx(t)}{dt} \qquad (12.24)$$

Similarly, the second derivative of $x(t)$ becomes

$$\frac{d^2x(t)}{dt^2} \rightarrow \frac{d^2x(T)}{d\left(\dfrac{T}{\gamma}\right)^2} = \gamma^2 \frac{d^2x(T)}{dT^2}$$

Hence

$$\frac{d^2x(T)}{dT^2} = \frac{1}{\gamma^2} \frac{d^2x(t)}{dt^2} = 0.01 \frac{d^2x(t)}{dt^2} \qquad (12.25)$$

Because we are only scaling the time-dependent terms in the system differential equation, we find that on the analog model only the input signals to integrators are affected by time scaling. For example, equation (12.24) would be represented on an analog computer diagram as shown in Figure 12-20.

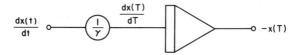

figure 12-20 time scaling an integrator

Consequently, in Figure 12-17, we can slow down the time scale by a factor of ten simply by multiplying all the inputs into the left-hand integrator by 0.01 [as from equation (12.25)] and multiplying the input into the right-hand integrator by 0.1 [from equation (12.24)]. The completed time and amplitude scaled computer diagram is shown in Figure 12-22.

We might ask at this stage how is it that this type of potentiometer scaling can affect the timing of the analog model. Consider the simple integrator illustrated in Figure 12-21.

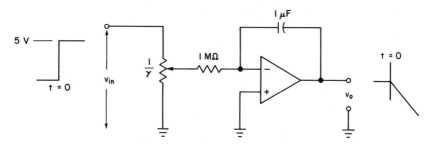

figure 12-21 circuit diagram of time scaling

If the wiper of the pot is initially at the top end of the pot, then the pot is effectively out of circuit. If a +5-volt step is applied at V_{in}, the output, V_0, is a negative-going ramp of 5 volts per second. If it is necessary to slow down the system by a factor of ten, then $1/\gamma$ would have a value of 0.1, and hence a step function of only 0.5 volts would be fed to the integrator (although V_{in} would still be a 5-volt step). The output rate would then be reduced to a negative-going ramp of only 0.5 volts per second. If, on the other hand, we wanted to speed up the response of the analog model, we would have to increase the gain of the integrator $(1/RC)$, which might involve changing either the fixed resistance (1 MΩ) or the feedback capacitor (1 μF).

The recorded response from the analog model of Figure 12-22, for a 1.3-volt step input at x_1, is shown in Figure 12-23, where the vertical axis has a scale of 1 volt = 10 cm and the horizontal scale shows both computer time and process time.

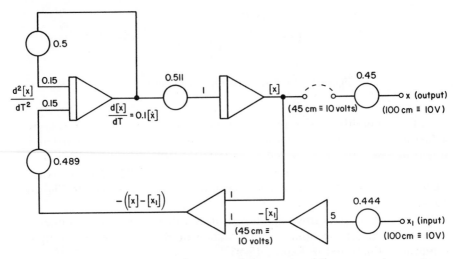

figure 12-22 the computer diagram of the automobile suspension system with amplitude and time scaling incorporated

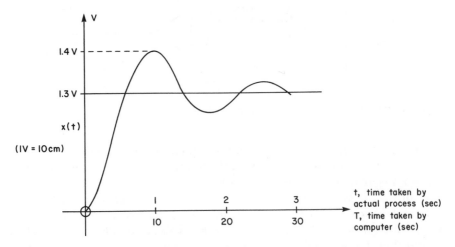

figure 12-23 the response of the completely scaled computer circuit to a 1.3-volt step input

12-4.3 a worked example—modelling a hydraulic control system

The Control System Equation The hydraulic control system of Figure 10-13 will be used to demonstrate modelling a control system on an analog computer, with a view to optimizing the circuit parameters. It will be remembered that the purpose of the control system is to control the action of a process control valve in order to regulate the flow of water to a hydraulic machine. The machine requires an "off-load" flow rate of 20 gpm and when switched to the "load" condition the demanded flow rate increases to 100 gpm. However, when solving the problem in Chapter 10, we assumed that the demanded flow rate changed from 0 to 80 gpm, and the final response curves were then adjusted by the addition of 20 gpm. For ease of modelling, this assumption will also be made here, without any loss of generality.

In Chapter 10 we employed a proportional-plus-derivative controller to improve the response of the control system; the circuit diagram for this type of controller is shown in Figure 10-25, and the resulting transfer function from equation (10.38) is

$$\frac{Q(s)}{R(s)} = \frac{8.33K_3 \ (s + 3K_1/K_3)}{s^2 + (4 + 0.833K_3)s + 2.5K_1} \qquad (10.38)$$

where K_1 is the proportional gain and K_3 is the derivative action gain. The differential equation corresponding to this transfer function is

$$\ddot{q} + (4 + 0.833K_3)\dot{q} + 2.5K_1q = 8.333K_3\dot{r} + 25K_1r \qquad (12.26)$$

(i.e., the initial values of q and $\dot{q}$ are zero).

Examination of this equation shows that the right-hand side of the equation contains a term that we have not encountered previously in our attempts at analog modelling; this term is $\dot{r}$, the first derivative of the forcing function (or demand) r. Since we obviously do not have $\dot{r}$ available, it would appear that this term is going to present us with a problem. This problem, however, is simply avoided by rearranging the equation. Instead of rewriting the equation with only $\ddot{q}$ on the left-hand side, as we have done when setting up earlier models, this time we rewrite the equation with both $\ddot{q}$ and $\dot{r}$ terms on the left-hand side; thus

$$\ddot{q} - 8.333K_3\dot{r} = -(4 + 0.833K_3)\dot{q} - 2.5K_1q + 25K_1r \qquad (12.27)$$

If we now start to build our analog model with the input to the first integrator being $\ddot{q} - 8.333K_3\dot{r}$, as given in equation (12.27), we know that the output of this integrator will be $-(\dot{q} - 8.333K_3r)$. Now by simply adding $-8.333K_3r$ to this output we obtain $-\dot{q}$, which can then be directly integrated to give q. Thus by adding a multiple of r ($-8.33 \ K_3r$, which we can easily generate) at a point *following* the first integrator, we have succeeded in synthesizing the presence of an $\dot{r}$ term within the analog model. The computer diagram for this arrangement can be seen in Figure 12-24.

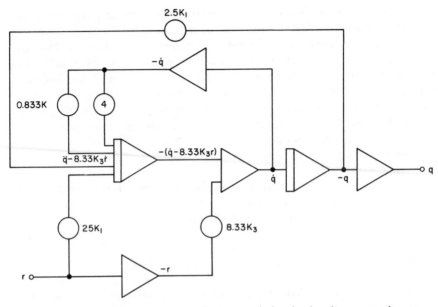

figure 12-24 analog computer diagram of the hydraulic control system shown in figure 10-25

Amplitude Scaling From the previously generated responses for this control system, we can see that time scaling of the analog model is not necessary and only amplitude scaling need be considered. Assuming that the maximum voltage allowed on the computer is ± 10 volts, we must scale r, q, and $\dot{q}$ so that the maximum range of the following values does not exceed zero to 10 volts:

$$r: 0 \text{ mA to } 16 \text{ mA}$$
$$q: 0 \text{ gpm to } 160 \text{ gpm}$$
$$\dot{q}: 0 \text{ gpm/s to } 500 \text{ gpm/s}$$

Rewriting equation (12.27) in scaled form, we find

$$\dot{q}_{max} \cdot \frac{d}{dt}\left(\frac{\dot{q}}{\dot{q}_{max}}\right) - 8.333 K_3 r_{max} \frac{d}{dt}\left(\frac{r}{r_{max}}\right)$$
$$= -(4 + 0.833 K_3)\dot{q}_{max}\left(\frac{\dot{q}}{\dot{q}_{max}}\right) - 2.5 K_1 q_{max}\left(\frac{q}{q_{max}}\right) \qquad (12.28)$$
$$+ 25 K_1 r_{max}\left(\frac{r}{r_{max}}\right)$$

which becomes

$$\frac{d}{dt}[q] - 8.333K_3 \cdot \frac{16}{500} \cdot \frac{d}{dt}[r]$$

$$= -(4 + 0.833K_3)[q] - 2.5K_1 \cdot \frac{160}{500} \cdot [q] + 25\,K_1 \cdot \frac{16}{500} \cdot [r] \quad (12.29)$$

which reduces to

$$\frac{d}{dt}[q] - 0.2667K_3 \frac{d}{dt}[r]$$

$$= -(4 + 0.833K_3)[q] - 0.8K_1[q] + 0.8K_1[r] \quad (12.30)$$

From Chapter 10, suitable experimental ranges for K_1 and K_3 are

$$0 \leqslant K_1 \leqslant 20 \quad (12.31)$$

$$0 \leqslant K_3 \leqslant 12 \quad (12.32)$$

Substituting the maximum values of K_1 and K_2 into equation (12.30) yields

$$\frac{d}{dt}[q] - 3.2\frac{d}{dt}[r] = -(4 + 10)[q] - 16[q] + 16[r] \quad (12.33)$$

The analog computer model for equation (12.29) is presented in Figure 12.25, where the potentiometer settings will provide the range of K_1 and K_2 described by expressions (12.30) to (12.33).

The equivalent value of each potentiometer is noted against the potentiometer symbol in Figure 12-25. Thus by using these expressions it is possible to calculate a particular value for a potentiometer setting for any value of K_1 and K_3, as the case may be. By adjusting the values of K_1 and K_3 in this way and applying a step input at $[r]$ of 5 volts ($\equiv 8$ mA), the response of q can be observed and subsequently optimized by finding the most satisfactory combination of K_1 and K_3.

If K_1 and K_3 are set at values

$$K_1 = 10 \quad (12.34)$$

$$K_3 = 5.5 \quad (12.35)$$

then the response observed at $[q]$ for a 5-volt step at $[r]$ will be identical to the response shown in Figure 10-27, after adjusting for the relevant scaling.

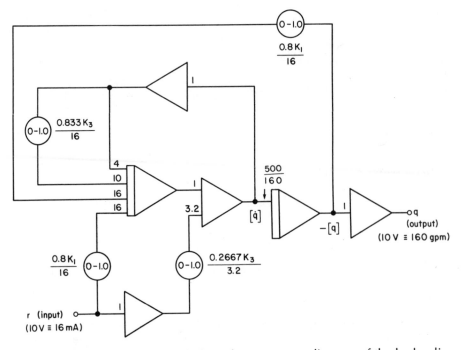

figure 12-25 completed, scaled, analog computer diagram of the hydraulic control system of Figure 10-25

12-5 SUMMARY

In this chapter we have talked about analog computers and analog modelling. We have described these machines and their capabilities and have described how to construct an analog model of a control system on such a machine. Finally, it was shown how to set up a model of a practical problem and optimize the parameters on the problem by adjusting potentiometer settings and observing the model's response.

PROBLEMS

Section 12-3   1. The steam-heated water system described in Chapter 3 has the following transfer function:

$$\frac{T(s)}{T_1(s)} = \frac{1}{\tau s + 1}$$

where $T(s)$ is the Laplace transform of water temperature (output) and $T_1(s)$ is the Laplace transform of steam temperature (input). Construct an analog computer diagram to solve this equation.

2. The transfer function of a dc motor field is represented by the block diagram below.

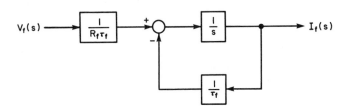

where V_f and I_f are the field voltage and current, respectively and $\tau_f = \dfrac{L_f}{R_f}$. Construct an analog computer diagram to solve for field current.

3. Construct an analog computer diagram for the system shown in Problem 8, Chapter 3.

4. Construct an analog computer diagram for the system shown in Figure 3-29.

5. If $K_1 = 5.0$, construct an analog computer diagram for the system shown in Figure 10-22.

Section 12.4

6. Provide amplitude and time scaling for Problem 1 above if $T_1 = 400°C$, $T_{max} = 600°C$, $R = 0.04°C$ s/joule and $C = 30 \times 10^3$ joules/°C.

7. Provide amplitude and time scaling for Problem 2 above if $L_f = 400$ mH, $R_f = 20\ \Omega$, $V_f = 12$ volts, and $I_{fmax} = 1$ amp.

8. In Problem 3 above the piston has a diameter of 10 cm and $K_v = 9.82 \times 10^{-3}$ m²/s. Scale the computer diagram for amplitude and time when $x_{0max} = 18$ cm and $x_{in} = 9.5$ mm. If the piston diameter can vary between 8 cm and 15 cm, how does this affect the diagram?

9. In Problem 4 above, x_3 is an input step of 13 cm. Scale the computer diagram if the system parameters are as follows:

$$M_1 = 400 \text{ Kg} \qquad M_2 = 30 \text{ Kg}$$

$$K = 15,000 \text{ N/m} \qquad K_2 = 75,000 \text{ N/m}$$

$$1460 \frac{\text{Ns}}{\text{m}} \quad < B \quad < 5100 \frac{\text{Ns}}{\text{m}}$$

$$x_{1 \text{ max}} = 46 \text{ cm} \qquad \dot{x}_{1 \text{ max}} = 1.4 \text{ m/s}$$

$$x_{2 \text{ max}} = 10 \text{ cm} \qquad \dot{x}_{2 \text{ max}} = 15 \text{ m/s}$$

10. In Problem 5 above the input is an 8 mA step. Properly scale the computer diagram. Obtain all maximum values from information provided in Chapter 10.

part 5

digital control

13

digital servo systems

Four other Oysters followed them,
And yet another four;
And thick and fast they came at last,
And more, and more, and more. . . .

Lewis Carroll, Through the Looking Glass

13-1 INTRODUCTION

In the previous twelve chapters we have dealt exclusively with the analysis and design of analog control systems that operate in the real, analog world; yet we are increasingly aware of the spread of digital technology into every corner of engineering. It would seem that the computer solution is the best way to do everything these days. But this is just not true; anyway, it is too vague a reason to invest a lot of effort into applying digital technology to the world of control. So why has it been done and why is its continued use on the increase?

A very important reason is that if the control signal can be digitized and processed throughout the system in a digital fashion, it can be protected from nearly all noise and can be made error-free. As a spectacular example of this, consider the color television pictures from Mars, Jupiter, and Saturn. This protection from noise errors can be maintained throughout any number of processings with absolutely no degradation of accuracy.

Second, the digital control signal can be remembered. However, although control systems use many different digital memory devices, they are essentially short-term memories. The huge memory capacities that would be required to remember the system response to past inputs, and hence allow the system to learn from such an experience, are impractical.

Notwithstanding this fact, neither of these two advantages are shared by analog signal processing methods.

Another advantage of the digital system is that it can be programmed. In addition, digital circuits are extremely reliable, comparatively simple, and very cheap, and they can be duplicated with great precision quite easily; again, these are not advantages shared by analog circuits.

In this chapter, therefore, we shall first of all examine the different control system configurations employing digital techniques, after which we shall consider in much greater detail the output end of the system and the digital devices used to move and monitor the load.

13-2 DIGITAL CONTROL SYSTEM CONFIGURATION

Systems having many inputs and outputs are called *multivariable systems,* and their complexity is beyond the level of our concern. It is worth noting, however, that the complexity of system design can be reduced, if, for example, a plant process having many inputs and outputs can be interfaced with a computer, thus taking advantage of its decision-making capability. At the present time a cheap microcomputer is capable of accommodating a couple of dozen input signals from transducers and monitoring devices. Alternatively, the larger, more powerful minicomputer can accept a couple of hundred input transducer and monitoring device signals.

Interfacing such a computer with a control system can be done in two ways. The computer can be located outside the closed-loop system; this is termed *off-line control* and is illustrated in Figure 13-1. In this type of control a human operator is always interposed between the control system and the computer. The operator monitors the system output and feeds data to the computer, which responds with the appropriate answers, enabling the operator to make the correct system input command. The computer can also be located inside the closed-loop system; this is termed *on-line* control and is illustrated in Figure 13-2. In this type of control the computer is effectively located at the summing junction of

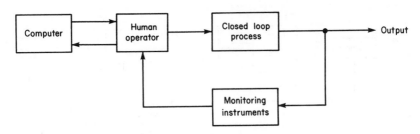

figure 13-1   off-line computer control

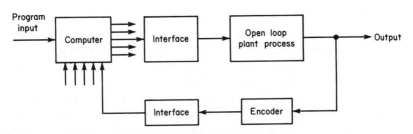

figure 13-2   on-line computer control

the overall system, where it compares the system output with the reference input and takes corrective action if necessary.

If we now restrict our interest to digital servo systems rather than considering the immense range of computer-controlled systems, we can more easily understand the signal conditioning necessary for control.

There are essentially two methods used when one is employing digital control: One method makes use of a hybrid (analog and digital) system such as that shown in Figure 13-3(a) and (b).

In Figure 13-3(a) the analog mechanical output from the servo motor is fed back as a digital signal by the use of a shaft encoder. In Figure 13-3(b) an analog transducer, such as a precision potentiometer, is used to sense the load position, the electrical output from which is converted to a digital signal by means of an analog to digital converter (ADC).

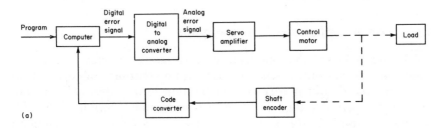

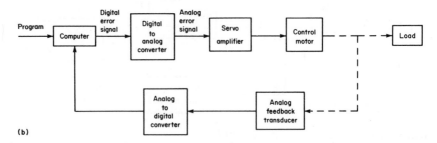

(b)

figure 13-3 a hybrid control system with output sensed by (a) shaft encoder, (b) analog transducer

In both systems the computer makes the decision to reduce the error and produces a digital error signal that is converted to an analog error signal by a digital to analog converter (DAC). The error is then used in the usual way to drive the load.

The second method of digital control uses no analog components and is accomplished entirely by digital techniques, as shown in Figure 13-4. Here the computer produces a digital error signal which is used as an input to the drive circuitry of the *stepper motor*. These circuits generate pulses at the desired rate, to enable the stepper to position the load properly. Should any steps be missed by the motor, causing a load position error, the shaft encoder, which senses the load position, will feed back the output position as a digital signal, permitting the computer to assess the correction necessary.

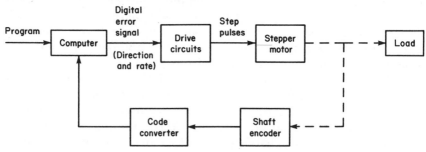

figure 13-4 a digital control system

13-2.1 open- and closed-loop operation

In Chapter 1 we saw that one feature which made the open-loop system unacceptable for control was the fact that no assurance of correct load position existed. But if correct load position is very desirable but not critical, an open-loop digital system such as the one shown in Figure 13-5 could be used. If this system were used in a computer peripheral

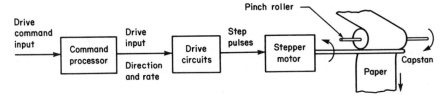

figure 13-5 open-loop digital drive system

device, such as a paper printer, it could maintain excellent output accuracy, and if it failed to do so, a simple adjustment could be made to rectify the problem. The failure of the system would not be critical to the computer operation and would, at the most, impose some short-term inconvenience. Also, because the paper drive presents a constant load, there are no problems due to load disturbances. The open-loop system is cheaper and simpler than a closed-loop system.

However, the same situation does not exist for a system required to position fuel rods in a reactor core at a nuclear power station. When each spent fuel rod is replaced by a new one, the loading machine must execute a motion similar to an *X-Y* plotter to seek the fuel rod location. If the stepper motor missed steps, the loading machine would arrive at the wrong place; if sufficient steps were lost, it would not be able to unlock the end fitting on the fuel channel, and therefore would be unable to load any new fuel. Consequently, for such a system a feedback loop is mandatory; its increased complexity and additional cost are far less important considerations than they would be in the design of a paper drive system.

13-3 DIGITAL SERVO OUTPUT DEVICES

Direct current and alternating current servo motors were first used extensively in digital control systems about 1954; with the rapid development of digital techniques, shaft position encoders were available three years later. It was not until 1960, however, that the stepper motor became a reliable commercial device. Since that time there have been many advances in its design, so today stepper motors provide the means of consistently achieving highly accurate position and speed control, in response to a digital pulse train. By controlling the total number of pulses delivered to the motor, it is possible to increment the motor position by a fixed distance or to drive the motor to a final known position. By controlling the rate at which pulses are delivered to the motor, we can control the acceleration, running speed, and deceleration of the motor.

13-1.1 the stepper motor

The theory of operation of the stepper motor is quite simple. The basic operation of such motors is illustrated in Figure 13-6. When the switch mechanism is at position 1, the rotor poles will align themselves with stator poles $N_1 - S_1$. If the switch is then moved to position 2, as shown, the rotor poles will rapidly move clockwise to align themselves with stator poles $N_2 - S_2$. It can be seen that with a little more sophistication the rotor could be made to rotate at 45 deg steps around the stator.

The relay switch in Figure 13-6 could be thought of as a drive circuit and is extremely important to the operation; it will be considered in some detail a little later.

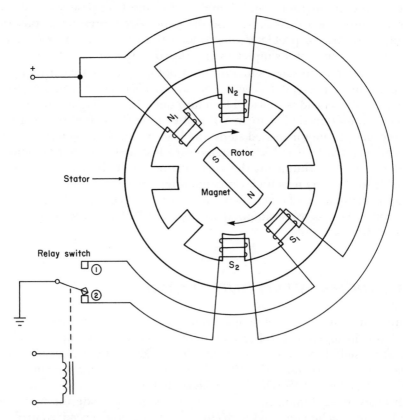

figure 13-6 the basic operation of a stepper motor

13-3.2 types of stepping motors

There are essentially two types of stepping motors commercially available, namely:

1. Permanent magnet
2. Variable reluctance

1. Permanent Magnet Stepper Motor Figure 13-7 shows the section of the type of permanent magnet stepping motor commonly found in digital drive systems where very fine step size is a requirement. The motor comprises a toothed cylindrical rotor which has an axially aligned permanent magnet set into it. The stator has each pole, doubly wound, with separate coils. The application of pulses from the drive circuitry will cause either one or two of the currents I_1, I_2, I_3, or I_4, to flow, as explained in more detail in Section 13-3.5

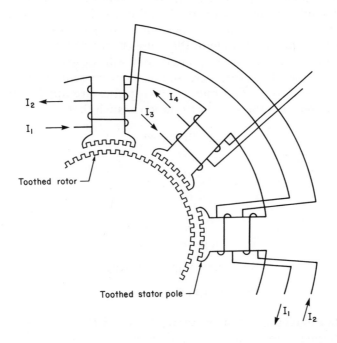

figure 13-7 typical configuration of a permanent magnet stepper motor

When I_1 or I_3 flows, it causes its respective poles to become north poles. I_2 and I_4 reverse the polarity of these poles so that they become south poles. The teeth on the stator poles and the teeth on the rotor are staggered so that only some of the poles have teeth aligned with the rotor teeth. This alignment which occurs when the coils of these poles are energized, represents a condition of magnetic equilibrium.

If other poles are now energized, the position of magnetic equilibrium is rotated and the rotor rotates to follow this position. Furthermore, the stator and rotor slots are staggered in such a way that by changing the sequence by which I_1, I_2, I_3, and I_4 are turned on and off, we can change the direction of rotation of the motor. For example, if the sequence I_1, I_3, I_2, I_4 causes a clockwise rotation, then changing the sequence to I_4, I_3, I_2, I_1 will cause the motor to turn in a counterclockwise direction.

In this type of stepping motor the number of stator and rotor teeth dictate the stepping angle so that the step angle is half the angle subtended by a rotor tooth, and there is usually a 1:1 correspondence between the applied pulses and the rotor steps (although half-step drives are possible).

Because there is a path of least reluctance, for the permanent magnet flux, at the position of magnetic equilibrium there is a "residual" or "detent" torque developed when the windings are not excited, although it is only 5 percent to 10 percent of the energized torque.

Increasing the magnitude of the pulses to the stator windings will not cause an increase in the step angle, nor will it cause a more rapid movement; it does, however, increase the stator current, which in turn produces more torque. The rate at which pulses are applied and the step angle determine the speed of rotation ω such that

$$\omega = \frac{(2\pi)(\text{SPS})(\text{SA})}{360}\text{rad/s} \qquad (13.1)$$

or

$$\omega = \frac{(2\pi)(\text{SPS})}{(\text{SPR})}\text{rad/s} \qquad (13.2)$$

where

SPS = steps per second

SA = step angle in degrees

SPR = steps per revolution

2. Variable Reluctance Stepper Motor Figure 13-8 shows a motor with a large toothed rotor made of magnetically soft iron. When a pulse energizes winding 1, the motor teeth a align themselves to the position of least reluctance in relation to the north poles (located, in this example, every 90 deg). The two intervening poles (whose windings are not excited) become south induced and act as the return path for the magnetic flux. If the next pulse in sequence energizes winding 3, the rotor will move one step clockwise to align its rotor teeth b to the new north pole configuration. The next pulse would energize winding 2, whence the rotor teeth a would once again step into alignment clockwise. For counterclockwise operation, the sequence would be 1, 2, 3. Again, there is a 1:1 correspondence between the applied pulses and rotor steps.

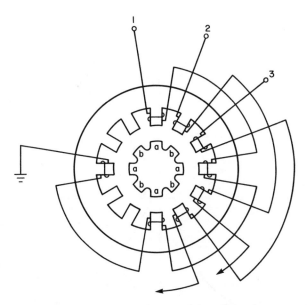

figure 13.8 typical configuration of a variable reluctance stepper motor

Since there is very little residual magnetism in the rotor iron, there is no "residual" or "detent" torque present when the stator is not excited.

This stepper happens to have a three-step sequence and eight rotor teeth. The product of the step sequence and the number of rotor teeth equals the number of steps that the rotor makes in one revolution, namely;

$$SPR = (SS)(T_r) \qquad (13.3)$$

where SS = number of step sequences

and T_r = number of rotor teeth.

13-3.3 stepper motor characteristics

If a stepper motor is steadily excited with a dc voltage to one step winding, a counter torque will be sensed if the shaft is rotated in either direction from the rest position. The relationship of torque versus angle is nearly sinusoidal and is shown in Figure 13-9. The maximum counter torque (T_{max}) measured at a deviation of one step angle either clockwise or counterclockwise is termed the *holding torque* and is always specified on manufacturers' data sheets, usually in units of ounce-inches. In fact,

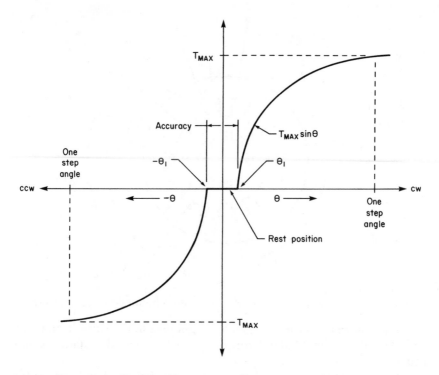

figure 13-9 torque versus rotation

one of the decided advantages of using a stepper motor as the drive in a position servo system is that the motor can maintain the holding torque under a stall condition indefinitely (e.g., up against an end stop) without suffering any damage. The motor "stiffness" is related to the holding torque by equation (13.4).

$$MS = \frac{HT}{SA} \qquad\qquad (13.4)$$

where   MS = motor stiffness (oz-in./degree)

                    HT = holding torque (oz-in.)

and   SA = step angle (degree)

      Figure 13-9 shows that as the rotor approaches the rest position, less and less torque is generated internally and at some point, θ_1 (or $-\theta_1$) motor friction torque equals rotor torque, and the rotor never achieves the precise rest position. This error in position is a measure of the accuracy of the motor and is typically 3 to 5 percent of the step angle.

It is also noncumulative, in that it remains 3 to 5 percent of one step angle, regardless of the number of steps taken. If, for example, a motor has an accuracy of ±5 percent and a step angle of 0.72 deg, the accuracy for one step is 0.72 deg ± 2.16 min or arc, and the accuracy for one revolution is also 360 deg ± 2.16 min of arc.

The dynamic step response to one input pulse is illustrated in Figure 13-10, where t_m is considered to be the time taken for one step.

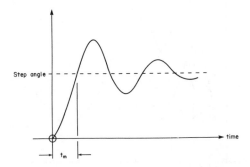

figure 13-10 typical angular step response

The value of t_m is determined in part by the ability of the motor to accelerate the rotor mass. The oscillations about the rest position indicate the underdamped condition of the motor dynamics. The degree of oscillation is determined by three factors: the load torque, T_L, the load inertia J_L, and the current rise time.

Load torque: Increasing the load torque reduces the oscillation and increases the time taken for one step. This is shown in Figure 13-11 when $T_{L_1} < T_{L_2} < T_{L_3}$

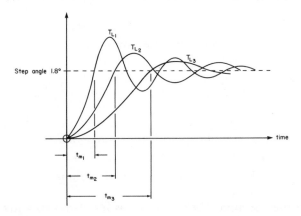

figure 13-11 effect of load torque on the step response

Load inertia: Increasing the load inertia causes increased oscillation, as well as increasing both the time taken for one step and the settling time, as illustrated in Figure 13-12, where $J_1 < J_2 < J_3$. It should be emphasized that an increase in the load inertia adversely affects the ability of the motor to stop, start, and restart quickly.

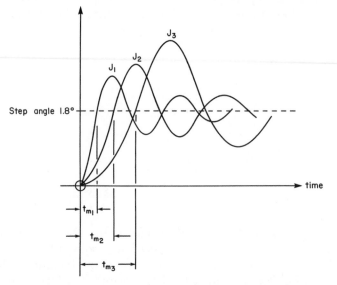

figure 13-12 effect of load inertia on the step response

Current rise time: The motor torque is developed by current in the stator windings. If the current takes a long time to rise to maximum, it will take the torque a long time to develop and thereby slow the motor response. The electrical time constant τ determines the current rise such that

$$\text{Current rise time} = 5\tau \qquad (13.5)$$

where $$\tau = \frac{L}{R}$$

L = motor phase inductance (henrys)

R = motor phase resistance (ohms)

To reduce the rise time, the resistance is increased by adding external circuit series resistance, R_s, so that

$$\text{Current rise time} = \frac{5L}{R + R_s} \qquad (13.7)$$

To overcome the current reduction due to increased circuit resistance, the excitation voltage must be increased. Consider the following example:

EXAMPLE 13.1

Stepper motor step size	1.8 deg steps
Inductance/phase	0.63 mH
Resistance/phase	0.395 Ω
Current/phase	3.8 amps
Voltage	1.5 Vdc

$$\text{Current rise time} = \frac{5 \times 0.63 \times 10^{-3}}{0.395} = 7.985 \text{ ms}$$

If the time for one pulse must be no longer than 1 ms, for example, then the current rise time would prohibit this. However, if 3 Ω (R_s) were added to each phase externally, then

$$\text{Current rise time} = \frac{5 \times 0.63 \times 10^{-3}}{3.395} = 0.928 \text{ ms}$$

To achieve the rated current of 3.8 A/phase, the dc voltage would have to be increased to (3.8 A) · (3.395 Ω) = 12.9 volts. The effect of current rise time on the rotor step response is illustrated in Figure 13-13.

If the rotor is required to run at a constant speed in response to a pulse train, it will do so in perfect synchronism with the incoming pulses, but only up to a certain maximum rate. *The maximum response rate* is the highest pulse rate that an unloaded motor can follow from a "standing start" without missing a step; it is indicated at *A* in Figure 13-14. In the figure J_0 represents zero load inertia and $J_0 < J_1 < J_2 < J_3 < J_4$. In the start-stop range, indicated by the value of the load inertia, the motor is capable of starting, stopping, restarting, or reversing, all without missing a step. For example, if the load inertia is J_4, the motor can perform these starting and stopping actions without missing a step, while developing a maximum torque of 600 oz-in., provided the pulse rate is less than 140 p.p.s. (pulses per second). However, once it is running, the motor can exceed the maximum response rate by entering the "slewing" range, where it will

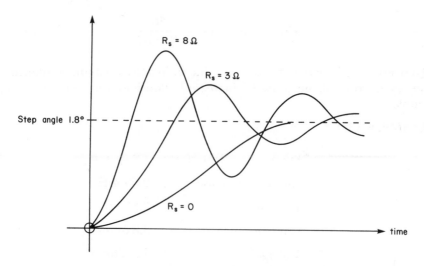

figure 13-13 effect of current rise time on the step response

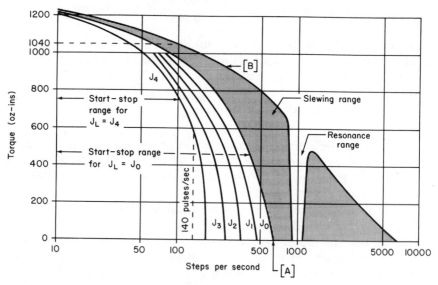

figure 13-14 speed/torque curves of a stepping motor

remain synchronized (one step per pulse) with the pulse train but cannot be made to stop and start without missing steps. A typical maximum slewing rate is shown by curve B in Figure 13-14. For most permanent magnet motors, there is a resonance range between 900–1200 p.p.s. in which a loss of torque exists, and this region should be avoided when designing the servo drive system.

13-3.4 the ramping mode operation

The usual start-stop sequence for a stepping motor is to start the motor at a speed in the start-stop range. (It will achieve this velocity in one step movement, of course.) When this speed, v_1 in Figure 13-15, has been reached, the motor is then "ramped" up into the slew range to a steady velocity v_2, where it will stay in synchronism, until it is required to stop. At that time it will be "ramped" down back to v_1 and stopped. In Figure 13-15 the ramp rise time is indicated by t_R and the ramp fall time by t_F. To accelerate up the ramp, the motor must develop enough torque to overcome the load inertia, its own rotor inertia, and the opposition presented by any load torque. The ramp down will often take less time, due to the aiding effect of the load torque.

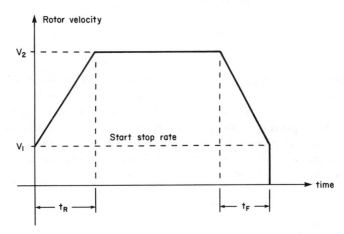

figure 13-15 velocity versus time for a stepping motor in the ramp mode

EXAMPLE 13.2

Calculate the available motor torque if a stepper is ramped from 100 to 500 steps per sec. in 30 ms when the load inertia is 0.104 oz-in. s^2 and the rotor inertia is 0.101 oz-in. s^2. The motor has a stepping angle of 1.8 deg, a slew curve represented by Figure 13-14, and negligible load friction. Also calculate the minimum ramp rise time.

solution:

Using equation (2.10), we obtain

$$T = J\alpha$$

or

$$T = J\frac{(\omega_2 - \omega_1)}{t_R} \tag{13.8}$$

where ω_2 = final angular velocity in rad/s

ω_1 = initial angular velocity in rad/s

t_R = ramp rise time

J = total inertia

To determine $(\omega_2 - \omega_1)$ we use equation (13.1) so that

$$\omega_2 - \omega_1 = \frac{(2\pi)(\text{SA})}{360°}(\text{SPS}_2 - \text{SPS}_1)$$

Hence $$T = \frac{(0.104 + 0.101) \times 2\pi \times 1.8 \times (500 - 100)}{360 \times 0.03}$$

Therefore, $T = 85.9$ oz-in.

Figure 13-14 suggests that the stepper could produce 800 oz-in. at 500 steps per second. However, because the motor is required to accelerate at a steady rate up to 500 SPS, we note that just before reaching this speed (at 499 SPS, say), 85.9 oz-in. of the motor torque is being used to accelerate the motor. Consequently, the available motor torque (i.e., the maximum load that the motor can accelerate in this way) is

$$800 - 85.9 = 714.1 \text{ oz-in.}$$

The minimum ramp rise time, $t_{R\,\text{min}}$, can be approximated quite accurately from the equation

$$t_{R\,\text{min}} = J\frac{(\omega_2 - \omega_1)}{T_{\text{ave}}}$$

where T_{ave} is the average value of the maximum motor torque as its speed changes from ω_1 to ω_2, i.e.,

$$T_{\text{ave}} = \frac{1040 + 800}{2} = 920 \text{ oz-in.}$$

Hence $$t_{R\,\text{min}} = \frac{(0.104 + 0.101) \times 2\pi \times 1.8 \times (500 - 100)}{360 \times 920}$$

or $$t_{R\,\text{min}} = 2.8 \text{ ms}$$

13-3.5 stepper motor drive circuit

The motor drive circuit is a very important part of the stepper motor system. In essence, it accomplishes four things:

1. Determines the direction of rotation.
2. Determines the winding excitation sequence.
3. Provides current to excite the winding.
4. Provides fast deenergizing of each phase winding after the pulse is removed.

Figure 13-16 shows a block diagram for a four-winding, permanent-magnet stepper motor.

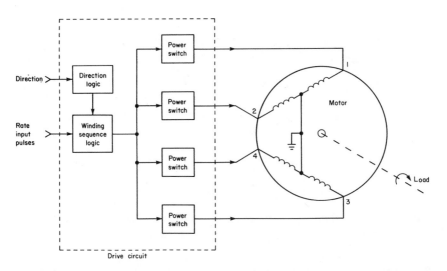

figure 13-16 block diagram of a drive circuit and a four-winding stepper motor

There are many different drive circuit configurations, the designs of which depend upon the demands made by the load and the type of stepper motor used. Since it would be unreasonable to cover each one, let us direct our attention to a limited number of the more common drive circuits. We will first examine the winding excitation sequence and the pulse train that determines the sequence.

Figure 13-17 illustrates two different excitation sequences for the permanent magnet stepper shown in the winding diagram of Figure 13-7 and shown schematically in the block diagram of Figure 13-16. In Figure 13-17(a) each winding is energized separately by a pulse train with a 25 percent duty cycle, but in Figure 13-17(b) pairs of windings

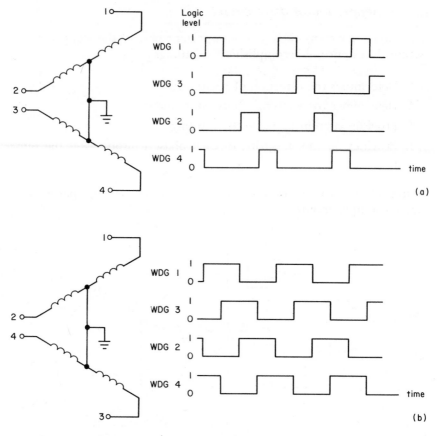

figure 13-17 ideal current logic pulse trains for: (a) single-phase, four-winding sequence, (b) two-phase, four-winding sequence

are energized alternately by overlapping pulses of 50 percent duty cycle. Logic level 1 means that current excitation is present, whereas logic level 0 means that current excitation is absent.

1. Single Phase Sequencing: In single-phase sequencing a 1-3-2-4 sequence, for example, would impart a clockwise rotation to the motor, and a 4-3-2-1 would cause counterclockwise rotation. To determine rotation, the input to the drive circuit must contain two kinds of information: direction and rate of stepping. A logic 1 from the processor is often used for the clockwise signal and a logic 0 for counterclockwise; these logic levels would be 5 V and 0 V, respectively. The MPU would then output a pulse train at a rate which controls the stepper speed and having the required number of pulses to control the output shaft po-

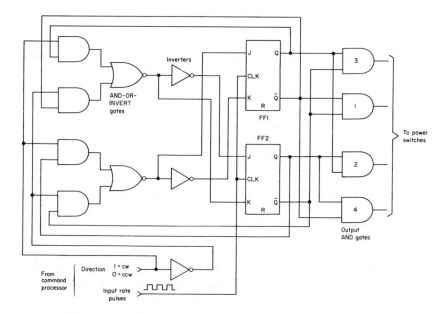

figure 13-18   single-phase sequencing logic

sition. A typical drive circuit providing these signals is shown in Figure 13-18. Fast negative-going pulses are required at the rate input connection to toggle the *J-K* flip-flops.

J-K flip-flops are used because their outputs can never be indeterminate and thereby cause the motor to miss steps. The sequencing operation of the circuit in Figure 13-18 takes place in the following way:

1. A logic 1 is applied to the "direction" inverter input and held there for the duration of the motor operation.

2. Flip-flop 1 (FF1) is initially conditioned such that $J = 1$ and $K = 0$; the output is reset to $Q = 0$ and $\bar{Q} = 1$.

3. FF2 is conditioned with $J = 0$ and $K = 1$; the output is reset to $Q = 0$ and $\bar{Q} = 1$.

4. This generates an output from "output gate 1" only, allowing motor phase 1 to be energized.

5. The negative-going edge on the first "input rate" pulse changes FF1 so that $Q = 1$ and $\bar{Q} = 0$.

6. The clock pulse cannot change FF2 because FF2 is already reset; i.e., for FF2, $Q = 0$ and $\bar{Q} = 1$.

7. The new state generates an output from "output gate 3" only, allowing motor phase 3 to be energized.

8. It also maintains the condition of FF1 so that $J = 1$ and $K = 0$.

9. It conditions FF2 so that $J = 1$ and $K = 0$.

10. The trailing edge on the next clock pulse cannot change the output of FF1, so FF1 remains at $Q = 1$ and $\bar{Q} = 0$.

11. FF2 changes state so that $Q = 1$ and $\bar{Q} = 0$.

12. The new state generates an output from output gate 2 only, allowing motor phase 2 to be energized.

The sequencing continues until the "input rate" pulses stop.

2. Two-Phase Sequencing: In two-phase sequencing, phases 1 and 3 are excited simultaneously for one step. Phase 1 is then deenergized, and phase 2 is excited with phase 3. After the same length of time has elapsed, phase 3 is deenergized and phase 4 is excited with phase 2, and so on. This could constitute a clockwise rotation perhaps, and conversely the counterclockwise rotation would be the reverse order of sequencing. Figure 13-19 shows such a circuit with a clockwise direction indicated and all the logic levels shown for excitation of phase 1 and 3. The circuit is ready for the negative clock trigger to excite phases 2 and 3.

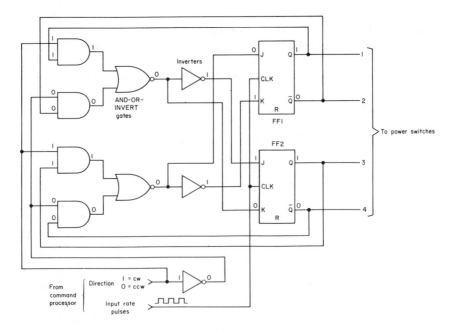

figure 13-19 two-phase sequencing logic

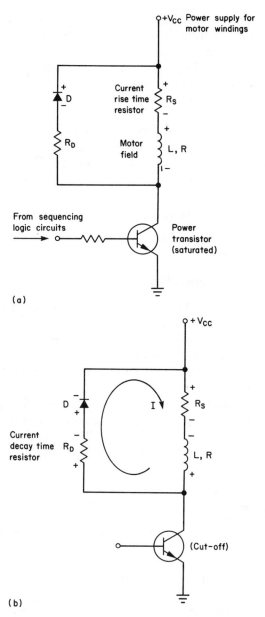

figure 13-20 (a) power switching circuit, (b) field deenergizing loop

3. Power Switches: Power switching of the individual field windings can be made by a circuit similar to the one shown in Figure 13-20(a).

When a voltage, representing a logic level 1, appears at the base of the power transistor, it conducts, allowing current to build up and energize the field. R_s limits the current if V_{cc} is large; it also reduces the current rise time as determined by equation (13.7). When the pulse at the transistor base terminates, the transistor turns off; however, the energy which has built up in the field has to be dissipated and a path is provided for the current to decay, through R_D, D, and R_s. When the field flux collapses, the induced emf produces the polarities shown in Figure 13-20(b). To permit the current decay to take place very quickly, an extra resistor, R_D, is added to the "deenergizing path." The current decay time may be calculated by

$$\text{Decay time} = \frac{5 \cdot L}{R + R_s + R_D} \qquad (13.9)$$

13-3.6 digital shaft encoders

Shaft encoders provide the simplest method of interfacing a feedback position or rate signal with a computer. The encoder is an electromechanical device that senses the rotation of the shaft and converts the angular movement to a digital signal. The signal can be just a series of pulses, or it can be either a weighted or unweighted binary code.

1. Incremental Shaft Encoders: Incremental encoders produce a train of pulses when the shaft is rotated. The pulses are produced by one of two methods, as illustrated in Figure 13-21(a) and (b). In both cases the pulse is produced at precise angular increments.

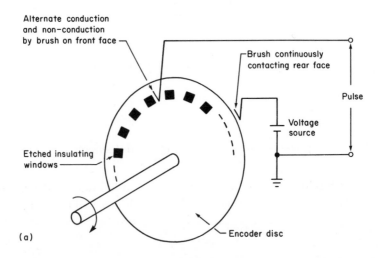

(a)

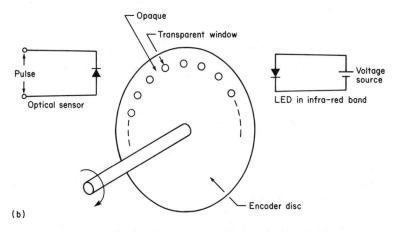

(b)

figure 13-21 incremental shaft encoders: (a) contacting type; (b) optical (noncontacting) type

The contacting disc has precious-metal conducting strips alternating with etched, epoxy-filled, insulating windows on one side and a precious-metal coated face on the other. Precious-metal wiping brushes touch both sides of the disc. The contact method is prone to all the problems inherent in such mechanical devices, e.g., brush wear, aging, noisy contacts, much smaller lifetime. The optical disc method, on the other hand, is much more reliable and costs less.

The pulses contain no information by themselves, but if counted they will be a measure of angular travel and if counted against a time base they will indicate the shaft speed.

If direction is to be determined, the encoder disc must contain two tracks of pulse-generating strips. Figure 13-22 shows a method used to determine direction, measured angle, and even speed. Figure 13-22(a) shows the disc having two tracks (which provide square wave outputs) staggered in such a way that the pulse trains are 90 deg out of phase.

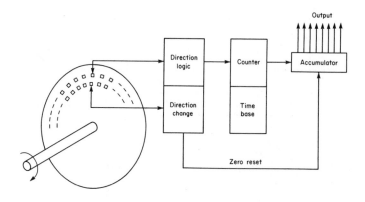

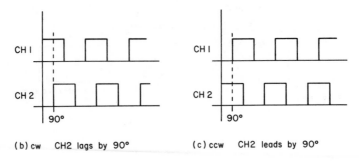

(b) cw   CH2 lags by 90°   (c) ccw   CH2 leads by 90°

figure 13-22   (a) angle/speed measurement and pulse train relationships, (b) clockwise sense, (c) counterclockwise sense

Figure 13-22(b) illustrates channel 2 lagging 90 deg behind channel 1, which indicates a clockwise sense of rotation. Figure 13-22(c) shows the pulse relationship for a counterclockwise sense. The direction logic connects the encoder to the counter and the accumulator and provides a reset signal if the direction changes.

Another technique is the generation of short-duration pulses and the multiplication of one channel by a factor of 2, 4, or 10 prior to the train's entering the direction logic. This method is shown in Figure 13-23; depending upon the direction, only one pulse train is fed from the direction logic to the counter.

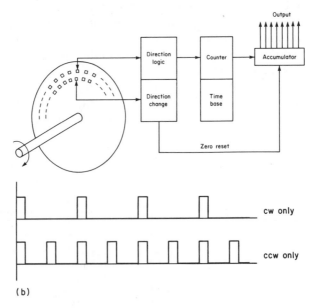

figure 13-23   (a) angle/speed measurement, (b) pulse train relationship for direction

Incremental encoders can produce up to 5000 pulses per revolution (PPR) and (for lower pulse repetition rates) be driven at 3500 RPM. The angular resolution in degrees of shaft rotation is determined by the relationship

$$\text{Resolution} = \frac{360°}{\text{PPR}} \qquad (13.10)$$

Because the pulses shown in Figure 13-22 have a 50 percent duty cycle, the duration of each pulse, or pulse width, is determined in the following way:

$$\text{PW} = \frac{30}{(\text{RPM}) \cdot (\text{PPR})} \qquad (13.11)$$

where
$$\text{PW} = \text{pulse width in seconds}$$
$$\text{RPM} = \text{speed of shaft in revs/min}$$
$$\text{PPR} = \text{pulses per revolution.}$$

In general, the pulse duration of an encoder, such as that shown in Figure 13-22, depends upon the duty cycle of the pulse train in the following way:

$$\text{PW} = \frac{\text{DC} \times 60}{(\text{RPM}) \cdot (\text{PPR})} \text{ seconds} \qquad (13.12)$$

where
$$\text{DC} = \text{duty cycle of the pulse train}$$

EXAMPLE 13.3

A two-channel, incremental shaft encoder is driven at 2400 rpm by a stepper motor in the slew mode. If the encoder has a square wave output (50 percent DC) and produces 600 pulses per revolution, calculate

1. The angular resolution of position
2. The pulse width
3. The pulse rate
4. The pulse delay between the two channels.

solution:

1. $\text{Resolution} = \dfrac{360°}{\text{PPR}}$

$$= \frac{360}{600} = 0.6°$$

$$= 36' \text{ of arc}$$

2. $PW = \dfrac{30}{(RPM) \cdot (PPR)}$

$$= \frac{30}{2400 \times 600}$$

$$= 20.8 \ \mu s$$

3. $\text{Pulse rate} = \dfrac{1}{2(PW)}$ *(13.13)*

$$= \frac{10^6}{2 \times 20.8}$$

$$= 24.038 \text{ kHz}$$

4. $\text{Pulse delay} = \dfrac{PW}{2}$

$$= \frac{20.8}{2}$$

$$= 10.4 \ \mu s$$

EXAMPLE 13.4

If the duty cycle of a $2\mu s$ pulse width train from an incremental encoder is 10 percent for a clockwise sense, what shaft speed does this represent when the encoder produces 3000 pulses per revolution in a clockwise sense? What is the pulse repetition rate?

solution:

Using equation (13.12), we can determine the speed thus

$$RPM = \frac{(DC) \cdot (60)}{(PW) \cdot (PPR)}$$

Hence $RPM = \dfrac{0.1 \times 60}{2 \times 10^{-6} \times 3000}$

or $\text{Speed} = 1000 \text{ rpm}$

$$\text{Pulse rate} = \frac{1}{T}$$

where

$$T = \frac{(PW)}{(DC)}$$

Hence

$$\text{Pulse rate} = \frac{0.1}{2\ \mu s} = 50\ \text{kHz}$$

2. Absolute Shaft Encoders: The absolute encoder produces a series of pulses on a number of lines simultaneously, in the form of a digital code. At any instant of time the code will measure the angular excursion from a reference position. If the output is referred to a time base, the speed of the shaft may be determined. The sensing of position is either contact or optical, as in the case of the incremental encoder; however, the method of encoding is to lay concentric tracks around the disc with a sensor for each track. The angular resolution is dependent upon the code used, the number of tracks, and the density of pattern laid down. Figure 13-24 shows a disc with a binary code pattern that would resolve a circle into 45 deg sectors (not very useful). In Figure 13-24 the code is laid down from the outside to the inside with each track representing an ascending order of binary, digit, i.e., 2^0, 2^1, 2^2, etc. Table 13-1 shows the relationships between the sector number, the angle, the binary code, and the etched pattern.

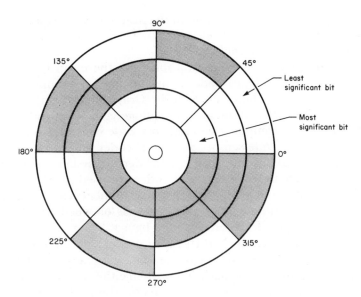

figure 13-24 a 3-track binary coded disc resolving 360 deg into 45 deg sectors

TABLE 13-1—CODE AND PATTERN RELATIONSHIPS
FOR THE DISC SHOWN IN FIGURE 13-24.

Angle (degrees)	Decimal number (sector)	Binary			Pattern on disc		
		MSB		LSB	MSB		LSB
0	0	O	O	O			
45	1	O	O	I			▓
90	2	O	I	O		▓	
135	3	O	I	I		▓	▓
180	4	I	O	O	▓		
225	5	I	O	I	▓		▓
270	6	I	I	O	▓	▓	
315	7	I	I	I	▓	▓	▓

It is obvious that the angular information from an asbolute encoder cannot be lost or changed due to power failure or noise transients, as is the case with the incremental encoder. However, the manufacture of the disc for an absolute encoder is much more expensive due to the increased complexity of the disc code and the necessity of mechanically locating the extra contacts or optical sensors.

Rotation sense is usually clockwise for an increasing count and counterclockwise for a decreasing count.

3. Codes: The digital codes used on the disc can be true binary, binary-coded decimal (BCD), BCD-excess-3, or Gray code. A sample of the four codes is shown in Table 13-2.

For the first three codes in Table 13-2 there can be as many as four simultaneous changes made at any one time for a simple 4-bit code. This does not occur, however, with the Gray code system, where only one bit changes each increment. Unfortunately, the Gray code is incompatible with most computer systems, so a code converter is interposed between the Gray code sensors and the encoder output terminals. Because they are a compromise between the decimal system and the binary system, the BCD and BCD-excess-3 codes are used nearly exclusively. The advantage of the excess-3 code is that numbers can be added and subtracted as for simple binary. The code is obtained by adding 3 to each decimal digit before converting to binary.

Because it is extremely difficult to precisely locate the contact or optical sensors on the disc, a code reading error can occur when the code is read. This is particularly likely when either the number of tracks or the code density is increased. To overcome this problem of ambiguity,

TABLE 13-2—CODE COMPARISONS

DECIMAL NUMBER	BINARY		BCD		BCD - Excess-3		GRAY
	32 16 8 4 2 1	2nd decade 8 4 2 1	1st decade 8 4 2 1	2nd decade 8 4 2 1	1st decade 8 4 2 1		unweighted
0	0 0 0 0 0 0	0 0 0 0	0 0 0 0	0 0 1 1	0 0 1 1	0 0 0 0 0 0
1	0 0 0 0 0 1	0 0 0 0	0 0 0 1	0 0 1 1	0 1 0 0	0 0 0 0 0 1
2	0 0 0 0 1 0	0 0 0 0	0 0 1 0	0 0 1 1	0 1 0 1	0 0 0 0 1 1
3	0 0 0 0 1 1	0 0 0 0	0 0 1 1	0 0 1 1	0 1 1 0	0 0 0 0 1 0
4	0 0 0 1 0 0	0 0 0 0	0 1 0 0	0 0 1 1	0 1 1 1	0 0 0 1 1 0
5	0 0 0 1 0 1	0 0 0 0	0 1 0 1	0 0 1 1	1 0 0 0	0 0 0 1 1 1
⋮						
22	0 1 0 1 1 0	0 0 1 0	0 0 1 0	0 1 0 1	0 1 0 1	0 1 1 1 0 1
23	0 1 0 1 1 1	0 0 1 0	0 0 1 1	0 1 0 1	0 1 1 0	0 1 1 1 0 0
24	0 1 1 0 0 0	0 0 1 0	0 1 0 0	0 1 0 1	0 1 1 1	0 1 0 1 0 0
25	0 1 1 0 0 1	0 0 1 0	0 1 0 1	0 1 0 1	1 0 0 0	0 1 0 1 0 1
26	0 1 1 0 1 0	0 0 1 0	0 1 1 0	0 1 0 1	1 0 0 1	0 1 0 1 1 1

the code is read by two sets of sensors displaced from one another by a very precise angle. Their outputs are then fed to antiambiguity logic circuits that resolve the actual angle of the shaft. Figure 13-25 illustrates an example of such a configuration. Because the sensors are mechanically displaced, one set will lead the angle to be read and one set will lag, and hence the code signals are called *lead* and *lag*.

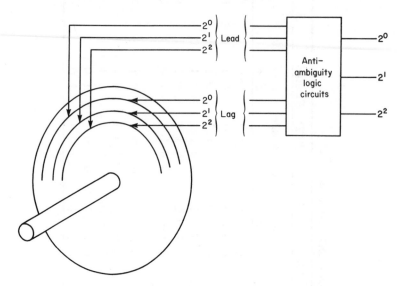

figure 13-25   two sets of sensors are used to resolve ambiguity

Regardless of the code used, the outputs are always compatible with DTL/TTL logic levels. The LSB (least significant bit) of the true binary code or the LSB of the first decade of the BCD and excess-3 codes are used as data-ready signals for computer interfacing. Also the same LSBs can be used in a time base counter to indicate the speed of the shaft rotation.

The angular resolution in degrees depends upon the number of bits if true binary is used and the number of decades if BCD is used, so that

$$\text{Resolution} = \frac{360°}{2^n} \quad \text{for binary} \qquad (13.14)$$

where $n = $ number of bits

and $\text{Resolution} = \dfrac{360°}{10^d} \text{ for BCD}$

where $\qquad d$ = number of decades

For a binary coded output the shaft speed can be determined from equation (13.16)

$$RPM = \frac{(LSBR) \cdot (60)}{2^{n-1}} \qquad (13.16)$$

where $\qquad$ LSBR = least significant bit rate

in pulses per second

n = number of bits

and the pulse width of the LSB is

$$PW = \frac{1}{(2) \cdot (LSBR)} \text{ seconds} \qquad (13.17)$$

For a BCD code output the shaft speed can be found by using equation (13.18) so that

$$RPM = \frac{(LSBR) \cdot (120)}{10^d} \qquad (13.18)$$

where $\qquad d$ = number of decades

The pulse duration of the LSB is found by using equation (13.17)

EXAMPLE 13.5

If (a) a true binary 8-bit code and (b) a BCD 3-decade code are used for an absolute encoder, what is

1. The angular resolution?
2. The output code for an angle of 40 deg?
3. The shaft speed if the LSB rate is 2500 pulses per second?
4. The pulse duration for the given shaft speed?

solution:

(a) Binary:

1.

$$\text{Resolution} = \frac{360°}{2^n}$$

$$\text{Resolution} = \frac{360°}{2^8} = 1.41°$$

2. The number of decimal angular increments in 40 deg is

$$\frac{40°}{1.41°} = 28.44$$

Rounded to the nearest whole number, the number is 28 and

$$28 \text{ (decimal)} = 00011100 \text{ (binary)}$$

3.

$$\text{RPM} = \frac{(\text{LSBR}) \times 60}{2^{n-1}}$$

$$\text{RPM} = \frac{2500 \times 60}{2^7}$$

$$\text{Speed} = 1172 \text{ rpm}$$

4.

$$\text{PW} = \frac{1}{2 \times (\text{LSBR})} = \frac{1}{2 \times 2500}$$

$$\text{PW} = 200 \ \mu s$$

(b) B.C.D.:

1.

$$\text{Resolution} = \frac{360°}{10^d}$$

$$\text{Resolution} = \frac{360}{10^3} = 0.36°$$

2. The number of decimal angular increments in 40 deg is

$$\frac{40°}{0.36°} = 111.11$$

Rounded to the nearest whole number, the number is 111 and

$$111 \text{ (decimal)} = 0001 \ 0001 \ 0001 \text{ (BCD)}$$

3.

$$RPM = \frac{(LSBR) \times 120}{10^d}$$

$$RPM = \frac{2500 \times 120}{10^3}$$

Speed = 300 rpm

4.

$$PW = \frac{1}{2 \times (LSBR)} = \frac{1}{2 \times 2500}$$

$$PW = 200 \ \mu s$$

13-4 SUMMARY

We have now seen just how accurate a digital system can be, particularly if an encoder is used in the feedback loop. It should be apparent that for interfacing with a computer the timing of pulses is very important, and, of course, the input and output voltage levels have to be compatible with the computer or microprocessor logic levels. At the present time the output power from the largest stepper motors is inadequate to meet the load demands made by some of the larger position control systems, but it is only a matter of time before such steppers are available.

PROBLEMS

Section 13-3.2

1. A 1.8 deg stepper can slew at a maximum rate of 10,000 steps per second, but only 500 steps per second in the start-stop range.
   (a) How many steps per revolution does the motor make?
   (b) What is the maximum shaft speed in both ranges?

2. If the motor in Problem 1 above is a permanent magnet type with a four-step sequence, how many rotor teeth does it have?

Section 13-3.3

3. Given the following specifications, which motor exhibits the greater "stiffness"?

	HOLDING TORQUE (oz-in.)	STEP ANGLE (deg)
motor 1	24	1.8
motor 2	96	9

4. A four-sequence stepper has a maximum slew rate of 4500 steps per second. If the current must rise to 63.2 percent of rated value in the pulse width, how much resistance must be added externally, and what is the maximum voltage applied if the stepper has the following characteristics?

Inductance/phase	2.8 mH
Resistance/phase	0.075 Ω
Current/phase	4 Amps
Voltage	3 volts

Section 13-3.4

5. The motor in Problem 4 above is run up to 200 steps per second in the start-stop mode, then ramped to 4500 steps per second in 800 ms. The rotor inertia is 0.106 oz-in. s^2, the step angle is 1.8 deg and the motor has a slew curve represented by Figure 13-14. What is the maximum value of load inertia that the motor can drive if the load torque is 64.55 oz-in. and the effects of resonance can be neglected? The load friction torque is negligible.

6. What is the minimum ramping time for the motor specified in Problems 4 and 5 above?

Section 13-3.5

7. Develop the gate, inverter, and flip-flop logic

sequences for CCW rotation for the two-phase stepper drive circuit logic shown in Figure 13-19.

Section 13-3.6

8. A two-channel incremental shaft encoder is driven at a rate of 200 rpm and has an angular resolution of 0.72 deg. What is the delay between the pulses if the encoder has a square wave output?

9. What are the pulse rate and pulse width from an incremental encoder if the duty cycle is 15 percent for a clockwise sense? The encoder produces 2500 pulses per revolution and the speed is 500 rpm.

10. A 4-decade BCD coded disc of an absolute encoder has an LSB rate of 8000 pulses per second. Calculate
    (a) The shaft speed.
    (b) The pulse duration.

11. What is the angular resolution for the absolute encoder described in Problem 10 above? Determine the output code for 135 deg of rotation.

converters: digital to analog and analog to digital

14

. . . "So you think you're changed, do you?" "I'm afraid I am Sir," said Alice.

Lewis Carroll, *Alice in Wonderland*

14-1 INTRODUCTION

If a control system is a hybrid of an analog actuator and digital controller, e.g., if the load is to be driven by a dc or ac servo motor, it is necessary that the digital error signal be changed to an analog voltage before being fed to the servo amplifier. This conversion is made by a digital to analog converter (DAC).

We have seen that the load position and rate can be sensed by a digital shaft encoder. However, if the feedback signal is developed by an analog transducer, it is necessary to convert the transducer output to a digital signal by means of an analog to digital converter (ADC). This signal can then be used by the computer to develop a digital error signal, which can then be acted upon according to the computer's control algorithm.

Sometimes the input to the ADC changes too rapidly. Under such conditions it is necessary to continually sample the analog signal at very small intervals; the sample is then held at a fixed value, for the duration of the sampling period, while the ADC converts the analog value of the sample to a digitally coded signal, after which the whole sampling process repeats itself. Sample and Hold (S/H) circuits such as this form an integral part of some ADC processes.

14-2 THE DIGITAL TO ANALOG CONVERTER (DAC)

Description: The DAC produces an output voltage or current proportional to the input digital code.

One type of input can be in the form of a pulse train, where each pulse contains no information by itself but represents one count. This method is of little concern to us when we are considering servo systems.

Alternatively, the digital input can be in the form of a binary code which is fed to the DAC either in serial or parallel fashion. The three input forms are shown in Figure 14-1. In the serial method [Figure 14-1(b)] only one input line is required, but to transfer the data into the DAC requires eight clock pulses to the input register. Only one clock pulse to the input register is needed for the parallel input illustrated in Figure 14-1(c). However, in this case, eight lines are required to transfer one byte. Because of its speed of operation, parallel data transfer to a DAC is the most common, and we will consider only this form of transfer from now on.

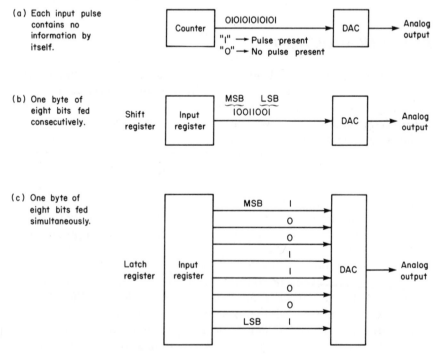

figure 14-1 different forms of input to a DAC: (a) count, (b) serial; (c) parallel

14-2.1 DAC Input Codes

Unipolar Codes: The output from a DAC is not smooth but is incrementally stepped due to the discrete form of the digital input. The resolution of the DAC (i.e., change in value of the analog output signal in one increment) must be such that, together with the gain of the servo amplifier, the resultant change in the motor input voltage is less than the dead-band of the motor.

The maximum possible output voltage of the DAC is referred to as the *full-scale voltage* (FSV); hence each input byte (8 bits) will represent some discrete fraction of the FSV. The resolution of the DAC is the smallest incremental change of the analog output that can be made with respect to the FSV. The resolution will, therefore, depend upon the number of bits if the input is true binary (and the number of decades if the input is BCD). The resolution may be determined as follows:

$$\text{Resolution} = \frac{1}{2^n} \text{ (true binary)} \qquad (14.1)$$

$$\text{Resolution} = \frac{1}{10^d} \text{ (BCD)} \qquad (14.2)$$

where n = number of bits

and d = number of decades

EXAMPLE 14.1

If a servo amplifier with unity gain drives a motor (such as the one described in Section 10-2.7) in a clockwise sense only and derives its input from a DAC with (a) an 8-bit true binary input, (b) a 2-decade 8421 BCD input, what is the resolution of the DAC?

solution:

The motor (Section 10-2.7) has an input of 48 volts dc.

(a) True binary:

$$\text{Resolution} = \frac{1}{2^n} \cdot 48 \ (n = 8)$$

$$\text{Resolution} = 187.5 \text{ mV}$$

(b) BCD:

$$\text{Resolution} = \frac{1}{10^d} \cdot 48 \ (d = 2)$$

$$\text{Resolution} = 480 \text{ mV}$$

The relationship between the input code and the output voltage for an 8-bit binary DAC whose FSV is 48 volts is shown in Table 14-1.

TABLE 14-1—INPUT/OUTPUT FOR AN 8-BIT BINARY DAC

BINARY NUMBER INPUT	FRACTION OF FSV	FACTOR	OUTPUT VOLTAGE (FACTOR) × (FSV)
00000000	0/256	0	0
00000001	1/256	0.0039	0.1875
00000010	2/256	0.0078	0.3750
00000011	3/256	0.0117	0.5625
⋮			
00001000	8/256 = 1/32	0.03125	1.5000
⋮			
00100000	32/256 = 1/8	0.125	6.0000
⋮			
10000000	128/256 = 1/2	0.5	24.0000
⋮			
11111111	255/256	0.9961	47.8125

Table 14-2 shows the input code and the output voltage for a DAC with a two-decade 8421 BCD format, whose FSV is 48 volts.

TABLE 14.2—INPUT/OUTPUT FOR A DAC HAVING A TWO-DECADE BCD FORMAT

BCD NUMBER INPUT	FRACTION OF FSV	FACTOR	OUTPUT VOLTAGE (FACTOR) × (FSV)
0000 0000	0/100	0	0
0000 0001	1/100	0.01	0.48
0000 0010	2/100	0.02	0.96
0000 0011	3/100	0.03	1.44
⋮			
0001 0000	10/100	0.1	4.80
0001 0001	11/100	0.11	5.28
⋮			
1001 1001	99/100	0.99	47.52

Three points should be apparent from Tables 14-1 and 14-2. The first is that a true binary input has better resolution than a BCD for the same number of bits. In this case, the resolution is improved by a factor of 2.56 (i.e., 0.48/0.1875). Second, the analog output never quite attains the FSV, but can only achieve the FSV less one LSB. If it is necessary to have FSV at the output, then the DAC reference voltage must be increased. The increased reference voltage can be determined by

$$\text{Reference voltage} = (\text{desired output}) \cdot \frac{2^n}{2^n - 1} \quad \text{(true binary)} \qquad (14.3)$$

and (14.4)

$$\text{Reference voltage} = (\text{desired output}) \cdot \frac{10^d}{10^d - 1} \quad \text{(BCD)}$$

In our example, if 48 volts is desired as the maximum achieved output voltage, then the reference voltage must be increased to

$$\text{Reference voltage} = 48 \times \frac{2^8}{2^8 - 1} = 48.1882 \text{ V} \quad \text{(binary)}$$

$$\text{Reference voltage} = 48 \times \frac{10^2}{10^2 - 1} = 48.4848 \text{ V} \quad \text{(BCD)}$$

Third, both negative and positive voltage is required by the servo-motor if it is to be bidirectional; hence, as both input codes represent positive numbers only, neither converter produces an output that would permit servo-motor rotation in both directions. The servo system, therefore, requires a DAC whose code represents both positive and negative numbers, in other words, a bipolar code.

Bipolar Codes: In these type of codes the MSB is used to express the sign such that a 0 is positive and a 1 is negative. The remaining digits are used for the magnitude of the number. For example:

$$\text{sign}$$
$$\downarrow$$

$$\text{binary } 01010110 = +86 \text{ decimal}$$
$$\text{binary } 10011001 = -25 \text{ decimal}$$
$$8421 \text{ BCD } 01010110 = +56 \text{ decimal}$$
$$8421 \text{ BCD } 10011001 = -19 \text{ decimal}$$

Table 14-3 shows the relationship between the input code and output voltage for an 8-bit binary DAC using a sign and magnitude code. The FSV available to drive the servo amplifier/motor combination is now ±48 volts.

TABLE 14-3—INPUT/OUTPUT FOR AN 8-BIT BINARY DAC USING A SIGN AND MAGNITUDE CODE

BINARY NUMBER INPUT (SIGN AND MAGNITUDE)	DECIMAL NUMBER	FRACTION OF FSV	FACTOR	OUTPUT VOLTAGE (FACTOR) × (FSV) FSV = ±48 volts
01111111	+127	+127/128	+0.9922	+47.625
01111110	+126	+126/128	+0.9844	+47.250
⋮				
00000001	+1	+1/128	+0.0078	+ 0.375
00000000	+0	+0	+0	+ 0
10000001	−1	−1/128	−0.0078	− 0.375
⋮				
11111110	−126	−126/128	−0.9844	−47.250
11111111	−127	−127/128	−0.9922	−47.625

TABLE 14-4—INPUT/OUTPUT FOR A DAC WITH A TWO-DECADE BCD FORM USING A SIGN AND MAGNITUDE CODE.

BCD NUMBER INPUT (SIGN AND MAGNITUDE)	DECIMAL NUMBER	FRACTION OF FSV	FACTOR	OUTPUT VOLTAGE (FACTOR) × (FSV) FSV = ±48 volts
0111 1001	+79	+79/80	+0.9875	+47.4
0111 1000	+78	+78/80	+0.985	+46.8
⋮				
0000 0001	+01	+1/80	+0.0125	+ 0.6
0000 0000	+00	+0	+0	+ 0
1000 0001	−01	−1/80	−0.0125	− 0.6
⋮				
1111 1000	−78	−78/80	−0.975	−46.8
1111 1001	−79	−79/80	−0.9875	−47.4

Table 14-4 shows the same relationship for a DAC using a sign and magnitude 8421 BCD code at the input.

It becomes apparent from Tables 14-3 and 14-4 that, although we are able to achieve either direction sense at the DAC output, the resolution has decreased for the same number of input bits. Of course, this problem could be overcome by increasing the number of bits for a true binary input or the number of decades for a BCD input. The resolution for the bipolar codes shown in Tables 14-3 and 14-4 can be determined in the following way:

$$\text{Resolution} = \frac{1}{2^{(n-1)}} \text{ (true binary)} \qquad (14.5)$$

or
$$\text{Resolution} = \frac{1}{8 \times 10^{(d-1)}} \text{ (BCD)} \qquad (14.6)$$

The maximum output range is now $\pm(48$ V less one LSB), and if once again it is necessary to produce the FSV at the output, the DAC reference voltage can be increased; thus:

$$\text{Reference voltage} = (\text{desired output}) \cdot \frac{2^{(n-1)}}{2^{n-1} - 1} \text{ (true binary)} \qquad (14.7)$$

and
$$\text{Reference voltage} = (\text{desired output}) \cdot \frac{8 \times 10^{(d-1)}}{[8 \times 10^{(d-1)} - 1]} (BCD) \qquad (14.8)$$

EXAMPLE 14.2

A digital to analog converter is required to produce a full voltage range of ± 28 volts at its output. If the input has sign and magnitude code in (a) 12-bit true binary form or (b) 3-decade 8421 BCD format, determine

1. The appropriate reference voltage.
2. The voltage resolution.
3. The actual output voltage and the equivalent input code for a desired output of -13 volts.

solution (a)

1.
$$\text{Reference voltage} = \frac{28 \times 2^{(12-1)}}{[2^{(12-1)} - 1]}$$

or   Reference voltage = 28.0137 volts

2.   Resolution voltage $= \dfrac{28.01}{2^{(12-1)}}$

hence   Resolution voltage = 0.01367 volt

3. The closest actual output voltage to -13 volts is

$$951(-0.01367) = -13.00195 \text{ volts}$$

Therefore, the input code = 101110110111, which represents decimal -951.

solution (b)

1.   Reference voltage $= \dfrac{28 \times 8 \times 10^2}{[8 \times 10^2 - 1]}$

or   Reference voltage = 28.035 volts

2.   Resolution voltage $= \dfrac{28.04}{8 \times 10^2}$

hence   Resolution voltage = 0.035 volt

3. The closest actual output voltage to -13 volts is

$$371 \, (-0.035) = -12.985 \text{ volts}$$

Therefore, the input code = 1011 0111 0001, which represents decimal -371.

14-2.2 digital to analog converter circuits

Weighted Resistor DAC: The block diagram of a DAC circuit is shown in Figure 14-2. A parallel shift register is located at the input of the DAC and receives the digital input data in parallel form. The register loads the instantaneous value of the input data when a clock pulse occurs and holds this value in the register until the next clock pulse, when a

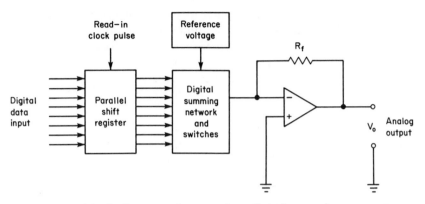

figure 14-2 a block diagram of a complete digital to analog converter

new piece of data can then be transferred. Each bit of data from the output of the shift register goes to an individual switch which, if activated, passes current through a weighted summing resistor. The current resulting from each data bit is then summed together by the op-amp circuit. The summing resistors are weighted in relation to the binary digits that they represent, so that they contribute the correct fraction of full-scale current at the summing junction of the op-amp.

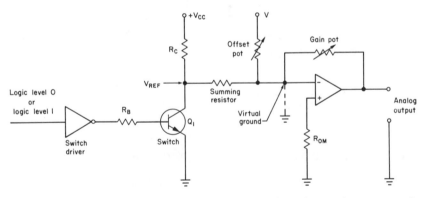

figure 14-3 simple circuit showing conversion of one binary bit to an analog signal

Figure 14-3 illustrates, by means of a very simple circuit, how a single binary digit (of logic level 0 or 1) can contribute to the analog output. If logic level 0 is passed through an inverter, then logic level 1 occurs at the base of Q_1, causing Q_1 to saturate and about 0.4 volt to appear at the collector. The offset potentiometer can be adjusted to give 0 volt at the op-amp input when Q_1 is saturated. If a logic level 1 (which is positive in this instance) is applied to the input inverter, then Q_1 is turned off and V_{cc} makes a contribution to the op-amp input.

To illustrate how a weighted resistor DAC works, consider the simple 4-bit true binary code below:

CODE	WEIGHT
0001	1/16
0010	1/8
0100	1/4
1000	1/2
1111	15/16

Four weighted resistors can now be used for summing, as indicated in Figure 14-4. If the input code is binary 1011, representing decimal

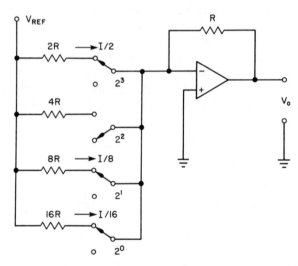

figure 14-4 simple binary DAC using weighted summing resistors

11 (or 11/16 FSV), the switch sequence would be $2^0 \rightarrow$ closed; $2^1 \rightarrow$ closed; $2^2 \rightarrow$ open, and $2^3 \rightarrow$ closed. Hence $V_0 = V_{Ref} \left[\dfrac{R}{16R} + \dfrac{R}{8R} + 0 + \dfrac{R}{2R} \right]$ or $V_0 = \dfrac{11}{16} V_{Ref}$. Naturally had we chosen a 12-bit binary input, we would have had 12 power switches, 2^{11}, 2^{10}, 2^9, . . . , 2^2, 2^1, and 2^0; and 12 summing resistors, $2R$, $4R$, $8R$, . . . , $1024R$, $2048R$, and $4096R$, respectively. If $R = 10$ KΩ (a reasonable value), the summing resistors would have had a range from 20 KΩ to 40.96 MΩ. The resistor values at the high end of this range prohibit the use of this type of weighted resistor circuit.

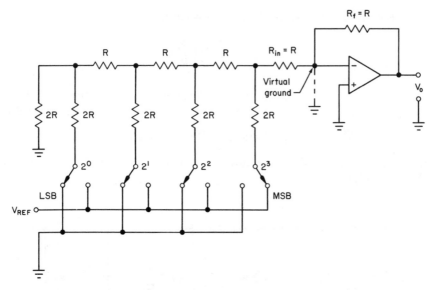

figure 14-5 an R-$2R$ ladder network and power switching for a 4-bit binary input.

R-$2R$ Ladder DAC: An alternative to the weighted resistor circuit is the R-$2R$ ladder network in which the largest resistor is only twice the value of the smallest. This type of circuit is illustrated in Figure 14-5 for a 4-bit binary input. If the transistor switches represent binary 1000, as shown in the figure, the voltage at the op-amp input resistor will be $\dfrac{V_{Ref}}{2}$. The reason for this is demonstrated with the help of the equivalent circuit in Figure 14-6.

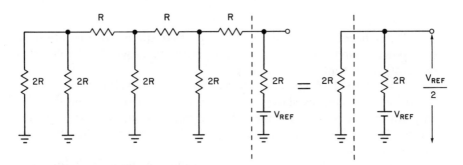

figure 14-6 the ladder network and its equivalent circuit

Simple series/parallel resistor addition results in one-half the reference voltage being applied to the op-amp input resistor.

Equation (11.3) shows that, in this case,

$$V_0 = -\frac{V_{\text{Ref}}}{2}\left(\frac{R}{R}\right)$$

If, on the other hand, the ladder network in Figure 14-5 had indicated binary 0001, the voltage at the op-amp input resistor would have been $\dfrac{V_{\text{Ref}}}{16}$. The application of Thevenin's theorem provides us with the means to reduce the ladder network to its equivalent circuit, as shown in Figure 14-7.

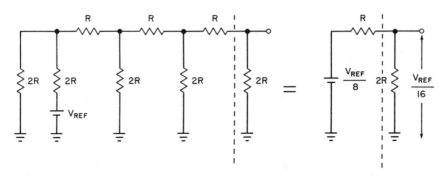

figure 14-7 the ladder network and its equivalent circuit

Once again, equation (11.3) allows us to determine the output of the op-amp, which now becomes

$$V_0 = -\frac{V_{\text{Ref}}}{16}$$

From here, it is not difficult to see that successive operation of each power switch will cause the appropriate binary weighted voltage to occur at the output of the op-amp.

If the DAC had a two-decade BCD input, the R-$2R$ ladder network could be connected as shown in Figure 14-8. Notice that the two op-amp input resistors now have a 10:1 ratio, so that the upper ladder represents the least significant decade and the lower ladder the most significant decade.

The very narrow range of resistor values in the R-$2R$ ladder network allows the use of input codes containing a much larger number of bits than is possible when one is using a weighted resistor network, such as that shown in Figure 14-4.

Although there are a number of different DAC's commercially available, accepting a variety of unipolar (positive values only) or bipolar (positive and negative values) codes, they all operate essentially in one of the two ways described above.

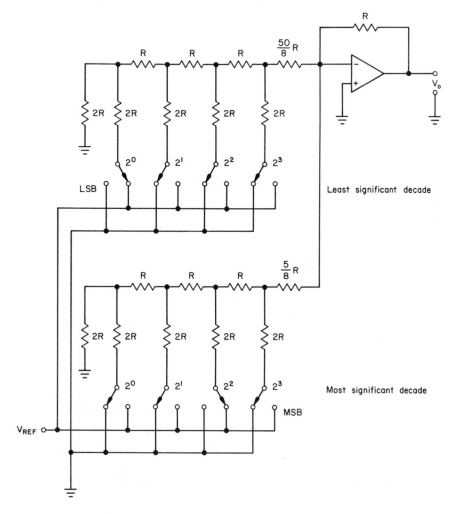

figure 14-8 a two-decade BCD input to an R-2R ladder network. the input code is 0100 0001; equivalent to decimal 41

14-2.3 converter characteristics

The major specifications used to evaluate the effectiveness of a DAC are explained below.

Resolution: This topic has already been addressed at some length in Section 14-2.1, but in summary, resolution can be defined as the smallest incremental change of the output referred to the reference voltage; therefore, it is equal to the weight of 1 LSB (see Figure 14-9).

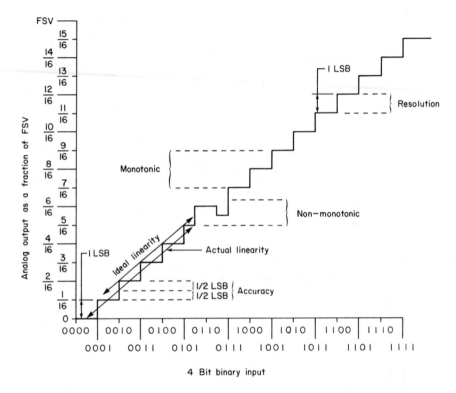

figure 14-9 the digital to analog relationship of a typical converter

Accuracy: This is a measure of the difference between the actual output voltage and the ideal output voltage (for the particular value of input code), expressed as a percentage of the FSV.

Linearity: As shown in Figure 14-9, when the output voltage of the DAC is plotted against the digital input, we obtain a staircase-type plot. If we plot the best straight line through the staircase, as shown by the "actual linearity" curve, the "linearity" of the DAC is measured by maximum deviation of the staircase curve from this straight line. "Standard linearity" is taken to be ±1/2 LSB.

Differential Linearity: Because the values of the resistors in the R-$2R$ ladder network (or weighted resistor network) can vary within their tolerance limits, the analog voltage produced by each bit has an error associated with it. Furthermore, the magnitude of this error changes from bit to bit, so that when two or more bits are combined (as in the code for 1010) the errors associated with each bit are added. Thus, for a particular input value, if the digital code contains a large number of bits, the accumulated output error (and, therefore, the deviation from linearity) can become quite large, possibly exceeding 1/2 LSB. If, at the next step, the input code changes, so that now only a small number of bits are included in the code, then an unusually large change in bit errors can occur between the two steps with a consequent change in the deviation from the line of linearity.

This change in linearity leads to the term *differential linearity*, which measures the amount by which any step deviates from the ideal step size of 1 LSB. If the differential nonlinearity is greater than 1 LSB and has negative polarity, then the output can actually decrease for an increasing input; this condition is referred to as *nonmonotonicity*.

Monotonicity: Monotonicity is the ability of the DAC to maintain an increasing (decreasing) incremental output for an increasing (decreasing) input code change. If, as shown in Figure 14-9, for example, the output "dips" for input code 0111, then the DAC is nonmonotonic.

Output Impedance: The output impedance is always low (a desirable feature) due to the fact that the output signal is usually derived from an op–amp and, therefore, from equation (11.32):

$$R_{\text{out}} = \frac{R_\text{o}}{1 + A_{\text{VOL}/2}} \qquad (11.32)$$

14-3 THE ANALOG TO DIGITAL CONVERTER (ADC)

Description: An ADC produces a digitally coded output having a digital value (binary, BCD, etc.) proportional to the analog voltage applied at its input. To do this conversion, the ADC senses the magnitude of the input and compares it to some reference level; if there is a difference in levels, it measures the difference digitally (by counting), and often a DAC is used within the ADC to make these comparisons.

All this comparison work takes time, so an ADC can not produce an output as fast as a DAC can. The speed of a DAC is usually limited only by the response of the output op-amp. In an ADC, although op-amps are used for level detection, they do not influence the speed of operation of the ADC to any extent; rather, it is the counting time required by the digital logic circuits when comparing voltage levels.

While digital to analog conversion uses only a limited number of circuits to execute a conversion efficiently, there are many methods used to make analog to digital conversions, as ADC's lend themselves so easily to the application of a wide variety of techniques. There are various methods of classifying ADC's, e.g., those which use a DAC as a feedback technique. We shall not concern ourselves here with the various classifications, but we shall discuss the features of one or two of the more common ADCs and examine the circuit techniques used to make the A/D conversion.

14-3.1 ADC GENERAL CHARACTERISTICS

The following explains the major characteristics and specifications used to evaluate the operation of the ADC:

Quantization: When we examined the DAC, we discovered that the output was not a smooth curve but a staircase curve of discrete, equal increments. Similarly, because the digital output of an ADC can only be incremental, each binary word of the output will represent a discrete value of the analog voltage input, over the range 0 to FSV. The operation of changing a smooth analog voltage signal to a series of fixed voltage levels is called *quantization*.

If, for example, we have an ADC whose input range is 0 to 12 volts (i.e., 0 to FSV) and the ADC has a 4-bit true binary output, then the correctly quantized transfer curve would resemble that illustrated in Figure 14-10. Because a 4-bit binary code is capable of changing 15 times, the input will be divided into 15 incremental steps, each having a step width of 0.8 V. As the input changes, in a continuous analog fashion from 0 V to 0.4 V, the output code remains at 0000. When the input reaches 0.4 V, the output code changes to 0001 and retains this value until the input reaches 1.2 V, whereupon the output changes to 0010 and so on. Under these conditions it can be seen from Figure 14-10 that the output of an ideal ADC has a quantizing error, whose magnitude cycles from $-\frac{1}{2}$ LSB to $+\frac{1}{2}$ LSB as the analog input changes. The form of the output error is plotted in Figure 14-11.

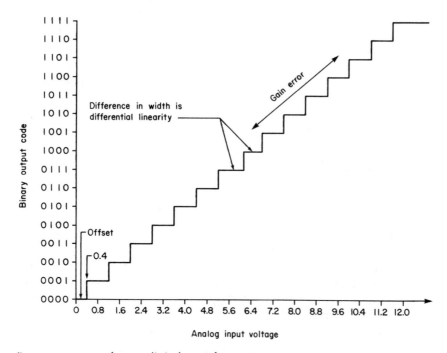

figure 14-10 analog to digital transfer curve

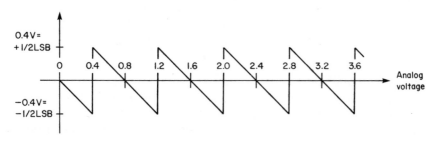

figure 14-11 the quantizing error

Resolution: The resolution of the ADC is the smallest change in the analog input that will cause a single bit change in the digital output. In the previous example, do not confuse the change from 0 to 0.4 V (½ resolution) as being the smallest change in the analog input, because the output code would remain at 0000 even if the input went negative by 0.4 V, which makes the smallest analog change again equal to 0.8 V. The ADC resolution is determined from

$$\text{Resolution} = \frac{1}{2^n - 1} \text{ (true binary)} \qquad (14.9)$$

or
$$\text{Resolution} = \frac{1}{10^d - 1} \text{ (BCD)} \qquad\qquad (14.10)$$

Equations (14.9) and (14.10) are only resolution factors, and if expressed as a voltage, by multiplying by the FSV, the resolution can be seen to be equivalent to the voltage weight of the LSB.

Differential Linearity: Because of the nonlinearities in switching and other circuit inaccuracies, there will always be a difference in the width of the individual steps of a transfer curve such as Figure 14-10. These differences are referred to as the differential linearity of the converter, as explained in Section 14-2.3.

Zero Offset Error: Another error that occurs in converters is the side-to-side displacement of the transfer curve (Figure 14-10) due to temperature. This error is termed the zero offset error and is measured by the position of the first step on the horizontal (analog) axis. Offset is usually adjustable and will be called up on the specification sheet as ±3 LSBs maximum, for example.

Gain Error: This type of error is known as either *gain error* or *scale factor error*. In our example, the input voltage range for full-scale digital output is 12 volts. If full-scale digital output is obtained with less than or more than the nominal 12-volt input, then the gain is in error. The gain error should be better than 1 percent of full scale.

Conversion Time and Conversion Rate: Conversion time and conversion rate are two very important characteristics of an ADC. Conversion time is the time taken to produce a digital output for a corresponding analog input and is usually specified as the conversion time per word. As a logical consequence, the conversion rate is the reciprocal of the conversion time and is used as a figure of merit in evaluating ADC's to denote the rate of measurements/second.

Linearity, Accuracy, and Monotonicity: These characteristics have the same meaning, when applied to an ADC, as they did when applied to the DAC in Section 14-2.3.

EXAMPLE 14.3
A hybrid servo system employs an ADC in the feedback loop to a microprocessor. The load position transducer produces a maximum

of 30 volts and its full-scale response time is 125 ms. If the ADC has
(1) an 8-bit binary output or (2) a 2-decade BCD output, determine

(a) The resolution of the ADC.

(b) The maximum available conversion time.

(c) The minimum conversion rate.

solution (a)

1. The input to the ADC is 30 volts; hence

$$\text{Resolution voltage} = \frac{\text{FSV}}{2^n - 1} = \frac{30}{255}$$

　　or　　Resolution voltage = 0.1176 V

2. For a BCD output,

$$\text{Resolution voltage} = \frac{\text{FSV}}{10^d - 1} = \frac{30}{99}$$

　　or　　Resolution voltage = 0.303 V

solution (b)

1. A change of 30 volts represents 255 binary word changes; hence

$$\text{Maximum available conversion time} = \frac{125}{255} \text{ ms}$$

　　or　　Maximum conversion time = 490.196 μs per word

2. Since 30 volts represents 99 word changes of BCD, then

$$\text{Maximum available conversion time} = \frac{125}{99} \text{ ms}$$

　　or　　Maximum conversion time = 1.263 ms per word

solution (c)

1.　　　　　　$$\text{Conversion rate} = \frac{1}{\text{conversion time}}$$

$$\text{hence} \quad \text{Binary conversion rate} = \frac{1}{490.196 \; \mu s} = 2040 \text{ Hz}$$

$$\textbf{2.} \; \text{For BCD the conversion rate} = \frac{1}{1.263 \text{ ms}} = 792 \text{ Hz}$$

14-3.2 analog to digital converter circuits

Because of the varying requirements of economy, speed of operation, and accuracy, in the application of ADC's, there are several different types of ADC to choose from. The most common types of ADC in current use are, in order of their conversion rate (slowest first),

1. Integrating type ADC.
2. Counter or Servo ADC.
3. Successive Approximation ADC.
4. Parallel or Simultaneous ADC.

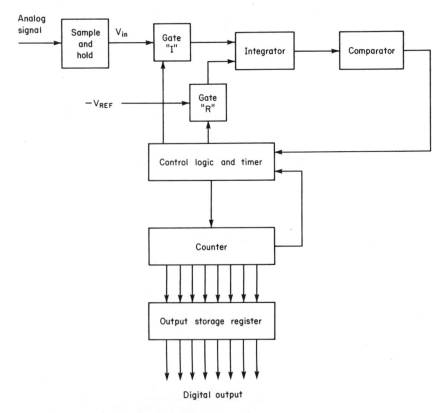

figure 14-12 an 8-bt dual slope ADC

Dual Slope ADC (Integrating Type): Because the dual slope ADC is of the integrating type, it has good noise rejection characteristics (Section 11-3.4); it also has a fairly slow conversion time, compared with other types of ADC, with conversion times typically of the order of 75 μs (10 kHz − 15 kHz). Although the resolution of this type of ADC is usually high (the use of 14 bits is not uncommon), its design is simple and the units are inexpensive to produce. For these reasons this type of ADC is usually found in measuring instruments such as digital voltmeters and panelmeters.

The dual slope ADC operates by comparing the input voltage value V_{in} against a known reference voltage $-V_{ref}$. In this type of ADC, the analog input signal is sampled by means of a sample and hold circuit in order that V_{in} is held constant for the duration of the conversion time. V_{in} and $-V_{ref}$ are compared in the following way: At the commencement of the conversion period a "start of conversion" pulse instructs the control logic to enable gate I, so that V_{in} appears at the input to the integrator. The integrator output now starts to decrease, in ramp fashion, as shown in Figure 14-13. As soon as the integrator starts to decrease, the comparator output goes high, which tells the control logic to start counting. The logic continues to count at a fixed clock frequency until the total number of pulses counted is P_1, a constant number determined by the number of bits used by the ADC. At this time the counter is reset, gate I is disabled, and gate R is enabled, thereby replacing V_{in} as the integrator input with the negative reference voltage $-V_{ref}$. The integrator output now starts to increase at a rate determined by the magnitude of V_{ref}, and the control logic starts to count once more. As soon as the integrator output crosses the zero volt line, the counter stops counting and transfers the number of counts P_2 to the storage register.

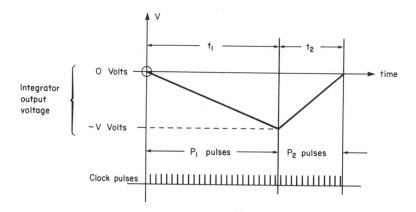

figure 14-13   integrator ramping and timing pulses in a dual slope ADC

If the time constant of the integrator is τ, then from equation (11.22a) and Figure 14-13,

$$-V = \frac{-V_{in} \cdot t_1}{\tau} \qquad (14.11)$$

and

$$0 = -\frac{-V_{ref} \cdot t_2}{\tau} - V \qquad (14.12)$$

Substituting equation (14.11) into equation (14.12) yields

$$V_{in} = V_{ref} \cdot \frac{t_2}{t_1} \qquad (14.13)$$

Since the rate of counting for both slopes is at the fixed clock frequency, then

$$\frac{t_2}{t_1} = \frac{P_2}{P_1}$$

and hence

$$V_{in} = \frac{V_{ref}}{P_1} \cdot P_2 \qquad (14.14)$$

V_{ref} and P_1 are constants of the ADC; thus the count P_2 varies directly as the magnitude of V_{in} hence V_{in} is obtained by simple scaling of P_2.

EXAMPLE 14.4

If the analog input to an 8-bit binary ADC is from 0 to 12 volts and the ADC uses a fixed count P_1 equivalent to its MSB, what is the reference voltage?

solution:

For an 8-bit binary:

$$MSB = 128 \text{ pulses}$$

therefore

$$P_1 = 128$$

If the maximum analog voltage V_{in} is 12 volts and the maximum output code is binary 11111111 (i.e., decimal 255), then

$$P_2 = 255$$

and since
$$V_{ref} = V_{in} \cdot \frac{P_1}{P_2}$$

we have
$$V_{ref} = 12 \cdot \frac{128}{255}$$

i.e.,
$$V_{ref} = 6.024 \text{ V}$$

EXAMPLE 14.5

If the ADC in Example 14.4 had a 3-decade BCD output, what would be the reference voltage?

solution:

For a 3-decade BCD code the MSB has a value of 800 (1000 0000 0000);

hence
$$P_1 = 800$$

and
$$V_{in} = 12 \text{ volts}$$

The maximum output code in BCD is 1001 1001 1001 or decimal 999.

Therefore,
$$P_2 = 999$$

Since
$$V_{ref} = V_{in} \cdot \frac{P_1}{P_2}$$

then
$$V_{ref} = 12 \cdot \frac{800}{999}$$

or
$$V_{ref} = 9.6096 \text{ V}$$

Counter or Servo ADC: This type of converter is one of the simplest and least expensive to implement. Although it is considered slow, with a conversion rate in the order of 15 kHz to 40 kHz, it is faster than the dual slope ADC, and in addition it does not require a sample and hold circuit input. The converter is called a servo type because of the feedback method used to control the counter. To understand its operation, consider the block diagram illustrated in Figure 14-14.

At the start of conversion, the clock is gated on and the control logic passes the clock pulses to the counter. As the clock pulses are accumulated in the counter, they are simultaneously fed to the input of a DAC (whose output, of course, is an analog voltage). The analog input voltage, V_{in}, and the DAC output are compared by the comparator. As long as V_{in} is greater than the DAC output, the comparator allows the

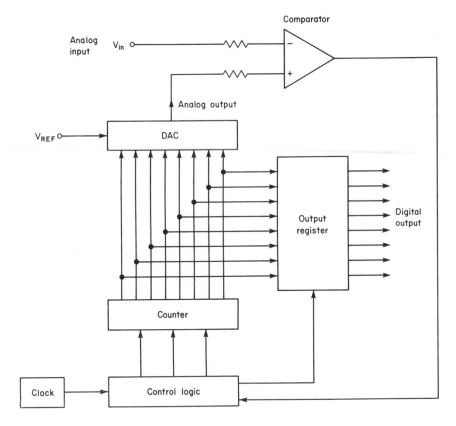

figure 14-14 counter or servo type ADC

control logic to pass the clock pulses to the counter. When the DAC output equals the analog input to the comparator, there is a fast change in the comparator output, which forces the control logic to inhibit the clock pulses. The control logic at this time also sends a pulse to the output register, which allows the contents of the counter to be transferred to the register after which the counter is reset to zero. Figure 14-15 shows the way in which the DAC output steps up to the level of the analog input.

The speed of this type of ADC can be improved by using an up-down counter, which allows the converter to count either up or down from its previous value, rather than resetting to zero prior to each conversion.

It is obvious, from this description and Figure 14-14, that the analog full-scale input is determined by the DAC output voltage; also, we learned in Section 14-2.1 that the DAC output is itself determined by

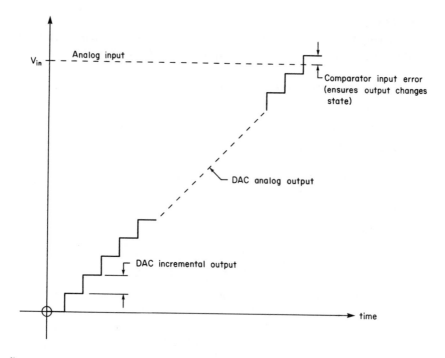

figure 14-15 comparator inputs

the DAC reference voltage. The use of equations (14.3) and (14.4) allows us to calculate the DAC output voltage and hence to determine the maximum analog input voltage possible:

$$\text{Analog full-scale input} = V_{ref} \cdot \frac{2^n - 1}{2^n} \quad \text{(binary)} \qquad (14.15)$$

or $$\quad \text{Analog full-scale input} = V_{ref} \cdot \frac{10^d - 1}{10^d} \quad \text{(BCD)} \qquad (14.16)$$

The longest conversion time is required when the counter counts from zero until every bit is a 1, for binary, or until every decade is a decimal 9 for BCD. The time to accomplish this is easily calculated by

$$\text{Conversion time} = \frac{2^n - 1}{\text{clock rate}} \quad \text{(binary)} \qquad (14.17)$$

or $$\quad \text{Conversion time} = \frac{10^d - 1}{\text{clock rate}} \quad \text{(BCD)} \qquad (14.18)$$

EXAMPLE 14.6

A binary, 10-bit counter-type ADC has a maximum analog input of 10 volts and an internal clock frequency of 10 MHz. Calculate

1. The internal reference voltage.
2. The voltage resolution.
3. The conversion time.
4. The conversion rate.

If the format is 3-decade BCD, recalculate 1, 2, 3, and 4.

solution

Binary Format

1.
$$V_{ref} = (\text{input}) \cdot \frac{(2^n)}{(2^n - 1)} = \frac{10 \times 2^{10}}{(2^{10} - 1)}$$

$$V_{ref} = 10.0098 \text{ V}$$

2.
$$\text{Resolution} = \frac{\text{maximum input}}{(2^n - 1)} = \frac{10}{1023}$$

hence $\text{Resolution} = 9.775 \text{ mV}$

3.
$$\text{Conversion time} = \frac{(2^n - 1)}{\text{clock rate}} = \frac{1023}{10 \text{ MHz}}$$

$$\text{Conversion time} = 102.3 \text{ } \mu s$$

4.
$$\text{Conversion rate} = \frac{1}{102.3 \text{ } \mu s}$$

or $\text{Conversion rate} = 9.775 \text{ kHz}$

3-Decade BCD Format

1.
$$V_{ref} = (\text{input}) \cdot \frac{(10^d)}{(10^d - 1)} = \frac{10 \times 10^3}{(10^3 - 1)}$$

$$V_{ref} = 10.01 \text{ V}$$

2.

$$\text{Resolution} = \frac{\text{maximum input}}{(10^d - 1)} = \frac{10}{999}$$

hence   Resolution = 10.01 mV

3.

$$\text{Conversion time} = \frac{(10^d - 1)}{\text{clock rate}} = \frac{999}{10 \text{ MHz}}$$

Conversion time = 99.9 μs

4.

$$\text{Conversion rate} = \frac{1}{99.9 \ \mu s}$$

or   Conversion rate = 10.01 kHz

Successive Approximation ADC: This type of ADC is probably the most popular because of its combination of very fast conversion time (as low as 1 μs for state of the art ADC's) and excellent resolution (up to 16 bits). Another advantage is that the successive approximation method ensures that conversion takes place at a fixed conversion time per bit, and is, therefore, independent of the value of the analog input. However, these advantages are reflected in the comparatively higher cost of this type of ADC.

Successive approximation ADC's can be quite accurate, but their accuracy is determined in large measure by the stability of the reference voltage and the precision of the DAC ladder network.

Figure 14-16 shows a simple diagram of a successive approximation ADC, where it is seen that a DAC is once again used as a feedback mechanism, for servoing the digital output until it corresponds to the analog input.

At the *start conversion* signal the MSB of the DAC is switched on. This produces $\frac{1}{2}$ FSV at the output of the DAC. If the output state of the comparator indicates that V_{in} is greater than the DAC output, the DAC MSB is left on and the programmer enables the next significant bit of the DAC. If, however, the output state of the comparator indicates that the DAC output is greater than V_{in}, the programmer turns off the DAC's MSB and enables the next most significant bit. This process continues, with the programmer making these successive decisions, until the input to the output storage register (i.e., the digital input to the DAC) is the digital equivalent of V_{in}.

Table 14-5 illustrates the successive decisions made by an eight-bit binary ADC having an FSV of 10 volts and an analog input of 8.7 volts.

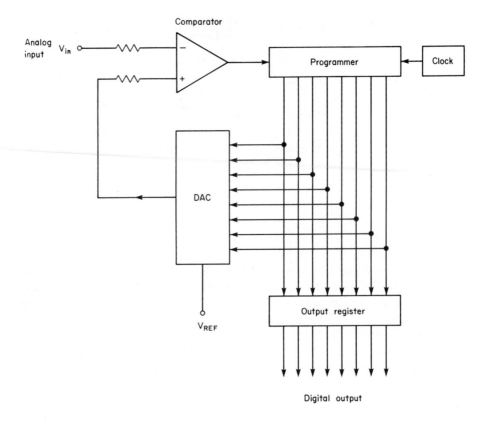

figure 14-16 successive approximation ADC

TABLE 14-5—SUCCESSIVE APPROXIMATION ADC-PROGRAMMER DECISIONS

	VOLTAGE CONTRIBUTION							
BIT WEIGHT	*MSB* *1/2*	*1/4*	*1/8*	*1/16*	*1/32*	*1/64*	*1/128*	*LSB* *1/256*
1st Decision	5							
2nd Decision	5	2.5						
3rd Decision	5	2.5	0					
4th Decision	5	2.5	0	0.625				
5th Decision	5	2.5	0	0.625	0.3125			
6th Decision	5	2.5	0	0.625	0.3125	0.15625		
7th Decision	5	2.5	0	0.625	0.3125	0.15625	0.07813	
8th Decision	5	2.5	0	0.625	0.3125	0.15625	0.07813	0
Binary	1	1	0	1	1	1	1	0

It will be noticed that the final output of the ADC represents an input voltage of only 8.6719 volts, rather than 8.7 volts, an error of 0.32 percent. If the ADC output also included the LSB, this would make the DAC output greater than the analog input; although this results in a smaller error, the comparator output state precludes the programmer from making this selection. The conversion time for this ADC is limited only by the time taken for the programmer to run through the ADC's full complement of bits, while making the appropriate decisions. This time can be calculated from

$$\text{Conversion time} = \frac{n}{\text{clock rate}} \qquad (14.19)$$

where n is the number of bits used by the ADC.

Parallel or Simultaneous ADC: This type of converter is so fast that the method of conversion is sometimes called the "flash technique." It is also the most expensive type of ADC. Conversion rates of 25 MHz are possible if only four bits are used. If more bits are required, the conversion rate is lowered. No feedback is used; hence no DAC is incorporated into the circuitry.

The analog input is fed to a string of input comparators (as shown in Figure 14-17), the inverting inputs of which are tied to a voltage divider network such that the bias voltage at each comparator differs by the weight of 1 LSB from the one below. When a voltage is applied at V_{in} all the comparators with a bias voltage less than V_{in} will turn on and all other comparators will remain off. The number of comparators required for an n-bit ADC is determined by

$$\text{Number of comparators} = 2^n - 1 \qquad (14.20)$$

The output of the comparators is not in any standard code and hence must be decoded before being transferred to the output storage register. To read a large number of bits requires a very large number of comparators, which is why this form of A/D conversion is so expensive. The resolution of the parallel ADC is given by

$$\text{Resolution} = \frac{V_{ref}}{2^n} \qquad (14.21)$$

The resolution of this type of ADC is usually poor, as the number of bits used tends to be limited in order to hold down the number of comparators required.

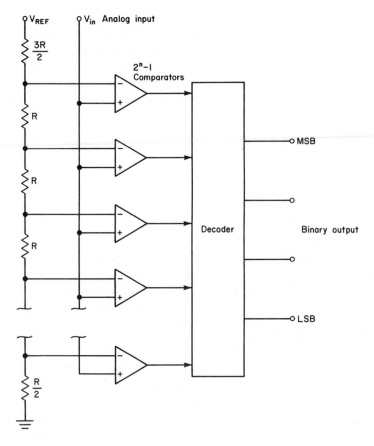

figure 14-17 block diagram of a parallel ADC

14-4 SAMPLE AND HOLD CIRCUITS

Description: Because the input to an ADC must be quantized, it is often necessary, due to the changing nature of the input, to periodically sample the input and to hold that value (or very nearly so) until the conversion is completed.

It is worthwhile at this point to briefly examine sampling theory; consider Figure 14-18. The illustration shows an analog voltage signal [A], which is sampled periodically by using a fixed frequency clock pulse train [B]. At each clock pulse the analog magnitude of [A] at that particular instant of time is allowed to pass to the next stage of conditioning.

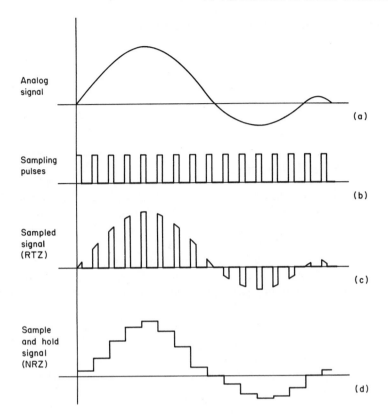

figure 14-18 the sampling process

Figure 14-18 [C] shows the waveform of the sampled signal. This type of waveform is called return to zero (RTZ) sampling and is the type of sampling performed by multiplexers. If, however, the value of [C] is somehow stored after each pulse (as a charge on a capacitor for example), these values can then be passed to the next conditioning stage as a nonreturn to zero (NRZ) sampled waveform, as shown in [D].

If the sampling frequency is too low, then some of the information contained in the input signal may be lost; also, for practical reasons, we want to set an upper limit on the frequency at which we do the sampling. How do we decide on what sampling frequency to use? The answer is provided by the sampling theorem, which says:

> The original signal can be completely recovered without distortion if it is sampled at a rate at least twice that of the highest frequency contained in the continuous bandwidth of the original signal.

Hence the sampling frequency should be chosen so that

$$\text{Sampling frequency} \geqslant 2f_{hf} \qquad (14.21)$$

where

f_{hf} = highest frequency component of the signal being sampled.

14-4.1 sample-hold characteristics

Figure 14-19 shows a block diagram of a S/H circuit. The sampling pulses close the switch, whereupon capacitor C begins to store the current value of the analog input. A specific period is allowed to charge the capacitor, after which the switch opens and the capacitor holds the last value of the input voltage until the next switch closure. The output circuit consumes only a very small amount of current and therefore passes the capacitor voltage, with very little decay, to the following ADC circuit.

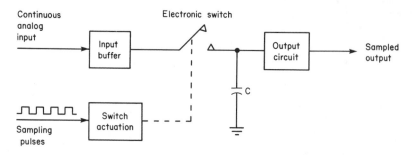

figure 14-19   a simplified S/H circuit

Figure 14-20 shows the sample and hold timing sequences. At time t_1 the command to sample is made by the sampling pulse. The switch begins to close and is fully closed at t_2. From t_2 to t_3 the capacitor voltage V_c rises to the magnitude of the input. The period from t_1 to t_3 is called the *acquisition time*, after which the capacitor voltage should be tracking the input. At time t_4 the command to hold is made by the removal of the sampling pulse; the switch begins to open and is completely open by time t_5. The period from t_4 to t_5 is called the *aperture time*. The aperture time is not precisely the same from pulse to pulse, and this difference is expressed as the *aperture uncertainty time*, wherein there is an uncertainty in the amplitude of V_c due to the rate of change of the

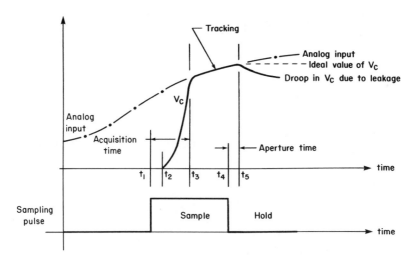

figure 14-20 sample-hold timing sequences

analog signal. The maximum rate of change of the input signal that can be followed by an ADC, with less than 1 LSB tracking error, is called the *tracking rate* and determines whether or not a S/H circuit will form part of an ADC.

For an ADC not employing a S/H circuit, in order for the tracking error not to exceed 1 LSB the maximum rate of change of the input signal must be less than 1 LSB per conversion time; i.e., for an n-bit ADC having a conversion time, T:

$$\text{Voltage weighting of 1 LSB} = \frac{\text{FSV}}{2^n}$$

Therefore, the maximum rate of change of the input, or tracking rate is

$$\text{Tracking rate} = \frac{\text{FSV}}{T \cdot 2^n} \qquad (14.22)$$

For an ADC employing a S/H circuit to maintain the tracking error below 1 LSB, the maximum rate of change of the input signal must be less than 1 LSB per period of aperture uncertainty. Therefore, the tracking rate for this type of ADC, if the aperture uncertainty time is t_u, is

$$\text{Tracking rate} = \frac{FSV}{t_a \cdot 2^n} \qquad (14.23)$$

EXAMPLE 14.7

A 10-bit dual slope ADC has a total conversion time of 80 ms. If tracking errors are to be less than 1 LSB, what is the maximum tracking rate permissible at its input, for an FSV of 12 volts? If the ADC is then used with a S/H circuit having an aperture uncertainty of 0.5 μs, what is the new maximum permissible tracking rate?

solution:

Converter only

$$\text{Tracking rate} = \frac{FSV}{T \cdot 2^n}$$

hence

$$\text{Tracking rate} = \frac{10^3 \times 12}{80 \times 2^{10}}$$

The maximum tracking rate possible = 146.5 mV/s

With S/H

$$\text{Tracking rate} = \frac{FSV}{t_a \times 2^n}$$

hence

$$\text{Tracking rate} = \frac{10^6 \times 12}{0.5 \times 2^{10}}$$

The maximum tracking rate possible = 23437.5 V/s

For the special case of sinusoidal input waveforms, since a sine wave has its maximum rate of change at the zero crossover, it is possible to determine the maximum input frequency that can be tracked by an ADC, in terms of the aperture uncertainty of the S/H circuit and the number of bits used in the ADC, as demonstrated in Figure 14-21.

14-4.2 types of S/H circuits

Figure 14-22 shows three typical S/H circuits. The circuit in Figure 14-22(a) has an open-loop configuration, whereas the other two examples have closed loops, making use of a feedback from the output buffer to the input stage. The first circuit uses an input buffer amplifier having a high input resistance and a fast response time. The FET switch, fol-

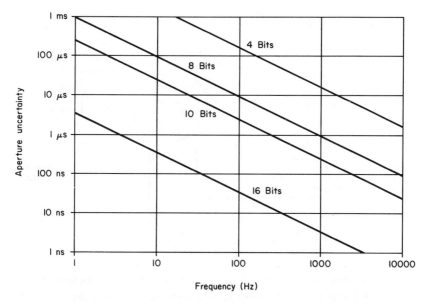

figure 14-21   aperture uncertainty vs frequency for various resolutions

lowing the input buffer, has an OFF resistance in the order of 10^{13} Ω and an extremely low ON resistance (which are, of course, the features of a good switch). The HOLD capacitor in this circuit is connected to the noninverting input of a fast-acting high impedance output buffer.

Figure 14-22(b) is essentially the same as the first circuit, except that the diode bridge possesses a very fast switching action; however, the OFF resistance of the diodes is only about 15×10^6 Ω. The HOLD

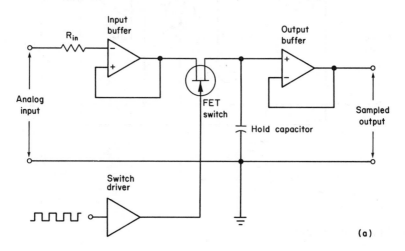

(a)

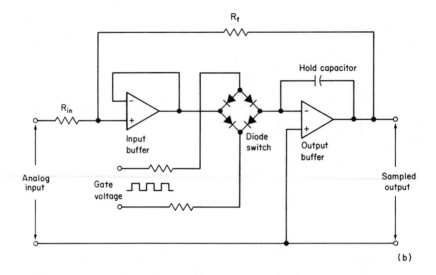

(b)

capacitor in this circuit is used in an integrator of the output buffer stage. This form of closed-loop configuration improves accuracy and linearity.

The circuit shown in Figure 14-22(c) is somewhat slower than the other two S/H circuits but has the advantages of extremely good accuracy and linearity.

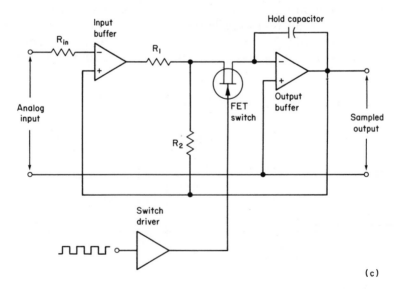

(c)

figure 14-22 S/H circuits

14-5 SUMMARY

In this chapter we have reviewed the equipment that forms the interface between analog components and the digital controller in a hybrid digital control system. This review included descriptions of the most common DAC, ADC, and S/H devices in current use and explanations of those device parameters significantly affecting their performance and capabilities.

PROBLEMS

Section 14-2.1

1. If the resolution for a 14-bit binary DAC is 610.352 μ volts,
   (a) What is the FSV?
   (b) What is the resolution for a 4-decade BCD input having the same FSV?

2. What is the internally generated reference voltage of a DAC using a 4-decade BCD format if the FSV is 6 volts?

3. If the DAC in Problem 2 above can accept a bipolar code and has an output of ± 6 volts,
   (a) What is the resolution?
   (b) What value is the internally generated reference voltage?

4. A DAC has a FSV of ± 10 volts with a 14-bit input. What is the input code for the following output voltages:
   (a) +3 volts, (b) −5.2 volts, (c) 0 volts,
   when the input code is sign and magnitude binary?

Section 14-2.2

5. If the DAC circuit shown in Figure 14-5 were modified to accept unipolar 8-bit binary and the input code was 10011010,
   (a) Determine the reference voltage for a FSV = +10 volts.
   (b) Determine the contribution at the output from each switch.
   (c) What is the output voltage?

Section 14-3.1

6. An ADC has a bipolar output of (i) 12-bit binary; (ii) 3-decade BCD. Determine
   (a) The resolution.
   (b) The conversion time per bit if the maximum conversion time is 5 ms.
   (c) The conversion rate.

Section 14-3.2

7. The analog input to the two ADCs in Problem 6 above is ± 10 volts, and the converter is a dual slope type.
   (a) What is the internal reference voltage?
   (b) The clock is running at 1 Mhz; calculate t_1 and t_2, if the ramp changes at the MSB voltage weighting.

8. A servo-type ADC has a bipolar output of ± 10 volts and either a (i) 12-bit binary or (ii) 3-decade BCD, input. If the conversion time is 60 μs what is
   (a) The input voltage resolution?
   (b) The clock frequency?

9. If the ADCs in Problem 8 above are of a successive approximation type, repeat the problem.

Section 14-4.1

10. A 12-bit counter-type ADC has an input of ± 10 volts and a conversion time of 9 ms. What is the maximum input sine wave frequency that it can accept?

11. A S/H circuit is connected at the input of the ADC in Problem 10 above. The aperture uncertainty is 0.2 μs. What is the new maximum sine wave frequency that can be accepted?

appendix I

si units

I-1 INTRODUCTION

Throughout the text we have tried to limit the use of non-SI units and deal only with internationally acceptable symbols and quantities. However, some of the worked examples were more comfortably expressed in the units in use, at present, in North America. By using the conversion factors listed below in Table I-1, it should be possible to convert most non-SI quantities into SI units as required.

Table I-1—SI SYMBOLS, UNITS, AND CONVERSION FACTORS

QUANTITY AND SYMBOL	SI UNIT AND SYMBOL	NON-SI EQUIVALENT
length, ℓ, L	metre, m	1 foot = 0.3048 m
volume, V	cubic metre, m³	1 gallon = 3.7854 × 10^{-3}m³ = 3.7854 L
time, t	second, s	
plane angle, θ, ϕ, ψ	radian, rad	1 degree = $\dfrac{\pi}{180}$ rad
angular velocity, ω	radian per second rad/s	

Table I-1—Continued

QUANTITY AND SYMBOL	SI UNIT AND SYMBOL	NON-SI EQUIVALENT
frequency, f	hertz, Hz	
mass, m, M	kilogram, kg	1 lb (mass) = 0.4536 kg
moment of inertia, J	kilogram metre squared, kg-m²	1 oz in.² = 18.29 × 10^{-6} kg-m²
force, F	newton, N	1 lb (force) = 4.4482 N
torque, T	newton metre, Nm	1 lb(f)ft = 1.3558 Nm 1 oz(f)in. = 7.0616 × 10^{-3} Nm
pressure, p, P	pascal, Pa	1 psi = 6.8948 kPa
flow, q	metre cubed per second, m³/s	1 ft³/min = 0.4719 × 10^{-3} m³/s
energy or work	joule, J	1 Btu = 1055 J
power, P	watt, W	1 HP = 745.6999W
thermodynamic temperature, T	kelvin, K	0° Celsius = 273.157K
quantity of heat or energy, E	joule, J	1 kWhr = 3.6 MJ
rate of flow of heat or energy, q	watt, W	
thermal resistance, R	watts per kelvin, W/K	
heat capacity, C	joules per kelvin J/K	
electric current, I	ampere, A	
electric charge, Q	coulomb, C	
electric potential, V	volt, V	
electromotive force, E		
capacitance, C	farad, F	
resistance, R	ohm, Ω	
self-inductance, L	henry, H	

appendix
roots of quadratic
and
cubic equations

II-1 THE QUADRATIC EQUATION

The general form of a quadratic equation is given by

$$as^2 + bs + c = 0 \qquad (II.1)$$

The solution is obtained by completing the square in the following manner:

Since
$$as^2 + bs + c = 0$$

then also,
$$4a(as^2 + bs + c) = 0$$

Hence
$$4a^2s^2 + 4abs + 4ac = 0 \qquad (II.2)$$

Adding b^2 to both sides of equation (II.2) yields

$$4a^2s^2 + 4abs + 4ac + b^2 = b^2 \qquad (II.3)$$

Now by substracting $4ac$ from both sides of equation (II.3), we obtain

$$4a^2s^2 + 4abs + b^2 = b^2 - 4ac \qquad (II.4)$$

Taking the square root of both sides of equation (II.4) yields

$$2as + b = \pm \sqrt{b^2 - 4ac}$$

Hence

$$s = \frac{-b \pm \sqrt{b^2 - 4ac}}{2a} \qquad (II.5)$$

Equation (II.5) can always be used to determine the two roots of a quadratic equation or expression. That is, if s_1 and s_2 are the two roots of equation (II.1), then

$$s_1 = \frac{-b + \sqrt{b^2 - 4ac}}{2a} \quad \text{and} \quad s_2 = \frac{-b - \sqrt{b^2 - 4ac}}{2a}$$

Also, if

$$4ac > b^2$$

then s_1 and s_2 will be complex conjugate roots.

II-2 THE CUBIC EQUATION

The cubic equations encountered in this text have constant, real (no imaginary parts) coefficients. Because this is so, then at least one of the roots of the cubic equation will always be real. This real root may be determined by using a technique called the Newton-Raphson method. It is an iterative approximation method, which means that if repeated enough times it yields a fairly accurate value of the real root. Once the real root is found, the two remaining roots can be determined by reducing the cubic equation to a quadratic form and solving the quadratic by the formula in equation (II.5).

The general form of a cubic equation is

$$as^3 + bs^2 + cs + d = 0 \qquad (II.6)$$

II-2.1 EXAMPLE II.1

It will be much easier to understand the procedure for solving cubic equations if we consider a specific example; e.g.,

$$2s^3 + 19s^2 + 54s + 45 = 0 \qquad (II.7)$$

Step 1. Specify equation (II.7) as $f(s)$.

Step 2. Differentiate equation (II.7) and specify it as $f'(s)$; namely

$$f'(s) = \frac{df(s)}{ds} = 6s^2 + 38s + 54$$

Step 3. Guess a value for s, say s_0, that looks as if it may make $f(s) = 0$ in equation (II.7) and introduce this value of s into the following equation; i.e.,

$$s_1 = s_0 - \frac{f(s_0)}{f'(s_0)} \qquad (II.8)$$

where s is the new, more accurate value of s. We now replace s_0 in equation (II.8) with s_1 and repeat the calculation to determine s_2, which will be an even better approximation of one of the roots of equation (II.7). This process is further repeated until the root is determined with acceptable accuracy.

For our example, let us begin with $s_0 = -1$ in equation (II.8):

Step 4. Substituting $s = -1$ into $f(s)$ and $f'(s)$, yields

$$f(s) = 2(-1)^3 + 19(-1)^2 + 54(-1) + 45$$

Therefore, $\quad f(s) = -2 + 19 - 54 + 45 = 8$

and $\quad f'(s) = 6(-1)^2 + 38(-1) + 54$

Therefore, $\quad f'(s) = 6 - 38 + 54 = 22$

Hence, $\quad s_1 = (-1) - \dfrac{(8)}{(22)} = -1.363$

Step 5. Repeat step 4, replacing s_0 by $s_1 = -1.363$. Having done this, we find that the new value for s is $s_2 = -1.5$.

Step 6. The procedure should be repeated until there is no significant difference between the new value and the old value of s.

Step 7. In our example, further iterations produce no change in the new value of s; thus, $s = -1.5$ is a real root of equation (II.7). If this equation is divided by $(s + 1.5)$, the result will be $2s^2 + 16s + 30$, i.e.,

$$(s + 1.5)\,(2s^2 + 16s + 30) = 0$$

and by using equation (II.5) we obtain all the roots of equation (II.7), i.e.,

$$s_1 = -1.5$$
$$s_2 = -3$$
$$s_3 = -5$$

Iterative methods such as this are somewhat laborious but provide a simple and reliable way of finding the roots of a cubic equation. We will summarize the above description by means of a second example.

II-2.2 EXAMPLE II.2

$$f(s) = s^3 + 10.42s^2 + 32.25s + 26.21 \qquad (II.9)$$

Following the same procedure as before, we have

$$\frac{df(s)}{ds} = 3s^2 + 20.84s + 32.25 = f'(s)$$

Now let
$$s = -1$$

Hence

$$f(s) = f(-1) = (-1)^3 + 10.42(-1)^2 + 32.25(-1) + 26.21$$
$$f(s) = 3.38$$

Also

$$f'(s) = f'(-1) = 3(-1)^2 + 20.84(-1) + 32.25$$
$$f'(s) = 14.41$$

Substituting the appropriate values into equation (II.8) yields

$$s_1 = s - \frac{f(s)}{f'(s)} = (-1) - \frac{(3.38)}{(14.41)}$$

Therefore, $s_1 = -1.23$, which is a new, more accurate solution of the initial equation. Repeating the procedure, we obtain

$$f(s_1) = f(-1.23) = (-1.23)^3 + 10.42(-1.23)^2 + 32.25(-1.23) + 26.21$$

Hence

$$f(s_1) = 0.39$$

Also,

$$f'(s_1) = f'(-1.23) = 3(-1.23)^2 + 20.84(-1.23) + 32.25$$
$$f'(s) = 11.09$$

from which

$$s_2 = s_1 - \frac{f(s_1)}{f'(s_1)} = (-1.23) - \frac{(0.39)}{(11.09)}$$

or $s_2 = -1.27$, which is again a new, more accurate solution of the initial equation. Successive repetition of the above procedure yields a root with no more accuracy than -1.27, so that $(s + 1.27)$ becomes a factor of the initial equation. To find the remaining factors, we must determine the quadratic expression which is a factor of the cubic equation by performing the simple division.

$$\frac{s^3 + 10.42s^2 + 32.25s + 26.21}{s + 1.27} = s^2 + 9.15s + 20.63$$

Employing equation (II.5), we find that

$$s^2 + 9.15s + 20.63 = (s + 5.12)(s + 4.03)$$

Therefore, the roots of the equation (II.9) are

$$s_1 = -1.27$$
$$s_2 = -4.03$$
$$s_3 = -5.12$$

appendix ▌▌▌

partial fraction expansion

III-1 INTRODUCTION

The analysis of a reasonably simple system may yield a Laplace transform of such complexity that its solution would be found only in a
comprehensive set of Laplace tables. However, it is often possible to
reduce the transform to the sum of a number of very simple Laplace
transform expressions by the use of partial fraction expansion. These
expressions can then either be inverse-transformed from memory or
looked up in an abbreviated table.

III-2 SINGLE FIRST-DEGREE FACTORS

Consider the transform described by

$$F(s) = \frac{3s + 2}{s(s^2 + 7s + 10)} \qquad (III.1)$$

This is not an easily recognizable transform, but it could probably
be solved if it could be reduced in complexity.

Step 1. The denominator must be factorized into its simplest factor form (by methods described in Appendix II, if need be). Equation (III.1) now becomes

$$F(s) = \frac{3s + 2}{s(s + 2)(s + 5)} \qquad (III.2)$$

Step 2. Separate the denominator factors and introduce a numerator constant thus:

$$\frac{3s + 2}{s(s + 2)(s + 5)} = \frac{A}{s} + \frac{B}{(s + 2)} + \frac{C}{(s + 5)} \qquad (III.3)$$

Step 3. Each constant may now be evaluated separately, by multiplying the left-hand side of equation (III.3) by the denominator term of the appropriate constant, and then evaluating the resulting expression when s equals a value which makes the denominator term zero; e.g., to determine the value of A multiply the left-hand side of equation (III.3) by s and let $s = 0$:

$$A = \frac{3s + 2}{\cancel{s}(s + 2)(s + 5)} \cdot \cancel{s} \Big|_{s=0}$$

Hence

$$A = \frac{1}{5}$$

To determine B, multiply by $(s + 2)$ and let $s = -2$, thus:

$$B = \frac{3s + 2}{s\cancel{(s + 2)}(s + 5)} \cdot \cancel{(s + 2)} \Big|_{s=-2}$$

Hence

$$B = \frac{2}{3}$$

Similarly, to find C, multiply by $(s + 5)$ and let $s = -5$, thus:

$$C = \frac{3s + 2}{s(s + 2)\cancel{(s + 5)}} \cdot \cancel{(s + 5)} \Big|_{s=-5}$$

Hence

$$C = -\frac{13}{15}$$

Step 4. Equation (III.1) now simplifies to

$$F(s) = \frac{1}{5s} + \frac{2}{3(s+2)} - \frac{13}{15(s+5)}$$

and the inverse Laplace transform is

$$\mathcal{L}^{-1}[F(s)] = f(t) = \frac{1}{5} + \frac{2}{3} e^{-2t} - \frac{13}{15} e^{-5t}$$

III-3 REPEATED FIRST-DEGREE FACTORS

Consider now the transform described by

$$F(s) = \frac{3s+2}{s(s^2 + 10s + 25)} \qquad (III.4)$$

Step 1. The denominator is once again factorized, and we see that it produces a repeated factor, namely,

$$F(s) = \frac{3s+2}{s(s+5)(s+5)} \qquad (III.5)$$

Step 2. The denominator of equation (III.5) is separated into distinct parts, as before, and constants are introduced. But this time the repeating factor is represented in increasing order; namely,

$$\frac{3s+2}{s(s+5)(s+5)} = \frac{A}{s} + \frac{B}{(s+5)} + \frac{C}{(s+5)^2} \qquad (III.6)$$

Step 3. Constants A and C may be evaluated as before, because they possess denominators clearly identified in the left-hand side of equation (III.6). However, constant B must be evaluated differently, which we will leave until Step 4. First let us determine A and C; A is given by

$$A = \frac{3s+2}{s(s+5)^2} \cdot s \Big|_{s=0}$$

Hence

$$A = \frac{2}{25}$$

and C is given by

$$C = \frac{3s + 2}{s(s + 5)^2} \cdot (s + 5)^2 \Big|_{s=-5}$$

Hence

$$C = \frac{13}{5}$$

Step 4. Constant B is found by multiplying the left-hand side of equation (III.6) by the highest order of repeating factor, i.e., $(s + 5)^2$, and then differentiating the result; when this is done, a substitution is made for s as before. Thus B is given by

$$B = \frac{d}{ds} \left[\frac{3s + 2}{s(s + 5)^2} \cdot (s + 5)^2 \right] \Big|_{s=-5}$$

$$\text{Hence} \quad B = \frac{3s - (3s + 2)}{s^2} \Big|_{s=-5}$$

$$\text{and} \quad B = -\frac{2}{25}$$

Step 5. Equation (III.4) now becomes

$$F(s) = \frac{2}{25s} - \frac{2}{25(s + 5)} + \frac{13}{5(s + 5)^2}$$

and the inverse transform is

$$\mathcal{L}^{-1}[F(s)] = f(t) = \frac{2}{25} - \frac{2}{25}e^{-5t} + \frac{13t}{5}e^{-5t}$$

III-4 COMPLEX FACTORS

A transfer function containing a denominator with a quadratic expression that will not factorize into real roots may be handled in a similar manner to that shown in Section III-2, with some slight modification. Consider the following transfer function:

$$F(s) = \frac{1}{s(s + 3)(s^2 + 2\zeta\omega_n s + \omega_n^2)} \tag{III.7}$$

where $\zeta = 0.6$ and $\omega_n = 5$ rad/s

Equation (III.7) simplifies to

$$F(s) = \frac{1}{s(s + 3)(s^2 + 6s + 25)} \qquad (III.8)$$

Step 1. Equation (III.8) must be separated into its usual factored form:

$$\frac{1}{s(s + 3)(s^2 + 6s + 25)} =$$

$$\frac{A}{s} + \frac{B}{(s + 3)} + \frac{C}{(s + 3 + j4)} + \frac{D}{(s + 3 - j4)} \qquad (III.9)$$

Step 2. The constants A and B may be evaluated in the usual manner:

$$A = \frac{1}{\cancel{s}(s + 3)(s^2 + 6s + 25)} \cdot \cancel{s} \Big|_{s=0}$$

Hence $A = \dfrac{1}{75}$

$$B = \frac{1}{s\cancel{(s + 3)}(s^2 + 6s + 25)} \cdot \cancel{(s + 3)} \Big|_{s=-3}$$

Hence $B = -\dfrac{1}{48}$

Step 3. Only one of the numerator constants having a complex denominator need be evaluated (either C or D), the reason being that ultimately we are trying to determine the time domain function of the Laplace transform. When two complex conjugate terms are involved, such as those found at the far right of equation (III.9), the form taken by the time function is

$$\mathscr{L}^{-1}\left[\frac{C}{(s + 3 + j4)} + \frac{D}{(s + 3 - j4)}\right] = Ce^{(-3-j4)t} + De^{(-3+j4)t} \qquad (III.10)$$

The right-hand side of equation (III.10) is a real quantity, and the only way that this may be demonstrated mathematically is for C and D to be complex conjugate quantities also.

Hence $|C| = |D|$, and the right-hand side of equation (III.10) can be rewritten as

$$Ce^{(-3-j4)t} + De^{(-3+j4)t} = [|C| + |D|]\, e^{-3t} \sin (4t + \phi) \qquad (III.11)$$

where $\phi = 90° - $ argument of C

or $\phi = 90° + $ argument of D

Hence it is only necessary to evaluate one of the quantities C or D, but not both.

Let us evaluate C.

$$C = \left.\frac{1}{s(s + 3)(s^2 + 6s + 25)} \cdot (s + 3 + j4)\right|_{s=-3-j4}$$

i.e., $$C = \left.\frac{1}{s(s + 3)(s + 3 - j4)}\right|_{s=-3-j4}$$

or $$C = \frac{1}{(-3 - j4)(-3 - j4 + 3)(-3 - j4 + 3 - j4)}$$

$$C = \frac{1}{160}\, \angle -53.13°$$

The right-hand side of equation (III.11) becomes

$$\frac{1}{80}\, e^{-3t} \sin (4t + \phi)$$

where $\phi = 90° - (-53.13°) = 143.13°$

Step 4. The inverse Laplace transform of equation (III.7) is, therefore,

$$f(t) = \frac{1}{75} - \frac{e^{-3t}}{48} + \frac{e^{-3t}}{80} \sin (4t + 143.13°)$$

appendix table of transforms IV

IV-1 INTRODUCTION

Listed below in Table IV-1 are some of the more common Laplace transforms. Although not as easily recognizable as those transforms listed in Table 5-1, the transform pairs in this appendix will prove quite useful in the solution of even the most difficult problems in the text and will also be of help in other circumstances.

TABLE IV-1—LAPLACE TRANSFORM PAIRS

NO.	$F(s)$	$f(t)$
1	$\dfrac{1}{(s + \zeta\omega_n)^2 + \omega_d^2}$	$\dfrac{e^{-\zeta\omega_n t}}{\omega_d}\sin\omega_d t$
2	$\dfrac{1}{s^2 + \omega^2}$	$\dfrac{1}{\omega}\sin\omega t$
3	$\dfrac{1}{(s^2 + \omega^2)^2}$	$\dfrac{1}{2\omega^3}(\sin\omega t - \omega t\cos\omega t)$
4(a)	$\dfrac{\omega_n^2}{(\tau s + 1)(s^2 + 2\zeta\omega_n s + \omega_n^2)}$	$\dfrac{\tau\omega_n^2}{\tau^2\omega_n^2 - 2\zeta\omega_n\tau + 1}e^{-t/\tau} + \dfrac{\omega_n e^{-\zeta\omega_n t}}{[(1 - \zeta^2)(\tau^2\omega_n^2 - 2\zeta\omega_n\tau + 1)]^{1/2}}\sin(\omega_n\sqrt{1 - \zeta^2}\,t - \phi)$ where $\phi = \tan^{-1}\left[\dfrac{\tau\omega_n\sqrt{1 - \zeta^2}}{1 + \zeta\omega_n\tau}\right]$
4(b)	$\dfrac{\omega_n^2}{(s + a)(s^2 + 2\zeta\omega_n s + \omega_n^2)}$	$\dfrac{\omega_n^2 e^{-at}}{\omega_n^2 - 2\zeta\omega_n a + a^2} + \dfrac{\omega_n e^{-\zeta\omega_n t}}{[(1 - \zeta^2)(\omega_n^2 - 2\zeta\omega_n a + a^2)]^{1/2}}\sin(\omega_n\sqrt{1 - \zeta^2}\cdot t - \phi)$ where $\phi = \tan^{-1}\left[\dfrac{\omega_n\sqrt{1 - \zeta^2}}{a - \zeta\omega_n}\right]$
5(a)	$\dfrac{\omega^2}{(\tau s + 1)(s^2 + \omega^2)}$	$\dfrac{\tau\omega^2 e^{-t/\tau}}{(\tau^2\omega^2 + 1)} + \dfrac{\omega\sin(\omega t - \phi)}{(\tau^2\omega^2 + 1)^{1/2}}$ where $\phi = \tan^{-1}(\tau\omega)$

TABLE IV-1—CONTINUED

NO.	$F(s)$	$f(t)$
5(b)	$\dfrac{\omega^2}{(s+a)(s^2+\omega^2)}$	$\dfrac{\omega^2 e^{-at}}{(a^2+\omega^2)} + \dfrac{\omega \sin(\omega_n t - \phi)}{[a^2+\omega^2]^{1/2}}$
		where $\phi = \tan^{-1}(\omega/a)$
6(a)	$\dfrac{1}{(\tau s + 1)^3}$	$\dfrac{t^2 e^{-t/\tau}}{2\tau^3}$
6(b)	$\dfrac{1}{(s+a)^3}$	$\dfrac{t^2 e^{-at}}{2}$
7	$\dfrac{1}{s^2(s^2+\omega^2)}$	$\dfrac{1}{\omega^2}[1 - \cos(\omega t)]$
8	$\dfrac{1}{s^2(s^2+\omega^2)}$	$\dfrac{t}{\omega^2} - \dfrac{1}{\omega^3}\sin(\omega t)$
9	$\dfrac{\omega^2}{s^2(s^2+\omega^2)}$	$t - \dfrac{1}{\omega}\sin(\omega t)$
10(a)	$\dfrac{\omega_n^2}{s^2(\tau s + 1)(s^2 + 2\zeta\omega_n s + \omega_n^2)}$	$t - \tau - \dfrac{2\zeta}{\omega_n} + \dfrac{\tau^3 \omega_n^2}{\tau^2\omega_n^2 - 2\zeta\omega_n\tau + 1} e^{-t/\tau}$
		$+ \dfrac{e^{-\zeta\omega_n t}\sin(\omega_n\sqrt{1-\zeta^2}\,t - \phi)}{\omega_n[(1-\zeta^2)(\tau^2\omega_n^2 - 2\zeta\omega_n\tau + 1)]^{1/2}}$

TABLE IV-1—CONTINUED

NO.	F(s)	f(t)
		$\left[\dfrac{\sqrt{1-\zeta^2}}{-\zeta}\right]+\tan^{-1}\left[\dfrac{\tau\omega_n\sqrt{1-\zeta^2}}{1-\zeta\omega_n\tau}\right]$ where $\phi=2\tan^{-1}$
10(b)	$\dfrac{\omega_n^2}{s^2(s+a)(s^2+2\zeta\omega_n s+\omega_n^2)}$	$\dfrac{1}{a}\left[t-\dfrac{2\zeta}{\omega_n}\right]-\dfrac{1}{a^2}\left[1-\dfrac{\omega_n^2 e^{-at}}{\omega_n^2+2\zeta\omega_n a+a^2}\right]$ $+\dfrac{e^{-\zeta\omega_n t}\sin(\omega_n\sqrt{1-\zeta^2}t-\phi)}{\omega_n[(1-\zeta^2)(\omega_n^2-2\zeta\omega_n a+a^2)]^{1/2}}$ where $\phi=2\tan^{-1}\left[\dfrac{\sqrt{1-\zeta^2}}{-\zeta}\right]+\tan^{-1}\left[\dfrac{\omega_n\sqrt{1-\zeta^2}}{a-\zeta\omega_n}\right]$
11	$\dfrac{1}{s^3}$	$\dfrac{t^2}{2}$
12	$\dfrac{\omega_n^2}{s^3(s^2+2\zeta\omega_n s+\omega_n^2)}$	$\dfrac{t^2}{2}-\dfrac{2\zeta}{\omega_n}t-\dfrac{1-4\zeta^2}{\omega_n^2}+\dfrac{e^{-\zeta\omega_n t}\sin(\omega_n\sqrt{1-\zeta^2}t-\phi)}{\omega_n^2\sqrt{1-\zeta^2}}$ where $\phi=3\tan^{-1}\left[\dfrac{\sqrt{1-\zeta^2}}{-\zeta}\right]$

TABLE IV-1—CONTINUED

NO.	$F(s)$	$f(t)$
13(a)	$\dfrac{1}{s^3(\tau_1 s + 1)(\tau_2 s + 1)}$	$\dfrac{t^2}{2} - (\tau_1 + \tau_2)t - \dfrac{\tau_2^3(1 - e^{-t/\tau_2}) - \tau_1^3(1 - e^{-t/\tau_1})}{\tau_1 - \tau_2}$
13(b)	$\dfrac{1}{s^3(s + a)(s + b)}$	$\dfrac{t}{ab}\left[\dfrac{t}{2} - \dfrac{1}{a} - \dfrac{1}{b}\right] - \dfrac{(1 - e^{-bt})/b^3 - (1 - e^{-at})/a^3}{(b - a)}$
14(a)	$\dfrac{1}{s^3(\tau s + 1)^2}$	$\dfrac{t^2}{2} - 2\tau t + 3\tau^2 - \tau(t + 3\tau)e^{-t/\tau}$
14(b)	$\dfrac{1}{s^3(s + a)^2}$	$\dfrac{t^2}{2a^2} - \dfrac{t}{a^3}[2 + e^{-at}] + \dfrac{3}{a^4}[1 - e^{-at}]$
15(a)	$\dfrac{(\alpha s + 1)\omega_n^2}{s^2 + 2\zeta\omega_n s + \omega_n^2}$	$\omega_n \sqrt{\dfrac{\alpha^2\omega_n^2 - 2\zeta\omega_n\alpha + 1}{1 - \zeta^2}}\, e^{-\zeta\omega_n t} \sin\left(\omega_n\sqrt{1 - \zeta^2}\, t + \phi\right)$ where $\quad \omega_n\phi = \tan^{-1}\left[\dfrac{\alpha\omega_n\sqrt{1 - \zeta^2}}{1 - \zeta\omega_n\alpha}\right]$
15(b)	$\dfrac{(s + c)\omega_n^2}{s^2 + 2\zeta\omega_n s + \omega_n^2}$	$\omega_n \sqrt{\dfrac{\omega_n^2 - 2\zeta\omega_n c + c^2}{1 - \zeta^2}}\, e^{-\zeta\omega_n t} \sin\left(\omega_n\sqrt{1 - \zeta^2}\, t + \phi\right)$

TABLE IV-1—CONTINUED

NO.	$F(s)$	$f(t)$
		$\phi = \tan^{-1}\left[\frac{\omega_n\sqrt{1-\zeta^2}}{c-\zeta\omega_n}\right]$
16(a)	$\frac{(\alpha s+1)\,\omega_n^2}{s(s^2+2\zeta\omega_n s+\omega_n^2)}$	$1+\sqrt{\frac{\alpha^2\omega_n^2-2\zeta\omega_n\alpha+1}{1-\zeta^2}}\,e^{-\zeta\omega_n t}\sin(\omega_n\sqrt{1-\zeta^2}\,t+\phi)$
		where $\phi = \tan^{-1}\left[\frac{\alpha\omega_n\sqrt{1-\zeta^2}}{1-\zeta\omega_n\alpha}\right] - \tan^{-1}\left[\frac{\sqrt{1-\zeta^2}}{-\zeta}\right]$
16(b)	$\frac{(s+c)\,\omega_n^2}{s(s^2+2\zeta\omega_n s+\omega_n^2)}$	$c+\sqrt{\frac{\omega_n^2-2\zeta\omega_n c+c^2}{1-\zeta^2}}\,e^{-\zeta\omega_n t}\sin(\omega_n\sqrt{1-\zeta^2}\,t+\phi)$
		$\phi = \tan^{-1}\left[\frac{\omega_n\sqrt{1-\zeta^2}}{c-\zeta\omega_n}\right] - \tan^{-1}\left[\frac{\sqrt{1-\zeta^2}}{-\zeta}\right]^*$
17(a)	$\frac{\alpha s+1}{s(\tau_1 s+1)(\tau_2 s+1)}$	$1+\frac{\tau_1-\alpha}{\tau_2-\tau_1}e^{-t/\tau_1}-\frac{\tau_2-\alpha}{\tau_2-\tau_1}e^{-t/\tau_2}$
17(b)	$\frac{s+c}{s(s+a)(s+b)}$	$\frac{c}{ab}+\frac{1}{ab(a-b)}[b(c-a)e^{-at}-a(c-b)e^{-bt}]$

* Equation (10.61a) is also a valid solution for ϕ

TABLE IV-1—CONTINUED

NO.	$F(s)$	$f(t)$
18	$\dfrac{\omega_n^2 s}{s^2 + 2\zeta\omega_n s + \omega_n^2}$	$\dfrac{\omega_n^2}{\sqrt{1 - \zeta^2}} e^{-\zeta\omega_n t} \sin\left(\omega_n\sqrt{1 - \zeta^2}\,t + \phi\right)$ where $\phi = \tan^{-1}\left[\dfrac{\sqrt{1 - \zeta^2}}{-\zeta}\right]$
19	$\dfrac{s}{(s^2 + \omega^2)}$	$\cos \omega t$
20(a)	$\dfrac{s\omega_n^2}{(\tau s + 1)(s^2 + 2\zeta\omega_n s + \omega_n^2)}$	$\dfrac{-\omega_n^2}{1 - 2\zeta\omega_n\tau - \tau^2\omega_n^2}e^{-t/\tau} + \dfrac{\omega_n^2 e^{-\zeta\omega_n t}\sin\left(\omega_n\sqrt{1 - \zeta^2}\cdot t - \phi\right)}{[(1 - \zeta^2)(\tau^2\omega_n^2 - 2\zeta\omega_n\tau + 1)]^{1/2}}$ where $\phi = \tan^{-1}\left[\dfrac{\tau\omega_n\sqrt{1 - \zeta^2}}{1 - \omega_n\tau\zeta}\right] - \tan^{-1}\left[\dfrac{\sqrt{1 - \zeta^2}}{-\zeta}\right]$
20(b)	$\dfrac{\omega_n^2 s}{(s + a)(s^2 + 2\zeta\omega_n s + \omega_n^2)}$	$\dfrac{-a\omega_n^2 e^{-at}}{a^2 - 2\zeta\omega_n a - \omega_n^2} + \dfrac{\omega_n^2 e^{-\zeta\omega_n t}\sin\left(\omega_n\sqrt{1 - \zeta^2}\cdot t - \phi\right)}{[(1 - \zeta^2)(\omega_n^2 - 2\zeta\omega_n a + a^2)]^{1/2}}$ $\phi = \tan^{-1}\left[\dfrac{\omega_n\sqrt{1 - \zeta^2}}{a - \zeta\omega_n}\right] - \tan^{-1}\left[\dfrac{\sqrt{1 - \zeta^2}}{-\zeta}\right]$

appendix

Fairchild linear

integrated circuits

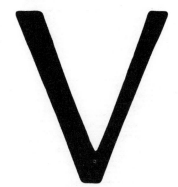

GENERAL DESCRIPTION — The μA741 is a high performance monolithic Operational Amplifier constructed using the Fairchild Planar* epitaxial process. It is intended for a wide range of analog applications. High common mode voltage range and absence of "latch-up" tendencies make the μA741 ideal for use as a voltage follower. The high gain and wide range of operating voltage provides superior performance in integrator, summing amplifier, and general feedback applications.

- NO FREQUENCY COMPENSATION REQUIRED
- SHORT CIRCUIT PROTECTION
- OFFSET VOLTAGE NULL CAPABILITY
- LARGE COMMON-MODE AND DIFFERENTIAL VOLTAGE RANGES
- LOW POWER CONSUMPTION
- NO LATCH UP

ABSOLUTE MAXIMUM RATINGS

Supply Voltage	
Military (741)	±22 V
Commercial (741C)	±18 V
Internal Power Dissipation (Note 1)	
Metal Can	500 mW
DIP	670 mW
Mini DIP	310 mW
Flatpak	570 mW
Differential Input Voltage	±30 V
Input Voltage (Note 2)	±15 V
Storage Temperature Range	
Metal Can, DIP, and Flatpak	−65°C to +150°C
Mini DIP	−55°C to +125°C
Operating Temperature Range	
Military (741)	−55°C to +125°C
Commercial (741C)	0°C to +70°C
Lead Temperature (Soldering)	
Metal Can, DIP, and Flatpak (60 seconds)	300°C
Mini DIP (10 seconds)	260°C
Output Short Circuit Duration (Note 3)	Indefinite

EQUIVALENT CIRCUIT

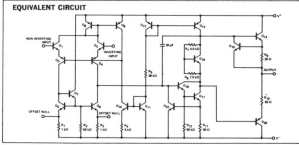

Notes on following pages.

CONNECTION DIAGRAMS

8-LEAD METAL CAN
(TOP VIEW)
PACKAGE OUTLINE 5B

Note: Pin 4 connected to case

ORDER INFORMATION

TYPE	PART NO.
741	741HM
741C	741HC

14-LEAD DIP
(TOP VIEW)
PACKAGE OUTLINE 6A

ORDER INFORMATION

TYPE	PART NO.
741	741DM
741C	741DC

10-LEAD FLATPAK
(TOP VIEW)
PACKAGE OUTLINE 3F

ORDER INFORMATION

TYPE	PART NO.
741	741FM

8-LEAD MINI DIP
(TOP VIEW)
PACKAGE OUTLINE 9T

ORDER INFORMATION

TYPE	PART NO.
741C	741TC

*Planar is a patented Fairchild process.

μA741 FREQUENCY -COMPENSATED OPERATIONAL AMPLIFIER — FAIRCHILD LINEAR INTEGRATED CIRCUITS

515

741

ELECTRICAL CHARACTERISTICS (V_S = ±15 V, T_A = 25°C unless otherwise specified)

PARAMETERS (see definitions)	CONDITIONS		MIN.	TYP.	MAX.	UNITS
Input Offset Voltage	$R_S \leqslant 10$ kΩ			1.0	5.0	mV
Input Offset Current				20	200	nA
Input Bias Current				80	500	nA
Input Resistance			0.3	2.0		MΩ
Input Capacitance				1.4		pF
Offset Voltage Adjustment Range				±15		mV
Large Signal Voltage Gain	$R_L \geqslant 2$ kΩ, V_{OUT} = ±10 V		50,000	200,000		
Output Resistance				75		Ω
Output Short Circuit Current				25		mA
Supply Current				1.7	2.8	mA
Power Consumption				50	85	mW
Transient Response (Unity Gain)	Risetime	V_{IN} = 20 mV, R_L = 2 kΩ, $C_L \leqslant$ 100 pF		0.3		μs
	Overshoot			5.0		%
Slew Rate	$R_L \geqslant 2$ kΩ			0.5		V/μs

The following specifications apply for −55°C ≤ T_A ≤ +125°C:

PARAMETERS	CONDITIONS	MIN.	TYP.	MAX.	UNITS
Input Offset Voltage	$R_S \leqslant 10$ kΩ		1.0	6.0	mV
Input Offset Current	T_A = +125°C		7.0	200	nA
	T_A = −55°C		85	500	nA
Input Bias Current	T_A = +125°C		0.03	0.5	μA
	T_A = −55°C		0.3	1.5	μA
Input Voltage Range		±12	±13		V
Common Mode Rejection Ratio	$R_S \leqslant 10$ kΩ	70	90		dB
Supply Voltage Rejection Ratio	$R_S \leqslant 10$ kΩ		30	150	μV/V
Large Signal Voltage Gain	$R_L \geqslant 2$ kΩ, V_{OUT} = ±10 V	25,000			
Output Voltage Swing	$R_L \geqslant 10$ kΩ	±12	±14		V
	$R_L \geqslant 2$ kΩ	±10	±13		V
Supply Current	T_A = +125°C		1.5	2.5	mA
	T_A = −55°C		2.0	3.3	mA
Power Consumption	T_A = +125°C		45	75	mW
	T_A = −55°C		60	100	mW

TYPICAL PERFORMANCE CURVES FOR 741

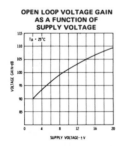

OPEN LOOP VOLTAGE GAIN AS A FUNCTION OF SUPPLY VOLTAGE

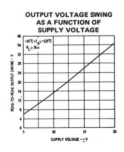

OUTPUT VOLTAGE SWING AS A FUNCTION OF SUPPLY VOLTAGE

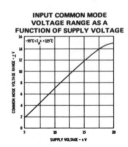

INPUT COMMON MODE VOLTAGE RANGE AS A FUNCTION OF SUPPLY VOLTAGE

741C

ELECTRICAL CHARACTERISTICS ($V_S = \pm 15$ V, $T_A = 25°$C unless otherwise specified)

PARAMETERS (see definitions)	CONDITIONS		MIN.	TYP.	MAX.	UNITS
Input Offset Voltage	$R_S \leqslant 10$ kΩ			2.0	6.0	mV
Input Offset Current				20	200	nA
Input Bias Current				80	500	nA
Input Resistance			0.3	2.0		MΩ
Input Capacitance				1.4		pF
Offset Voltage Adjustment Range				± 15		mV
Input Voltage Range			± 12	± 13		V
Common Mode Rejection Ratio	$R_S \leqslant 10$ kΩ		70	90		dB
Supply Voltage Rejection Ratio	$R_S \leqslant 10$ kΩ			30	150	μV/V
Large Signal Voltage Gain	$R_L \geqslant 2$ kΩ, $V_{OUT} = \pm 10$ V		20,000	200,000		
Output Voltage Swing	$R_L \geqslant 10$ kΩ		± 12	± 14		V
	$R_L \geqslant 2$ kΩ		± 10	± 13		V
Output Resistance				75		Ω
Output Short Circuit Current				25		mA
Supply Current				1.7	2.8	mA
Power Consumption				50	85	mW
Transient Response (Unity Gain)	Risetime	$V_{IN} = 20$ mV, $R_L = 2$ kΩ, $C_L \leqslant 100$ pF		0.3		μs
	Overshoot			5.0		%
Slew Rate	$R_L \geqslant 2$ kΩ			0.5		V/μs

The following specifications apply for $0°$C $\leqslant T_A \leqslant +70°$C:

		MIN.	TYP.	MAX.	UNITS
Input Offset Voltage				7.5	mV
Input Offset Current				300	nA
Input Bias Current				800	nA
Large Signal Voltage Gain	$R_L \geqslant 2$ kΩ, $V_{OUT} = \pm 10$ V	15,000			
Output Voltage Swing	$R_L \geqslant 2$ kΩ	± 10	± 13		V

TYPICAL PERFORMANCE CURVES FOR 741C

OPEN LOOP VOLTAGE GAIN AS A FUNCTION OF SUPPLY VOLTAGE

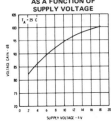

OUTPUT VOLTAGE SWING AS A FUNCTION OF SUPPLY VOLTAGE

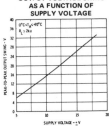

INPUT COMMON MODE VOLTAGE RANGE AS A FUNCTION OF SUPPLY VOLTAGE

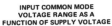
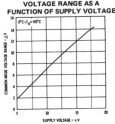

NOTES:
1. Rating applies to ambient temperatures up to 70°C. Above 70°C ambient derate linearly at 6.3 mW/°C for the Metal Can, 8.3 mW/°C for the DIP, 5.6 mW/°C for the Mini DIP and 7.1 mW/°C for the Flatpak.
2. For supply voltages less than ±15 V, the absolute maximum input voltage is equal to the supply voltage.
3. Short circuit may be to ground or either supply. Rating applies to +125°C case temperature or 75°C ambient temperature.

517

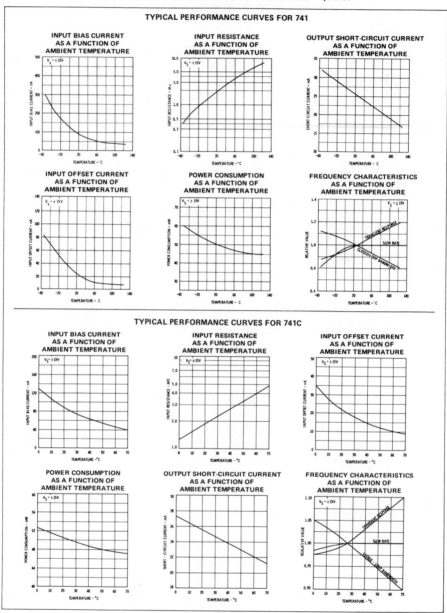

TYPICAL PERFORMANCE CURVES FOR 741

INPUT BIAS CURRENT
AS A FUNCTION OF
AMBIENT TEMPERATURE

INPUT RESISTANCE
AS A FUNCTION OF
AMBIENT TEMPERATURE

OUTPUT SHORT-CIRCUIT CURRENT
AS A FUNCTION OF
AMBIENT TEMPERATURE

INPUT OFFSET CURRENT
AS A FUNCTION OF
AMBIENT TEMPERATURE

POWER CONSUMPTION
AS A FUNCTION OF
AMBIENT TEMPERATURE

FREQUENCY CHARACTERISTICS
AS A FUNCTION OF
AMBIENT TEMPERATURE

TYPICAL PERFORMANCE CURVES FOR 741C

INPUT BIAS CURRENT
AS A FUNCTION OF
AMBIENT TEMPERATURE

INPUT RESISTANCE
AS A FUNCTION OF
AMBIENT TEMPERATURE

INPUT OFFSET CURRENT
AS A FUNCTION OF
AMBIENT TEMPERATURE

POWER CONSUMPTION
AS A FUNCTION OF
AMBIENT TEMPERATURE

OUTPUT SHORT-CIRCUIT CURRENT
AS A FUNCTION OF
AMBIENT TEMPERATURE

FREQUENCY CHARACTERISTICS
AS A FUNCTION OF
AMBIENT TEMPERATURE

TYPICAL PERFORMANCE CURVES FOR 741 AND 741C (Cont'd)

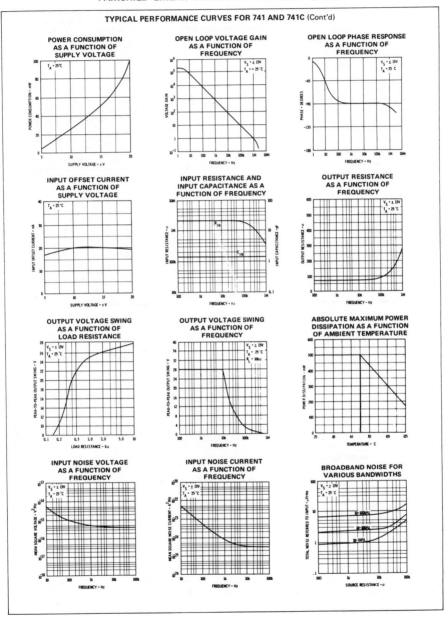

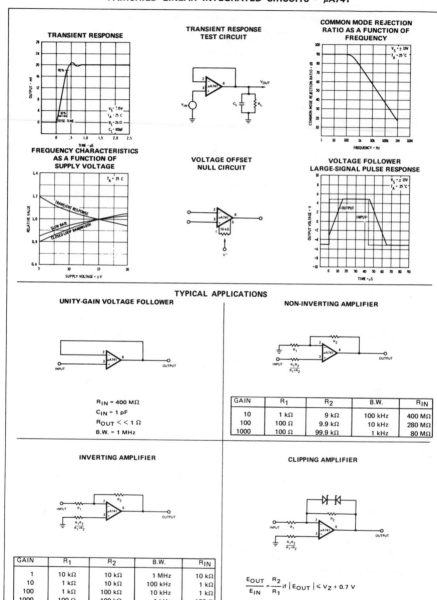

TRANSIENT RESPONSE

TRANSIENT RESPONSE TEST CIRCUIT

COMMON MODE REJECTION RATIO AS A FUNCTION OF FREQUENCY

FREQUENCY CHARACTERISTICS AS A FUNCTION OF SUPPLY VOLTAGE

VOLTAGE OFFSET NULL CIRCUIT

VOLTAGE FOLLOWER LARGE-SIGNAL PULSE RESPONSE

TYPICAL APPLICATIONS

UNITY-GAIN VOLTAGE FOLLOWER

R_{IN} = 400 MΩ
C_{IN} = 1 pF
R_{OUT} << 1 Ω
B.W. = 1 MHz

NON-INVERTING AMPLIFIER

GAIN	R_1	R_2	B.W.	R_{IN}
10	1 kΩ	9 kΩ	100 kHz	400 MΩ
100	100 Ω	9.9 kΩ	10 kHz	280 MΩ
1000	100 Ω	99.9 kΩ	1 kHz	80 MΩ

INVERTING AMPLIFIER

GAIN	R_1	R_2	B.W.	R_{IN}
1	10 kΩ	10 kΩ	1 MHz	10 kΩ
10	1 kΩ	10 kΩ	100 kHz	1 kΩ
100	1 kΩ	100 kΩ	10 kHz	1 kΩ
1000	100 Ω	100 kΩ	1 kHz	100 Ω

CLIPPING AMPLIFIER

$$\frac{E_{OUT}}{E_{IN}} = \frac{R_2}{R_1} \text{ if } |E_{OUT}| \leqslant V_Z + 0.7 \text{ V}$$

where V_Z = Zener breakdown voltage

TYPICAL APPLICATIONS (Cont'd)

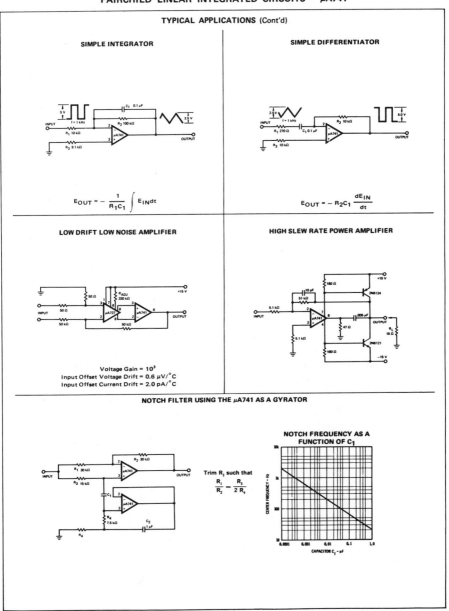

SIMPLE INTEGRATOR

$$E_{OUT} = -\frac{1}{R_1 C_1} \int E_{IN}dt$$

SIMPLE DIFFERENTIATOR

$$E_{OUT} = -R_2 C_1 \frac{dE_{IN}}{dt}$$

LOW DRIFT LOW NOISE AMPLIFIER

Voltage Gain = 10^3
Input Offset Voltage Drift = 0.6 µV/°C
Input Offset Current Drift = 2.0 pA/°C

HIGH SLEW RATE POWER AMPLIFIER

NOTCH FILTER USING THE µA741 AS A GYRATOR

Trim R_3 such that
$$\frac{R_1}{R_2} = \frac{R_3}{2R_4}$$

NOTCH FREQUENCY AS A FUNCTION OF C_1

521

PHYSICAL DIMENSIONS

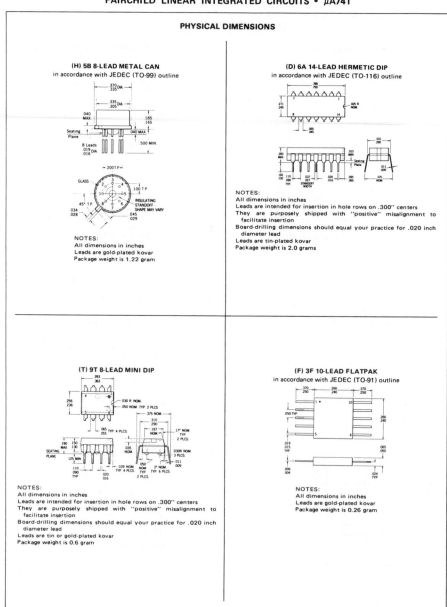

(H) 5B 8-LEAD METAL CAN
in accordance with JEDEC (TO-99) outline

NOTES:
All dimensions in inches
Leads are gold-plated kovar
Package weight is 1.22 gram

(D) 6A 14-LEAD HERMETIC DIP
in accordance with JEDEC (TO-116) outline

NOTES:
All dimensions in inches
Leads are intended for insertion in hole rows on .300" centers
They are purposely shipped with "positive" misalignment to facilitate insertion
Board-drilling dimensions should equal your practice for .020 inch diameter lead
Leads are tin-plated kovar
Package weight is 2.0 grams

(T) 9T 8-LEAD MINI DIP

NOTES:
All dimensions in inches
Leads are intended for insertion in hole rows on .300" centers
They are purposely shipped with "positive" misalignment to facilitate insertion
Board-drilling dimensions should equal your practice for .020 inch diameter lead
Leads are tin or gold-plated kovar
Package weight is 0.6 gram

(F) 3F 10-LEAD FLATPAK
in accordance with JEDEC (TO-91) outline

NOTES:
All dimensions in inches
Leads are gold-plated kovar
Package weight is 0.26 gram

By kind permission of Fairchild Camera and Instrument Corporation, 464 Ellis Street, Mountain View, CA 94042.

appendix VI

bibliography

Barwell, F. T., *Automation and Control In Transportation*. London, England: Pergamon Press, Ltd., 1973.

Bateson, Robert, *Introduction to Control System Technology*. Columbus, Ohio: Charles E. Merrill, 1973.

Bell, David A., *Fundamentals of Electric Circuits*. Reston, Virginia: Reston Publishing Company, Inc., 1978.

Bell, David A., *Solid State Pulse Circuits*. Reston, Virginia: Reston Publishing Company, Inc., 1976.

Buckstein, Ed., *Basic Servomechanisms*. New York: Holt, Rinehart & Winston, 1963.

Bulliet, L. J., *Servomechanisms*. Reading, Mass: Addison Wesley, 1967.

Canadian Standards Association, *Metric Practice Guide*. Rexdale, Ontario, Canada, 1979.

Cooper, William D., *Electronic Instrumentation and Measurement Techniques*. Englewood Cliffs, N.J.: Prentice-Hall, Inc., 1978.

D'Azzo, John D. and Constantine H. Houpis, *Feedback Control System Analysis*. New York: McGraw Hill, 1966.

Dempsey, John A., *Basic Digital Electronics with MSI Applications*. Reading, Mass: Addison Wesley, 1977.

DiStefano, J. T., A. R. Stubberud, and I. J. Williams, *Theory and Problems of Feedback and Control Systems*. New York: McGraw Hill, 1967.

Eckman, Donald P., *Automatic Process Control*. New York: John Wiley & Sons, Inc., 1958.

Garrett, Patrick H., *Analog Systems for Micro Processors and Mini Computers*. Reston, Virginia: Reston Publishing Company, Inc., 1978.

Humphrey, William M., *Introduction to Servomechanism System Design*. Englewood Cliffs, N.J.: Prentice-Hall, Inc., 1973.

Hunter, Ronald P., *Automated Process Control Systems*. Englewood Cliffs, N.J.: Prentice-Hall, Inc., 1978.

Institution of Electrical Engineers, *Distributed Computer Control Systems*. London, England, 1977.

Johnson, Clarence, *Analog Computer Techniques*. New York: McGraw Hill, 1963.

Johnson, Eric, *Servomechanism*. Englewood Cliffs, N.J.: Prentice-Hall, Inc., 1963.

Kuo, Benjamin C., *Automatic Control Systems*. Englewood Cliffs, N.J.: Prentice-Hall, Inc., 1975.

Lenk, John D., *Handbook of Simplified Solid State Circuit Design*. Englewood Cliffs, N.J.: Prentice-Hall, Inc., 1971.

Lenk, John D., *Manual for Integrated Circuit Users*. Reston, Virginia: Reston Publishing Company, 1973.

MacFarlans, A. G. J., *Engineering System Analysis*. London, England: George Harrap & Co., 1964.

Manera, Anthony S., *Solid State Electronic Circuits for Engineering Technology*. New York: McGraw Hill, 1973.

Miller, Richard W., *Servomechanisms, Devices and Fundamentals*. Reston, Virginia: Reston Publishing Company, Inc., 1977.

Morris, N. M., *Control Engineering*. Maidenhead, England: McGraw Hill Co., 1968.

Nixon, Floyd E., *Handbook of Laplace Transformation*. Englewood Cliffs, N.J.: Prentice-Hall, Inc., 1965.

Philco Labs, *Servomechanism Fundamentals and Experiments*. Englewood Cliffs, N.J.: Prentice-Hall, Inc., 1964.

Pitman, R. J. A., *Automatic Control Systems Explained*. London: McMillan & Co., 1966.

Ricci, Fred J., *Analog/Logic Computer Programming and Simulating*. New York: Spartan Books, Inc., 1972.

Savant, C. J., *Control System Design*. New York: McGraw Hill, Inc., 1964.

Stout, David F. and Milton Kaufman, *Handbook of Operational Amplifier Design*. New York: McGraw Hill, 1976.

Streitmatter, Gene and Vito Fiore, *Microprocessors: Theory and Applications*. Reston, Virginia: Reston Publishing Company, Inc., 1979.

Tocci, Ronald J., *Fundamentals of Pulse and Digital Circuits*. Columbus, Ohio: Charles E. Merrill, 1977.

Warfield, John, *Introduction to Electronic Analog Computers*. Englewood Cliffs, N.J.: Prentice-Hall, Inc., 1959.

Weyrick, Robert C., *Fundamentals of Automatic Control*. New York: McGraw Hill, 1975.

Wiebelt, John A., *Radiation Heat Transfer Engineering*. New York: Holt, Rinehart and Winston, 1966.

Wiberg, Donald, *State Space and Linear Systems*. New York: McGraw Hill, 1971.

Woolvet, G. A., *Tranducers in Digital Systems*. Stevenage, England: Peter Pereginus, Ltd., 1977.

Young, R. E., *Supervisory Remote Control Systems*. Stevenage, England: Peter Pereginus, Ltd., 1977.

Index